## PART II: EUTROPHICATION EFFECTS ON BIOGEOCHEMISTRY OF NITROGEN AND SULFUR IN COASTAL LAGOONS

# Preface

The **C**oastal **L**agoon **E**utrophication and **AN**aerobic processes research programme (CLEAN) was one of the projects funded under the European Community Environment Research Programme 1991–1994. Coastal lagoons constitute an important transition zone between the terrestrial environment and the open ocean. They comprise 13% of the world's coastline from the tropics to the polar regions. Their productivity is derived from interactions between the land, oceans and atmosphere. They are enriched by the oceanic and continental inputs and are amongst the most productive ecosystems in the biosphere.

The effect of human activity on these coastal ecosystems, whether direct or indirect, has accelerated considerably in recent years and the impact of such changes on these sensitive and economically important environments is now a cause of major concern. Along the European coasts, many shallow coastal lagoons receive large amounts of organic and mineral nutrients which has led to a rapid increase in eutrophication over the two last decades.

The objective of this project was to study different lagoon systems subject to increasing levels of eutrophication in order to identify the key processes controlling these changes and how they impact on the species composition and population dynamics of the indigenous biological communities.

The papers in this special volume provide a comprehensive set of data on the effects of eutrophication on the population dynamics of the indigenous biota, biodiversity, trophic relationships, biogeochemical cycling, nutrient inputs and fluxes in two constrasting lagoon systems, the highly eutrophic Etang du Prevost and the moderately eutrophic Bassin d'Arcachon. By adopting a multidisciplinary approach it has been possible to identify the key processes regulating eutrophication in these lagoons and their impact on the community structure of the primary and secondary producers as well as the nutrient exchanges which occur at the sediment-water interface. These data provide a valuable source of new information for scientists working on related problems in similar ecosystems as well as those involved in the management of these important natural resources.

The editors wish to acknowledge the assistance given by the numerous referees who carefully reviewed the manuscripts and to thank Professor Henri Dumont, editor-in-chief of *Hydrobiologia* for his invaluable assistance in the production of this special volume. We also wish to acknowledge the support and encouragement given by Dr Canice Nolan of the European Commission Directorate DG XII to publish the data obtained during the course of this project as a single special volume of *Hydrobiologia.*

We are grateful to the Conservatoire du Littoral (Mrs Musson) and to the Conseil Général of Gironde (Mr Boutet) for field work facilities and permission to enter and work on the site of Certes in Arcachon Bay and the Company of Maguelone, and their director Mr Sallas for help during sampling work in the Prevost lagoon.

Finally, it is a pleasure to thank the staff of the Centre d'Océanographie et de Biologie Marine, Arcachon, for the unstinting support that they have given to all the participants during the course of this project. Particular thanks are due to Françoise Truong who provided the administrative support for the project.

*Arcachon, October 1995*

P. Caumette
J. Castel
R.A. Herbert

## PARTICIPATING LABORATORIES

| | | |
|---|---|---|
| 1 | FRENCH GROUP<br><br>Pierre Caumette<br>Coordinator<br>Jacques Castel<br>Université Bordeaux I | Centre d'Océanographie et de Biologie marine<br><br>2 rue Jolyet<br>33120 Arcachon |
| 2 | DUTCH GROUP<br>Lucas Stal<br>University of Amsterdam<br>(proposer)<br><br>Hans van Gemerden<br>University of Groningen<br>(associate proposer) | Laboratory for Microbiology<br>Nieuwe Achtergracht 127<br>NL–1018 WS Amsterdam<br><br><br>Dept. of Microbiology<br>P.O. Box 14<br>Kerklaan 30<br>8750 NN Haren |
| 3 | DANISH GROUP<br>Niels Peter Revsbech<br>University of Aarhus | Institute of Biological Sciences<br>Dept. of Microbial Ecology<br>Ny Munkegade<br>Building 540<br>DK 8000 Aarhus C |
| 4 | U.K. GROUP<br>Rod Herbert<br>University of Dundee | Dept. of Biological Sciences<br>Dundee DD1 4HN |
| 5 | ITALIAN GROUP<br>Victor Ugo Ceccherelli<br>Università di Ferrara<br>(proposer)<br><br>Ireneo Ferrari<br>Università di Parma<br>(associate proposer) | Dipart. Biologia Evolutiva<br>Via L. Borsari 46<br>44100 Ferrara<br><br><br>Instituto di Ecologia<br>Viala della Scienze<br>43100 Parma |
| 6 | SPANISH GROUP<br>Francisco Rodriguez-Valera<br>Universidad de Alicante | Departamento de Genetica<br>Molecular y Microbiologia<br>Campus de San Juan<br>Apartado 374<br>03080 Alicante |

*Hydrobiologia* **329**: ix–xxviii, 1996.
*P. Caumette, J. Castel & R. Herbert (eds), Coastal Lagoon Eutrophication and ANaerobic Processes (C.L.E.AN.).*



# Eutrophication gradients in coastal lagoons as exemplified by the Bassin d'Arcachon and the Étang du Prévost

Jacques Castel[1], Pierre Caumette[1] & Rodney Herbert[2]
[1] *Laboratoire d'Océanographie Biologique, Université Bordeaux I, 33120 Arcachon, France*
[2] *Department of Biological Sciences, University of Dundee, Dundee DD1 4HN, Scotland*

*Key words:* Coastal lagoons, eutrophication, macrophytes, microorganisms, nitrogen cycle, sulphur cycle

## Abstract

The conditions of eutrophication are described in three lagoon systems differing by their structure, their catchment area and their connection with the sea: the Bassin d'Arcachon on the Atlantic coast, SW France, the semi-artificial fish ponds of the Bassin d'Arcachon, and the Étang du Prévost on the Mediterranean. The Bassin d'Arcachon is a shallow semi-enclosed bay, strongly influenced by climatic factors and tidal currents. The Bassin receives significant inflow of freshwater and the waters are only partially renewed. The greatest part of the primary production is due to the seagrass *Zostera noltii*. Although the ecosystem remains on the whole in steady state, some evidence of potential eutrophication are visible. For instance, the flux of nitrogen into the Bassin d'Arcachon has increased by more than 50% during the last 25 years. The most significant change among primary producers is the massive development since 1988 of the green alga *Monostroma obscurum*. The fish 'réservoirs' of the Bassin d'Arcachon are man-made enclosures designed for extensive aquaculture and where the water renewal is only possible during certain periods of time. Thus, because of the shallowness and the confined nature of these fish ponds, acute eutrophication is sometimes observed in summer. The Étang du Prévost is extremely eutrophic due to agricultural and urban run-off. Red waters occur periodically during the warm summer months as a consequence of ecological events beginning in the early spring with a bloom of green macroalgae (*Ulva* sp.). In summer, the algal biomass is degraded by aerobic heterotrophic bacteria; the oxygen demand encompasses the oxygen production, leading to the predominence of anaerobic processes and dystrophic crisis. From the comparison of the selected sites, three stages of eutrophication are recognized according to the conceptual model of Nienhuis (1992) describing the relation between the relative dominance of primary producers connected to the availability of nutrients. Such macroscopic observations should, now, be explained by the study of microbiological processes including meiofauna, protozoa, bacteria and all the components of the microbial loop.

## Introduction

Littoral ecosystems such as lagoons, estuaries, and salt marshes are highly productive environments but how they function is still not fully understood (Lasserre, 1979a; Lasserre & Postma, 1982). Most of them are subjected to continuous natural modifications. Lagoons, in particular, may undergo periods of disorder due to excessive eutrophication which can lead to dystrophic crisis, the so-called malaïgue in Mediterranean lagoons (Caumette & Baleux, 1980). The control of production and consequently the control of eutrophication are amongst the major problems faced by those responsible for the management of these sensitive ecosystems.

Lagoons have been historically important as sites for human settlement providing access to both the land and the sea. Not only they are important for transportation, they also provide natural food resources, such as oysters, clams, shrimps, fish, as well as providing a convenient place to dump urban and industrial waste. Some of these multiple uses are compatible; others are not. It is a key objective of the management of these systems that the maximum benefit can

*Table 1*. Comparison of the characteristics of seawater and lagoonal water. Example from the Bassin d'Arcachon, SW France (August 1986, after Escaravage, 1990). Seawater = open water in the Bassin d'Arcachon, fish ponds = shallow lagoonal enclosures.

| | Seawater | Fish ponds |
|---|---|---|
| $PO_4$ ($\mu$mol $l^{-1}$) | 0.25 | 1 |
| $NH_4$ ($\mu$mol $l^{-1}$) | 0.78 | 5.62 |
| $NO_3$ ($\mu$mol $l^{-1}$) | 1.14 | 0.30 |
| $NO_2$ ($\mu$mol $l^{-1}$) | 0.26 | 0.15 |
| pH | 8.5 | 7.3 |
| $O_2$ (% saturation) | 87 | 35 |

be derived from these exploitations without jeopardising their long-term future.

## Ecological structure and biogeochemical cycles of eutrophic coastal lagoons

### *Basic model of lagoon ecosystem*

Coastal lagoon ecosystems are directly related to the physical and chemical environment, i.e. coastal lagoons are dynamic, open systems where functions are dominated and controlled by physical processes (Figure 1).

The driving forces of these systems are characterized by:

- flux vectors (currents, tide, solar energy, rain),
- marine inputs (sediment, coastal waters and associated elements such as nutrients, plankton),
- continental inputs (rivers, groundwater, nutrients, sediment, organic matter).

Because of their position as an interface between the terrestrial and marine environments, they are subject to both continental and marine influences. Thus they are highly dynamic environments.

The lagoon ecosystem receives from the marine environment oxidized forms of inorganic nitrogen and phosphorus which are restored as reduced compounds. Ammonia is a preponderant form of nitrogen (Table 1). Its concentration in the sediment is high, due to the mineralization of large amount of organic matter. In contrast, nitrate concentrations are generally low. Low phosphate concentrations recorded in lagoon waters are related to adsorption by particulate matter and the sediments. However, frequent release of $PO_4$ leads to high variation in the measured concentrations.

The continental inputs in the coastal lagoons are mainly characterized by river waters and, sometime, by ground water or rain water that drain the surrounding soils. These waters not only carry large amounts of particulate material in the form of clay particles and organic detritus but also dissolved material in the form of dissolved organic matter and nutrients arising from human activity in the vicinity of the lagoons (fertilizers, domestic and industrial effluents...). Most of these materials are deposited and concentrate in the lagoons.

Nutrient and organic inputs, together with shallow water conditions, good light penetration and good mixing, lead to high primary production. It is widely accepted that the rates of primary production in coastal lagoons are among the highest measured for natural ecosystems (Lasserre, 1979a). The bulk of the primary production is due to macrophytes: phanerogams (*Ruppia, Zostera*) and macroalgae (*Enteromorpha, Ulva*). However, significant production by microphytes (epiphytic and/or epibenthic) have also been recorded.

The pelagic compartment (phyto- and zooplankton) is not the most important component in lagoon ecosystems. The shallowness of the water column associated with the stagnation of the water normally does not allow the planktonic community to develop to any great extent. Most of the micro-invertebrates encountered in the water column are phytophilous or epibenthic.

In lagoons, the benthic compartment largely controls the functioning of the ecosystem. The superficial film of sediment on the lagoon bed is generally covered by a blanket of unicellular organisms (cyanobacteria, peridinians, dinoflagellates, diatoms). Through the intense browsing action of herbivorous fauna such as small gastropods and crustaceans, this vegetal material is converted into increasingly fine debris (Figure 2) which is quickly colonized by a large number of microorganisms (microalgae, bacteria) and small-sized invertebrates (meiofauna, crustaceans, annelids). This colonized detritus constitutes an important food source for higher trophic levels (shrimps, fish, especially mullet).

The excessive primary production leads to high rates of production in the rest of the biological food web in these ecosystems. Thus eutrophication is characteristic of many coastal lagoons which should be considered as frontal ecosystems, which whilst very productive are unstable and sensitive to changes in physical and chemical conditions. Recent research, however, suggests that their biological communities are well adapted to such perturbations and are able to function as apparently stable ecosystems.

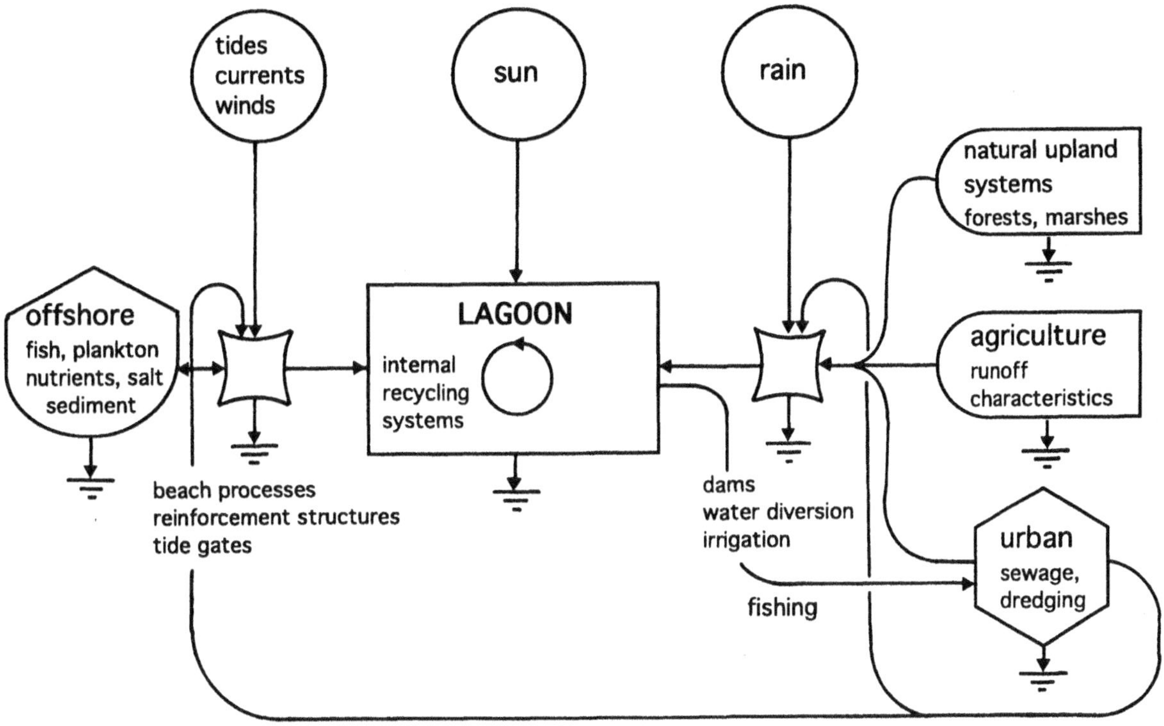


*Figure 1.* The lagoon as concentrating mechanism and controllable interface between land and sea (redrawn from Lasserre, 1979a).

*Microbial communities in coastal lagoons*

Many shallow coastal lagoons exhibit well defined salinity gradients from seawater salinity near the connection with the sea to low salinity in the continental areas of the lagoon receiving river inputs, or to brines in solar evaporation lagoons. Such variable conditions of salinity have a very important impact on the distribution and the selection of the components of the biological communities of coastal lagoons.

In the water column bacterial communities are dominated by aerobic heterotrophic bacteria, either with strictly oxidative metabolism or facultatively fermentative metabolism. These bacteria are halotolerant organisms and are mainly of terrestrial origin (Carmouze & Caumette, 1985). The distribution of marine bacteria depends on the influence of tides (Erchenbrecker & Stevensen, 1975). Although many authors have reported on their distribution and physiology (see references in Caumette, 1989a), little is known about their activity and role in coastal lagoons. They mineralize organic material either in free living conditions, or associated with clays or particulate organic matter. They contribute to the release of inorganic nutrients. It is known that fast growing plants such as phytoplankton, filamentous algae and *Ulva* sp. benefit from the increased levels of nutrients and will grow in volume and dominance at the expense of flowering plants and other algae. They commonly dominate in many coastal lagoons.

In the sediment, the microbial community is more complex depending on the gradients of oxygen penetrating the sediment and subsequently on the interface between oxic and anoxic phases. This interface is generally found within the first 2 mm of the sediment. Therefore many different kinds of semi-aerobic and anaerobic bacteria coexist, living by fermentation or anaerobic respiration. In most coastal lagoons, it is known that the anaerobic bacterial community is dominated by sulphate-reducing bacteria that can transform up to 50% of the organic material (Jørgensen, 1983) but produce $H_2S$, a toxic compound that accumulates as $FeS_2$ (pyrite) in the anoxic environment. If light reaches the anoxic layers, phototrophic sulphide-oxidizing bacteria can grow and sometimes populate these environments by forming dense purple to brown colored masses (Caumette, 1989b; Van Gemerden, 1983). In addition to sulphur bacteria, anoxic layers of coastal lagoons are densely populated by nitrogen bacteria and methanogenic bacteria (Caumette, 1989a). However,

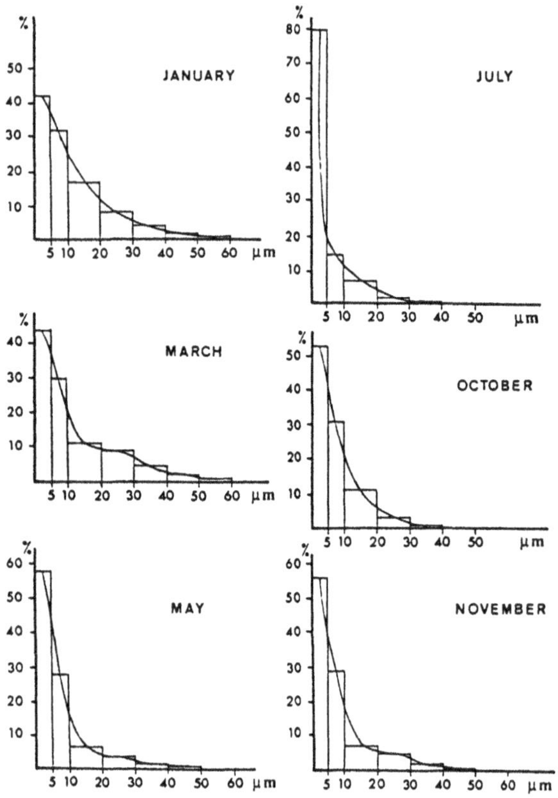


*Figure 2.* Seasonal variation of the particle size (first mm of the sediment) in a semi-enclosed lagoon ecosystem: the fish ponds of the Bassin d'Arcachon (Castel, unpublished).

although many papers describe these bacteria and their role in marine or coastal estuarine environments (Herbert, 1975; McFarlane & Herbert, 1984; Marty et al., 1990), little is known on their distribution and role in coastal lagoons.

Lagoon environments display very high densities of meiofauna composed mainly of nematodes, harpacticoid copepod, oligochaetes, ostracods, turbellarians, and polychaetes (Castel, 1992). These meiofauna components may consume much more energy in the form of organic matter (detritus + microflora + bacteria) than do the macrofauna (Lasserre et al., 1976). Furthermore respiration rates when related to biomass are very high, indicating the importance of the micro- and meiofaunal communities in coastal lagoons (Table 2). One gram of meiofauna respires a rate at least ten times greater than an equivalent macrofaunal biomass. In shallow eutrophic areas, the ratio macrofauna, meiofauna, microfauna, in term of biomass is approximately 100, 10, 1; whereas from a metabolic point of view it is 4, 2, 1.

In coastal lagoons, the trophic structure is characterized by a number of different primary food types and a highly connected food web of generalist feeders. Much of the emergent and submergent plant material enters the food web as detritus. Because of the net nutrient uptake which occurs during the fermentation of detritus, this material may have a higher nutritional value than the original plant material. For both detritus and epiphytes, it is not just plant material, but an entire community which is consumed, including bacteria, fungi, microalgae and protozoans.

The increased pool of autochthonous particulate matter results in intense microbial activity in the surface sediment. This might be further enhanced by benthic-pelagic coupling in the topmost layer of the sediment. Due to the good light conditions, the high primary production resulting from eutrophication exceeds the capacity of the heterotrophic decomposers to completely mineralize the biomass. As a consequence, the pool of particulate organic matter in the sediment is increasing due to the imbalance between production and mineralisation.

In coastal lagoons, the excess detritus is accumulated from late autumn through winter. The eutrophic conditions prevailing in spring and summer contribute to a rapid decomposition of detritus. As a result decomposer production and the microbial loop are also stimulated. This provides an abundant food source on which luxurious development of the meiofauna can occur. In this case, the ratio of detritus to biomass results in an imbalance if the energy released from the detritus by the meiofauna is dissipated too rapidly. These high decomposition rates result in an increased oxygen demand and may lead to a temporary disappearance of dissolved oxygen in the water column (Lasserre et al., 1976). This phenomenon causes mass mortality of the benthic macrofauna and fish.

These summer dystrophic crises lead to the formation of high bacterial biomasses in the lagoons as a consequence of high bacterial activity involved in mineralization (both aerobic and anaerobic bacteria). Such biomass can further be used as a food source by the meiofauna. In many coastal lagoons, particularly at the sediment surface, bacteria play an important role as a food source for higher organisms via the detritic food chain (Coull, 1973). However the quantitative importance of bacteria was not appreciated until relatively recently (Fenchel & Jørgensen, 1977) although some early studies (see Zobell, 1946) emphasized the importance of bacteria in marine food chains.

*Table 2*. Total *in situ* oxygen consumption and relative uptake by micro- and meiofauna (in percent) in man-made lagoons (fish ponds) and salt marshes in the Bassin d'Arcachon (from Lasserre et al., 1976 and Lasserre, 1979b).

| Biotope | Compartment | Oxygen consumption ($ml\ O_2\ m^{-2}h^{-1}$) | | % Micro- and meiofauna | |
|---|---|---|---|---|---|
| | | July | December | July | December |
| Fish ponds | Phytal | 218 | 44 | 17.9 | 20.5 |
| | Benthos | 108 | 28 | 58.3 | 42.9 |
| Salt marshes | Phytal | 99 | 23 | 11.1 | 21.7 |
| | Benthos | 81 | 16 | 16.0 | 25.0 |

Recent studies have shown very high bacterial productivity in coastal lagoons (references in Torreton, 1991) but few describe accurately the interrelationships between such biomass and the consumers in the water column and benthos (Gophen et al., 1974; Caumette et al., 1983; Rieper, 1982; Carman & Thistle, 1985; Decho & Castenholz, 1986; Souza-Santos et al., 1996). Fenchel & Jørgensen (1977) calculated that bacteria in the detritic food chain provided between 60–80% of the diet of the meiofauna. On average, meiofauna graze at a rate of 1% of the standing stock of both heterotrophic bacteria and autrotrophic microalgae per hour (Montagna, 1995). Meiofauna grazing is therefore broadly in equilibrium with microbial production. However, new more accurate data are now required to clarify our understanding of the functioning of coastal lagoons, in terms of carbon turnover, organic matter fluxes via the biogeochemical cycles, and the mineralization and the production of organic material through the food chain.

*Biogeochemical nitrogen and sulphur cycles in coastal lagoons*

The nitrogen and sulphur cycles include a diverse range of oxidation and reduction reactions (Figures 3 & 4). These can be divided into two categories: dissimilatory reactions that are found principally amongst prokaryotes and assimilatory reactions that occur in both prokaryotes and eucaryotes (Ferguson, 1988).

In both cycles, the production of reduced compounds such as $NH_3$, $NO_2^-$, $H_2S$, $S^0$ are either the end products of dissimilatory anaerobic respiration or derive from aerobic and anaerobic degradation of organic material (Herbert, 1982; Caumette, 1986; Herbert & Nedwell, 1990). These compounds accumulate in the anoxic layers mainly in the sediment of eutrophic coastal lagoons. Their oxidation is dependent on the redox state of the sediments and overlaying water column. It is well established that their oxidation is mainly mediated by chemotrophic prokaryotes that use oxygen as terminal electron acceptor.

Such organisms live around the redoxcline of the lagoon ecosystem, depending on reduced compounds from the anoxic layer and oxygen from the overlying oxic layer. Thus reduced nitrogen compounds are generally oxidized via the activity of ammonia- or nitrite-oxidizing bacteria. In contrast, reduced sulphur compounds can be oxidized under anoxic conditions. Chemotrophic bacteria that perform anaerobic respiration such as *Thiobacillus denitrificans* generate a proton electrochemical gradient as electron flow from reduced sulphur compounds to nitrate, nitric or nitrous oxides. Moreover, in the upper part of the anoxic layers reached by light, phototrophic sulphur oxidizing bacteria are able to oxidize reduced sulphur compounds to sulphate as a consequence of their anoxygenic photosynthesis. Thus, in coastal lagoons, both cycles are functioning in similar conditions: reduction of nitrogen or sulphur compounds occurs under anoxic conditions whereas oxidation of the reduced compounds mostly occur at the interface between oxic and anoxic zones, i.e. the redoxcline. Almost all eutrophic shallow water coastal lagoons have a redoxcline occurring at the sediment surface. Therefore most of these metabolic processes coexist within the first millimeters of the sediment. In eutrophic coastal lagoons, ammonia is the major nitrogen compound produced in anoxic layers (Koike & Hattori, 1978; Nishio et al., 1982, 1983; Marty et al., 1990). However, production of reduced nitrogen compounds is lower compared to that of reduced sulphur compounds which are derived from sulphate reduction which is the most important anaerobic respiratory process occurring in coastal lagoons (Caumette, 1986) and coastal marine environments (Jørgensen,

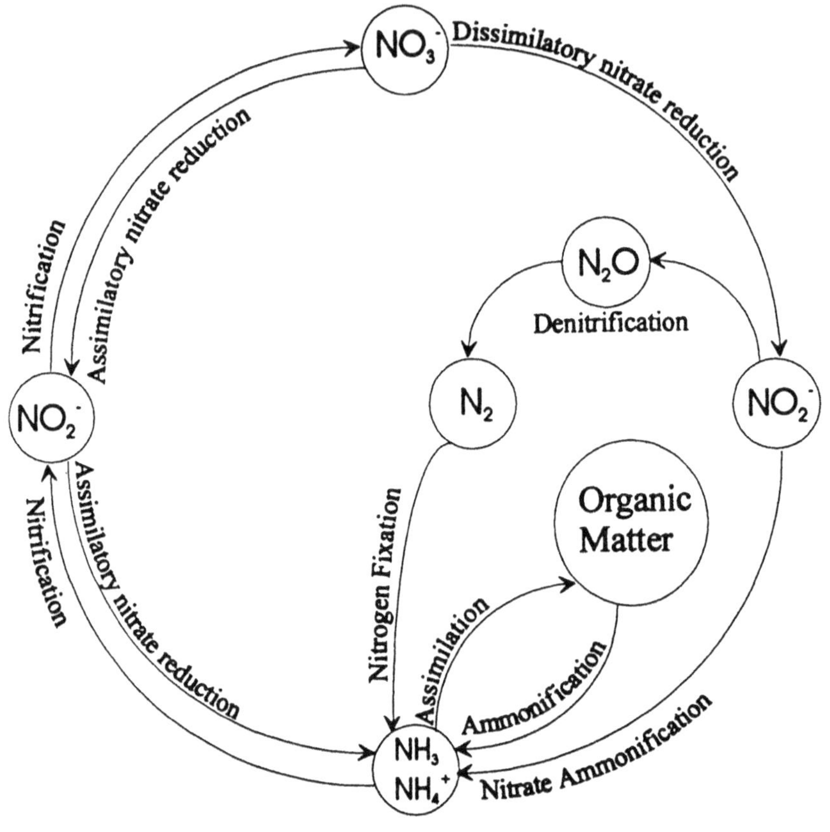


*Figure 3.* Nitrogen cycle.

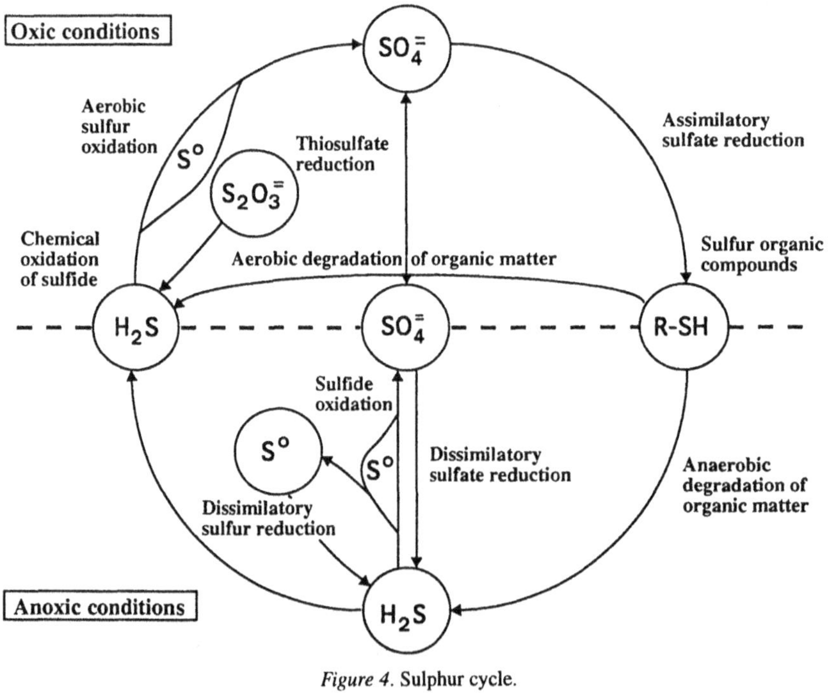


*Figure 4.* Sulphur cycle.

1982, 1983). The bacterial oxidation of reduced nitrogen or sulphur compounds at the sediment surface prevents their diffusion into the overlying oxic layer (the water column). Little is known on the oxidation of nitrogen compounds in coastal lagoons and field work as well as laboratory work are required to determine and quantify the oxidative pathways of reduced nitrogen compounds. In contrast, it is established that in most of the coastal sediments studied so far, between 50 to 95% of the $H_2S$ produced is reoxidized, at the interface between the oxic and anoxic zones. However, sulphide does not usually reach the oxic zone unless sulphate reduction is very intensive.

Thus, it appears that the abiotic and biotic equilibrium in coastal lagoons is primarily dependent on the balance between oxidation and reduction activities in the biogeochemical cycles. When reduction of nitrate or sulphate in the anoxic layers is enhanced, the reduced compounds produced can diffuse into the oxic layers leading to the establishment of anoxic conditions and the release of reduced nitrogen and/or sulphur compounds to the atmosphere. In recent years, attention has been paid to volatile methylated sulphur compounds and their metabolism at the sediment surface of coastal environments. In coastal lagoons, their intensive production and transformations may have important consequence on the global sulphur cycle and the atmospheric behaviour of sulphur. It is known that these conditions develop when organic matter accumulates and the subsequent activity of mineralization processes are stimulated in the water column and sediments. However, detailed field experiments on the oxidation and reduction processes taking place in both cycles at the sediment surface are required in order to understand the development of such drastic events in coastal lagoons.

In many shallow coastal lagoons, much of the surface light irradiance reaches the sediment surface leading to the development of photosynthetic benthic communities which in turn may lead to the development of microbial mats. These mats, which are characteristics of many coastal lagoons, are composed of different laminated layers of oxygenic and anoxygenic phototrophic bacteria (cyanobacteria and different kinds of purple or green sulphur oxidizing bacteria) depending on oxygen and sulphide microgradients in the upper sediment. It has been recently shown that methylated sulphur compounds are very important in the sulphur cycle that takes place in microbial mats, at the oxygen sulphide interface (Visscher et al., 1991).

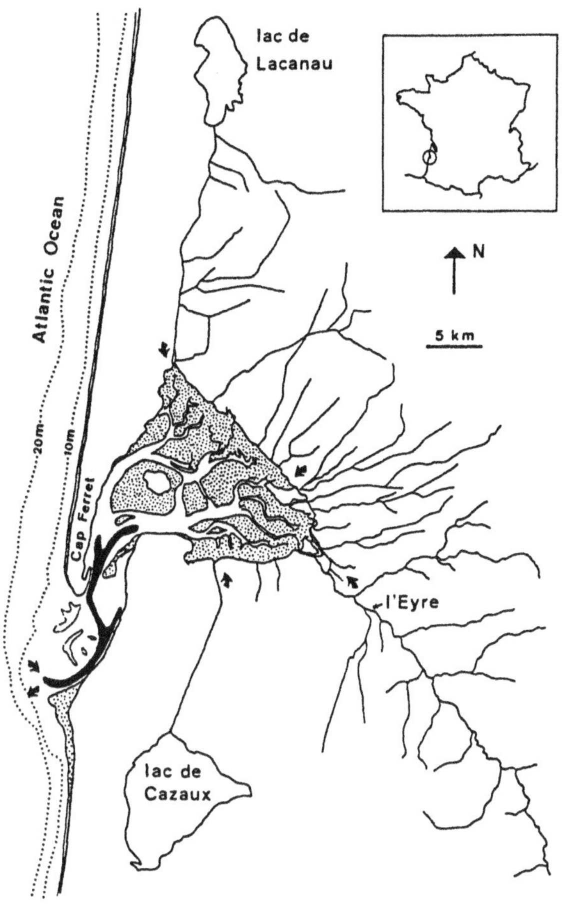


*Figure 5.* Location of the Bassin d'Arcachon along the Atlantic coast, showing the drainage basin and the most important inputs of freshwater.

## Examples of eutrophication gradients in coastal lagoons

### *The Bassin d'Arcachon, a moderately eutrophic lagoon*

#### *General hydrography*

The Bassin d'Arcachon (44°40′ N, 1°10′ W) is a triangular shaped embayment on the South-West Atlantic coast of France (Figure 5). Channels and intertidal areas cover 155 km$^2$ but only 40 km$^2$ of the bay are subtidal. Seventy percent of the lagoon is composed of intertidal flats called 'crassats' which are used for oyster farming. The intertidal areas, especially in the eastern half of the Bay are covered by *Zostera noltii* Hornem. whilst *Zostera marina* L. is found in the channels. The oyster beds cover an area of 10 km$^2$. The decaying eelgrass and detritus from the oysters provides a rich supply of organic nutrients for the intertidal sediments leading to an enhancement of the abundance of the small invertebrate fauna (Castel et al., 1989). The Bassin d'Arcachon is connected to the Atlantic Ocean at the South-West end by two narrow channels (4–5 m deep at low water). The presence of sandbanks at the entrance to the lagoon have a major effect on water exchange between the bassin and the Atlantic. On a spring tide $370 \times 10^6$ m$^3$ of water are exchanged whilst on a neap tide this is reduced to $130 \times 10^6$ m$^3$. These volumes are similar to those calculated by Caspari in 1863 (*cit. in* Labourg, 1985). This shows how stable the tidal influence has been over more than a century. The exchange of such large volumes of water induces considerably high water flow: the mean discharge through the channels is $17 \times 10^3$ m$^3$s$^{-1}$ (comparable to the average discharge of the St Lawrence). These waters carry a great amount of sand in suspension particularly on the ebb during spring tides. Sediment transport has been estimated to be as much as 11 500 t through the channels on a spring tide. In contrast, in the inner bay, resuspension is much less; on average the concentration of suspended matter is around 3–7 mg l$^{-1}$. At Arcachon Eyrac, the tidal range is 4.9 m on a spring tide and 1.1 m on a neap tide. Tidal currents are strong, reaching velocities of 2 m s$^{-1}$ in the channels.

The oceanic water entering the bassin is diluted by freshwater inflow, mainly from the northern and eastern parts of the bassin ($1.8 \times 10^6$ m$^3$d$^{-1}$). Furthermore, ground waters provide approximately $10^6$ m$^3$d$^{-1}$. As a consequence, salinity and temperature variations are directly proportional to the distance 'upstream' from the inlet. In summer the water temperature reaches 21–22 °C in August, falling to 6–8 °C in mid-winter. Salinity ranges from 30–33‰ at high water but occasionally decreases to 20‰ after periods of heavy rain. In spring and summer the temperatures are notably higher in the inner bay than in the outer channels: the converse is observed in autumn and winter. Temperatures are uniform in the bassin in February-March and in September-October. During these periods water temperatures of the Bassin d'Arcachon and the Bay of Biscay tend to be similar. This has strong ecological implications, especially in spring when migrating species (cephalopods, fish) enter the bassin.

Three distinct water masses have been recognised in the Bassin d'Arcachon (Bouchet, 1968; Figure 6):

- External neritic waters: temperature: 9.5–21 °C, salinity: 34–35‰
- Intermediate neritic waters: temperature 6–22.5 °C, salinity: 26.8–33.2‰

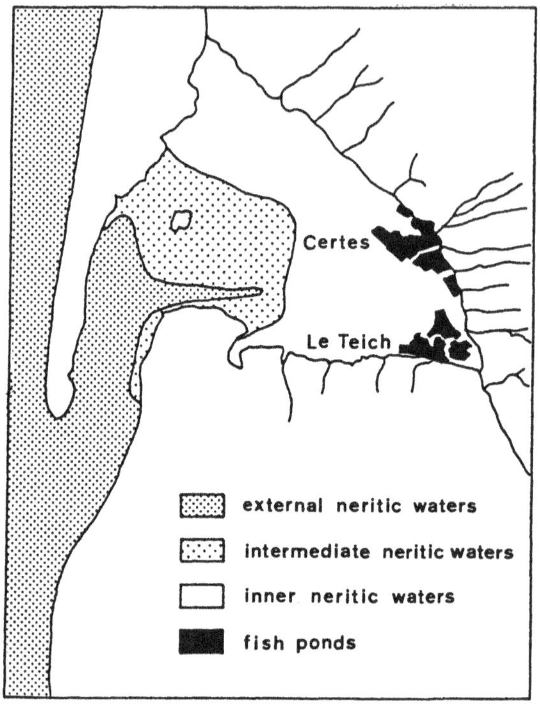


*Figure 6.* Distribution of the water masses in the Bassin d'Arcachon according to salinity (modified from Bouchet, 1968). Fish ponds are shallow brackishwater enclosures.

- Inner neritic waters: temperature: 1–25 °C, salinity: 22–32‰

These water masses move according to the tidal state and their renewal is only partial. In the North-East and eastern sectors of the lagoon, water exchange with the Atlantic is occasional, probably once or twice a year. The inner waters are eliminated from the Bassin as a laminar flow or nappes at the surface. The intermediate body of water oscillates and tend to mix with the inner waters whilst the water mass associated with the deep channel is well mixed with Atlantic water during each tidal cycle.

Whilst the Bassin d'Arcachon is continuously changing, the system as a whole remains in a steady-state. The lagoon is subject to erosion but this is balanced by sedimentation. The most important changes occur near the inlet and the adjacent coast area:

- slow advance and sporadic erosion of the Cap Ferret spit
- migration of the navigation channel and associated sand banks to the South
- erosion of the southern coast.

These changes tend to reduce exchange between the Bassin and the Atlantic. In the inner bay the sed-

*Table 3.* Annual flux of mineral nitrogen (in tonnes) originating from the drainage basin in the Bassin d'Arcachon (Auby et al., 1994).

| Origin | Years | | |
|---|---|---|---|
| | 1970 | 1980 | 1990 |
| Forests | 263 | 257 | 252 |
| Agriculture | 281 | 415 | 575 |
| Lakes | 17 | 66 | 27 |
| TOTAL | 561 | 738 | 854 |

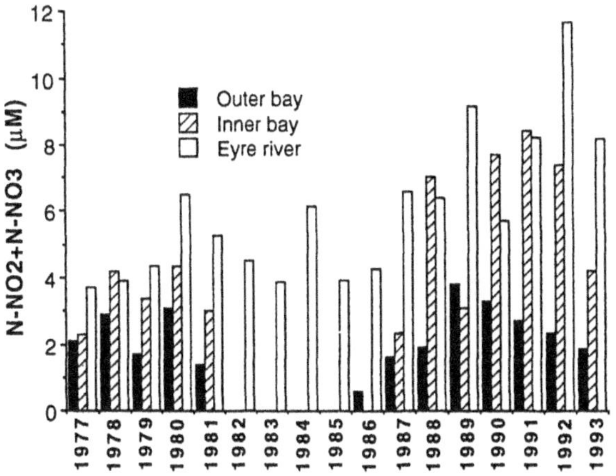


*Figure 7.* Long-term distribution of nitrite + nitrates in the Bassin d'Arcachon (outer and inner water masses) and in the river Leyre (modified from Auby et al., 1994).

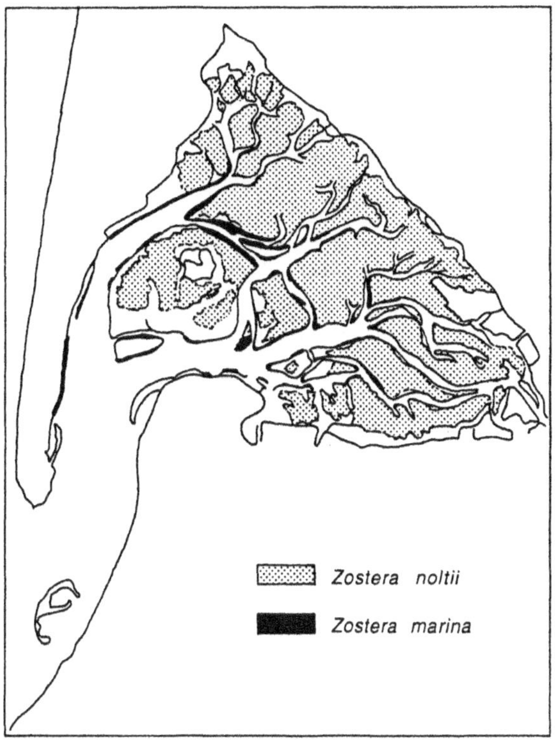


*Figure 8.* Location of the seagrass beds of *Zostera noltii* and *Z. marina* in the Bassin d'Arcachon (redrawn from Auby, 1993).

imentation rate is quite low (10 cm per century today cf. 1 m per century 1700 years ago).

*Evidence of eutrophication*

The catchment of the Bassin d'Arcachon covers an area of 4140 $km^2$ (Figure 5). Approximately a quarter of the catchment can be considered to indirectly contribute nutrients to the Bassin through coastal lakes. Since the 1970's intensive agriculture in the region (especially maize) has increased. In 1970 agriculture was responsible for 50% of the annual flux of nitrogen to the Bassin d'Arcachon; at present it is responsible for 66% of this flux (Table 3). The annual flux of total phosphorus remains quite stable (25–30 t $yr^{-1}$). Urban sewage has significantly decreased during the last twenty years (from 127 t total N $yr^{-1}$ to 40 t total N $yr^{-1}$).

Most of the freshwater flux comes from the river Leyre (see Figure 5) thus it is not surprising that nutrient concentrations have increased during in recent years (Figure 7). Although there is a dilution effect from the river input to the outer region of the Bassin, the nutrient concentrations have clearly increased in a large part of the Bassin d'Arcachon (by a factor 2 in the inner Bassin from 1977–1981 and 1990–1993). Such increases could have stimulated the primary production and thus eutrophication. Measurements made since 1976 (Castel, unpublished; Robert et al., 1987) indicate however that the standing stock of phytoplankton (expressed in chlorophyll *a*) is not particularly high and remains at a relatively constant level from one year to another: the baseline value is around 2 $\mu g\ l^{-1}$ and the spring bloom does not exceed 15 $\mu g\ l^{-1}$.

The most abundant primary producer in the Bassin d'Arcachon is the seagrass *Zostera noltii* (Table 4). The seagrass beds occupy 7000 ha in the intertidal zone (Figure 8) which represents the largest area in Europe. In contrast to other places in Europe where the biomass of *Z. noltii* has declined in recent years, the seagrass meadows in Arcachon have remained almost constant over the last 30 years. Mean annual biomass is 70–100 g DW $m^{-2}$ for the leaves and 70–160 g DW $m^{-2}$ for the roots and rhizomes (Auby, 1991).

*Table 4.* Estimated annual production of the different primary producers in the Bassin d'Arcachon (from Auby et al., 1993).

| Taxa | Total production (t d.w. $yr^{-1}$) | Carbon (t $yr^{-1}$) | Nitrogen (t $yr^{-1}$) | Phosphorus (t $yr^{-1}$) |
|---|---|---|---|---|
| Halophilous phanerogams | 7612–9098 | 3045–3636 | 537–686 | 73–93 |
| *Zostera noltii* | 30, 790–43,700 | 9275–13,300 | 660–960 | 70–100 |
| *Zostera marina* | 6213 | 2003 | 157 | 15 |
| *Monostroma obscurum* | 7600 | 2508 | 342 | 23 |
| Other macroalgae | unknown | unknown | unknown | unknown |
| Microphytobenthos | | 4930–12,270 | 860–2140 | 120–290 |
| Phytoplankton | | 3540 | 625 | 85 |

The subtidal eelgrass *Zostera marina* occupies 4.26 $km^2$ in the channels (Figure 8). It constitutes a unique refuge for invertebrates as well as fish.

The most significant change among primary producers is the development, since 1988, of the green alga *Monostroma obscurum* (= *Ulvaria obscura* Kützing). This species was first described in 1843 on the Basque coast. It is a cosmopolitan species found in Northern Europe as well as in the Pacific. Since the early 1980's, fishermen from Arcachon have exploited the natural deposits of oysters near the Adour estuary, close to the site where this species was first described. The oysters were transported to Arcachon for growth, without cleaning the shells. It is probable that some thalli were imported together with the oysters. The rapid development of *Monostroma* shows that this alga finds environmental conditions in the Bassin ideal for growth. Although no causal relationship can be demonstrated, it is interesting to note that the development of *Monostroma* coincided with the increased nutrient input into the Bassin d'Arcachon. In spring, the total biomass of *Monostroma* has been estimated to be between 18 000 to 21 000 t. The alga mainly colonizes the inner region of the Bassin, in the intertidal zone as well as in the channels (Figure 9). The maximum density is around 6 kg W W $m^{-2}$, which is comparable to other macroalgal blooms (Étang du Prévost, *Ulva*: 5 kg W W $m^{-2}$; Baie of St Brieuc, *Ulva*: 8 kg W W $m^{-2}$) although it is less than the massive development of *Enteromorpha* and *Ulva* observed in the lagoon of Venice (30 kg W W $m^{-2}$). Arcachon is the only place in the world where a bloom of *Monostroma* has been observed.

Changes in animal populations and other algae have also been observed (Labourg, 1985). It is not always possible to differentiate between the impact of natural variation of the environment and the effect of

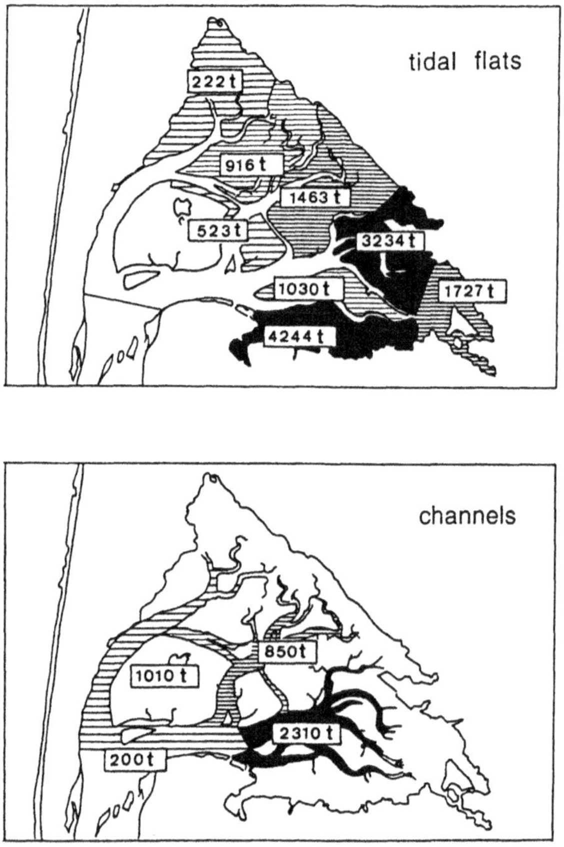


*Figure 9.* Total biomass (tonnes wet weight) of the green alga *Monostroma obscurum* in the Bassin d'Arcachon in June 1993 (redrawn from Auby et al., 1994).

the human activities, however, it is clear that some species have disappeared and some appeared during the last century. The barnacle *Elminius modestus* Darwin, originating from Australia, after colonizing the Mediterranean coasts, appeared in Arcachon in 1960. The Mediterranean balist *Balistes capriscus* L. was reported in the bassin in 1962. These are example of species having extended their area of distribution. Other species have appeared since 1968 as a consequence of the importation of the Pacific oyster *Crassostrea gigas* Thunberg, including the tunicate *Styella clava* Herdman, some Bryozoans and Annelids. More recently, the large brown alga *Sargassum muticum* (Yendo) Fensholt accidentally introduced in Great Britain with the Pacific oyster, invaded the coast of Britanny and appeared some years ago in Arcachon. Even small organisms (< 1 mm) such as the Harpacticoid Copepod *Stenhelia latioperculata* Itô (originally from Japan) are likely to have been introduced with oysters. Conversely, some species have disappeared due to either modifications of hydrological conditions (e.g. the clam *Venus verrucosa* Linné) or to overexploitation (e.g. the bivalve *Chlamys varia* Linné).

*Fish ponds of the Bassin d'Arcachon, a moderately eutrophic lagoon system*

*General description*

The 'fish reservoirs' of the Bassin d'Arcachon (Figure 10) are man-made enclosures created in the lagoonal marshes (wetlands) and where a number of euryhaline fish are farmed: grey mullet (*Chelon labrosus* Risso, *Liza ramada* Risso), sea bass (*Dicentrarchus labrax* Linné), eels (*Anguilla anguilla* Linné) and gilt head bream (*Sparus aurata* Linné). Such fish reservoirs are also known from the Charente and Vendée along the Atlantic coast. These structures, designed for traditional extensive (without food supply) aquaculture, are comparable to the 'valli', situated along the Adriatic coast, and also to the tropical 'tambaks' in Indonesia. They constitute mixohaline and shallow (0.2–1.5 m depth) environments where the rich input of detritus plays a prominent role in the food chain. Originally they were salt-pans exploited since the end of the Middle Ages. Progressively, during the late eighteenth century the exploitation of the salt decreased and the salt-pans were converted into fish ponds. These fish reservoirs schemes flourished all through the nineteenth century up to the end of the Second World War but they have since gone into decline because of the labour costs and the relatively low yield (50 kg $ha^{-1}$) of this type of extensive aquaculture. These semi-enclosed lagoonal systems cover a surface area of approximately 940 ha in the Bassin d'Arcachon. The old salt-evaporation areas and the salt-pan runoff ditches are separated by embankments from the sea and through which inflow and outflow are regulated by sluice gates (Figure 11). The fish ponds or fish reservoirs have a characteristic shape: channels or ditches, 1.5 to 2 m in depth, feed large expanses known as 'flats', each of which covers an area ranging from 1000 to 10 000 $m^2$. In the canals the fish are 'penned' during the winter months to counteract the colder temperatures prevailing in the shallow areas. These flats, 20–50 cm deep, are separated from each other by ridges formed from the mud removed during the digging of the ditches. All inflows and outflows can be regulated by sluices located at intervals along these embankments.

At high tide, the sluices are manoeuvred in such a way that a current is created and the fish, both immature and adults, are drawn into the reservoirs. The fish are prevented from returning to the sea by a system of mesh frames. These operations are usually carried out during the cooler seasons, from March to November. Every two weeks, when high tides occur at the time of the full or new moon, water from the sea is allowed to flow into the reservoir through the sluices. This operation lets in the young fish and natural mineral salts, and replenishes the reservoir by exchanging the water held in the ponds and fresh seawater. This operation is carried out at low tide. The young fish are free to move and to grow inside the complicated network of basins, in which the salinity may vary from very dilute brackish water to full seawater.

Because of the shallowness and the confined nature of the fish ponds, the salinity regime is extremely variable both in time and space. The salinity ranges from almost freshwater to hypersaline (60‰). The salinity regime of the fish reservoirs is strongly linked to (1) the relative location of the ponds between the sea and the continental freshwater inputs, and (2) the renewal of the water through the sluices. The man-induced renewal of water is supposed to maintain salinity compatible with biological activities. However, the maintenance of a given salinity requires a careful and periodic operating of the sluices since, due to the shallowness of the ponds, the salinity tends to vary widely (Figure 12) between periods of renewals. The same situation can also be observed for nutrients (Escaravage, 1990). In such ponds, the greatest part of the regulation of nutrient concentrations is controlled by *in situ* biochemical processes, especially at the benthic level. These mechanisms are hardly affected by the renewal of the overlaying water.

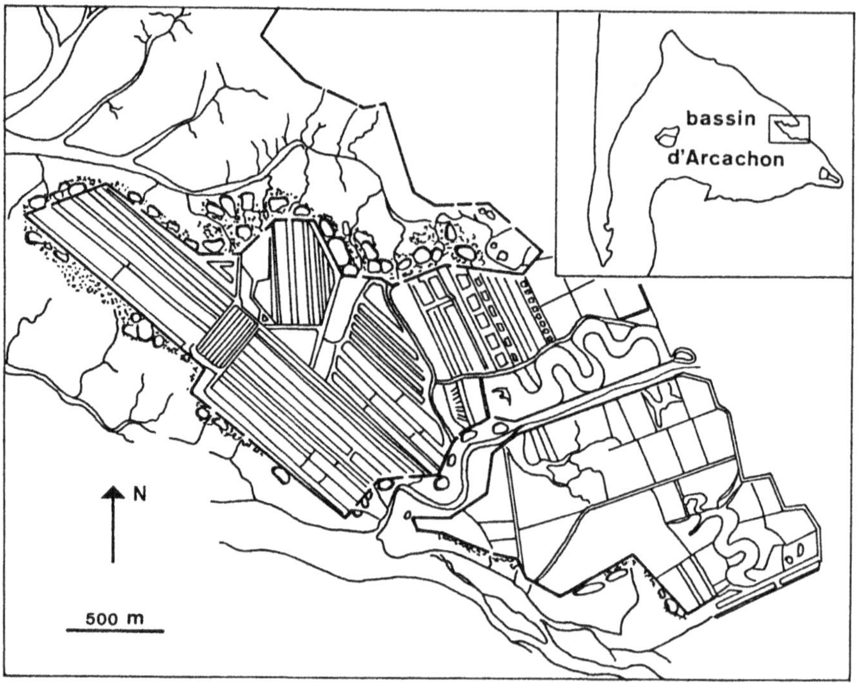


*Figure 10.* Map of the fish ponds of the Bassin d'Arcachon (Certes reservoirs).

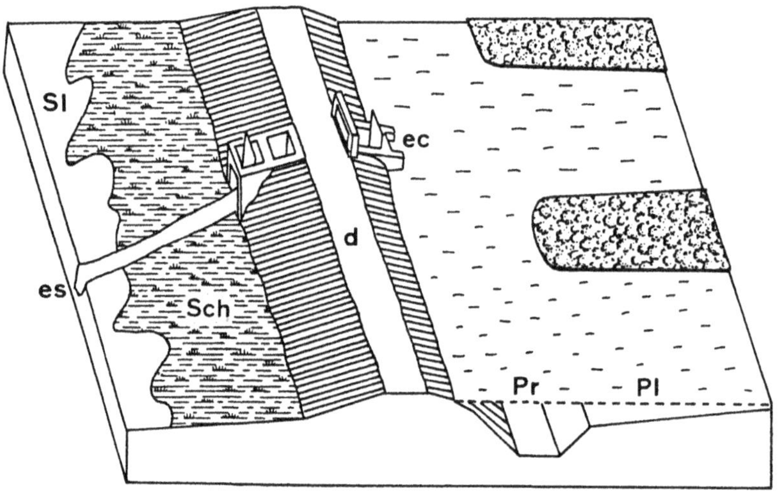


*Figure 11.* Schematic diagram of the fish ponds of the Bassin d'Arcachon. Left: salt marsh comprising the 'slikke' (Sl) = mudflat, the 'schorre' (Sch) = high marsh and 'estey' (es) = small channel. The ponds are separated from the salt marsh by a dyke (d) along which a sluice (ec) is established. Right: typical structure of the reservoirs, Pr = channel (width: 3–4 m, depth: 0.8–1.5 m), Pl = shallow basins (width: 10–40 m, length: 100–800 m, depth: 0.2–0.5 m). Redrawn from Lasserre (1979a) and Castel (1989).

*From eutrophy to dystrophy*

As in many inshore areas, the rich input of detritus, organic matter, bacteria and benthic microflora play a prominent role in the food chain of the fish ponds. The most important primary producers are sea grasses (*Ruppia cirrhosa* Petag), and filamentous green algae: *Cladophora vadorum* (Aresch.) Kütz at the water surface, and mats of green algae (*Lamprothamnium papulosum* J. Groves, *Chaetomorpha aerea* (Dillwyn) Kütz, or cyanobacteria at the sediment surface. The biomass of *Ruppia cirrhosa* has been estimated to 126 g W W $m^{-2}$ for the above-ground and to 51 g W W $m^{-2}$ for the

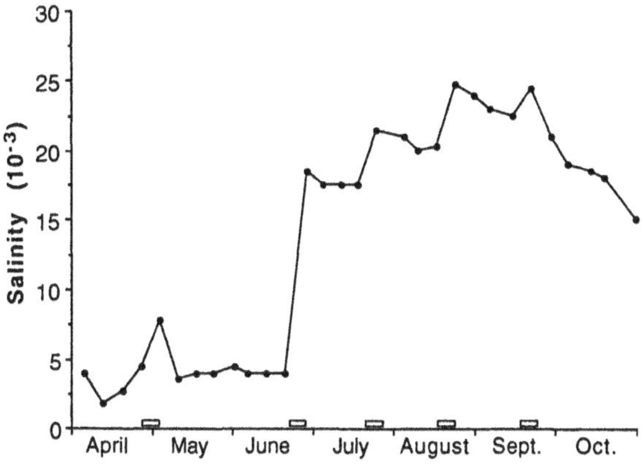


*Figure 12.* Temporal variation of salinity in a fish pond of the Bassin d'Arcachon (Le Teich in Figure 6, April-October 1974). Measurements were made 100 m far from a sluice. The bars on the x-axis indicate the periods of water renewal (Castel, unpublished).

below-ground structures (Soriano-Sierra, 1988). This biomass apperas to be relatively stable from year to year and has on average standing stock of 149–186 g W W $m^{-2}$ (Labourg, 1979). Green macroalgae are strongly dominated by *Chaetomorpha area* with an average biomass of 79 g W W $m^{-2}$. Macro-epiphytes (*Cladophora vadorum*, *Rhizoclonium kernerii* Stockmayer) are found on the stalks and blades of *R. cirrhosa*. Their average biomass has been estimated at 68 g W W $m^{-2}$ (Soriano-Sierra, 1988).

Temperature is probably one of the principal factors conditioning the eutrophication processes and further the dystrophic crisis. In the fish ponds, the variations in temperature depend both on the sun light and renewal of the waters. Each time seawater is allowed to enter the pond the temperature significantly decreases (Figure 13). In contrast, on a neap tide, when there is no entry of seawater, the temperature rapidly increases, reaching sometimes values close to 30 °C. Generally, production of hydrogen sulfide is observed for temperature reaching 24–25 °C, at the end of the neap tide period, when the water is stagnant. Concurrently (even before the production of $H_2S$) the oxygen concentration drops to very low values. pH values also tend to decrease due to a hyper production of acids (organic acids, sulfate, sulphur). A biological indication of eutrophication is the presence of Peridinians in the plankton, usually dominated by diatoms in the fish reservoirs (Castel, 1978).

In some years, during warm summer, true dystrophic crises may occur, with the formation of white waters due to the precipitation of carbonates. During such periods, high mortalities of fish may be observed

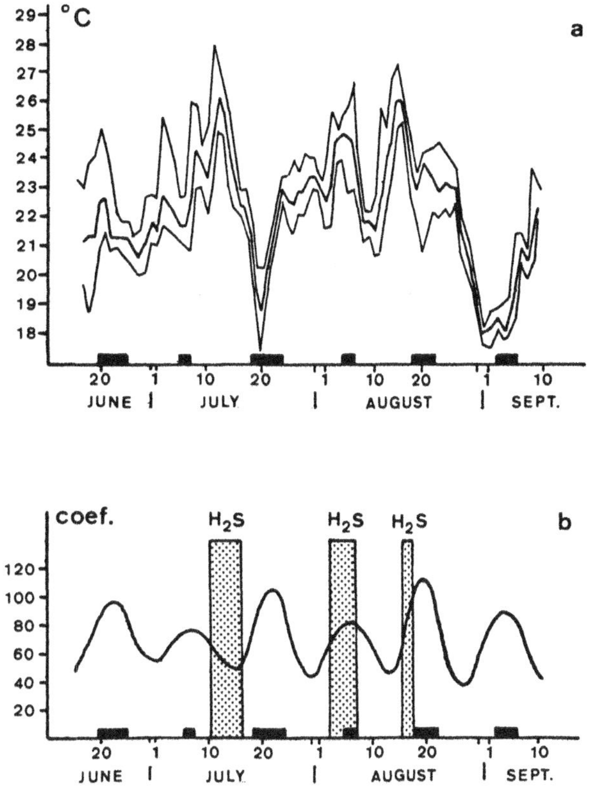


*Figure 13.* a) Daily variation of temperature (mean, maximum, minimum) during the period 19 June – 11 September 1974 in a fish pond of the Bassin d'Arcachon (Certes in Figure 6). Measurements were made 20 m far from a sluice at 0.40 m depth. b) Daily variation of the tide coefficient (proportional to the height of water) during the same period. Periods of $H_2S$ production are indicated by vertical shaded bars. The horizontal bars on the x-axis indicate the period of water renewal (redrawn from Castel, 1978).

(Labourg, 1975; Castel, 1978). Several tonnes of dead fish have been collected during particularly acute crises. However, such dramatic events do not occur every year. Generally, dystrophic crises in the fish ponds of the Bassin d'Arcachon are less acute and less extensive in space than in the Mediterranean lagoons.

### *The Étang du Prévost, a strongly eutrophic lagoon*

#### *General hydrography*

The Prévost Lagoon (43°30′ N, 3°54′ E), located on the French Mediterranean coast, belongs to a lagoon system that extends along the coast, between Sète and Montpellier. This littoral zone is formed by a succession of coastal brackish ponds ('étangs'), surrounded by marsh and separated from the sea by a low, sandy beach allowing only limited amount of

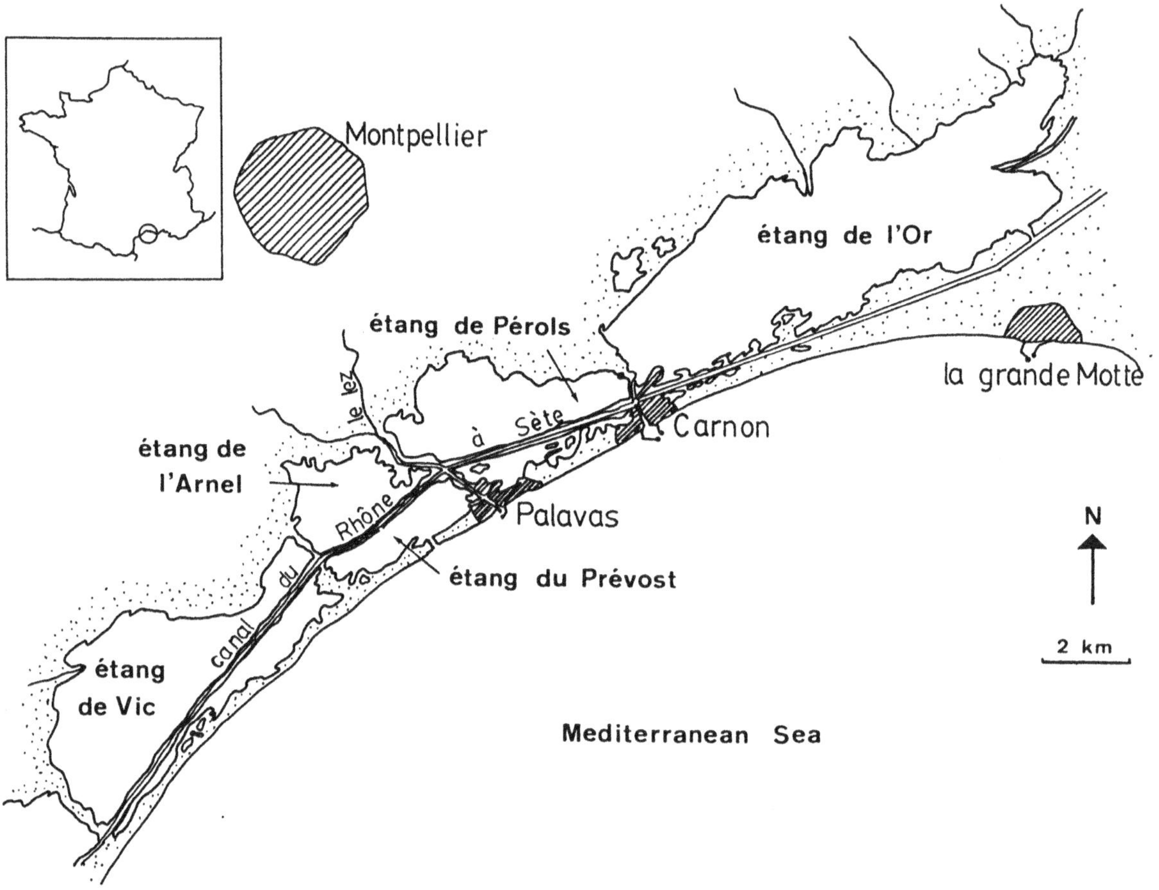


*Figure 14.* Location of the principal lagoons along the Languedocian Mediterranean coast.

water exchanges. The Étang du Prévost is situated in a lagoonal complex delimited to the South-West by the Étang de Thau and the Sète promontory and to the North-East by the wetlands of the Camargue and Rhône delta (Figure 14). These ponds are of recent geological formation. The sandy and detrital zones in this region explain their continental formation. The creation by the currents of offshore bars ('cordons littoraux') has allowed the transformation of the coastline and the formation and evolution of these coastal lagoons since the quaternary period.

The water in these lagoons is slowly renewed, via small channels ('graus') communicating with the Mediterranean sea. These shallow bodies of water are subjected to continental inputs. They are directly influenced by urban or industrial effluents.

In contrast to the Bassin d'Arcachon, the Étang du Prévost is a shallow lagoon with an average depth of about 0.8 m and a surface area of 380 ha (Figure 14). It receives fresh water and sewage carried by a small river 'Le Lez', and seawater through the only connection with the sea 'Le Grau du Prévost'. Although the lagoon is situated in a nontidal zone, the water level can vary from –0.30 m to +0.40 m relative to the mean level because of the strong winds. In the meantime, the current speed can reach 0.40 m $s^{-1}$ in the lagoon and 0.80 m $s^{-1}$ in the 'grau'. When the wind is blowing, 20 to 25% of the water can be renewed in one day. In contrast, during calm periods, the water remains stagnant. This is particularly true in summer. The wind action may induce very strong and rapid variations of the salinity (Figure 15). In summer, because of the evaporation, the salinity can reach 40‰.

The system is extremely eutrophic due to agricultural and urban run-off. For the Étang du Prévost, 85–90% of the input of N and P are of domestic origin (Table 5). The input per volume unit of lagoonal water is approximately 24.6 t $10^6$ $m^{-3}$ for the total nitrogen and 4.3 t $10^6$ $m^{-3}$ for the total phosphorus (Anony-

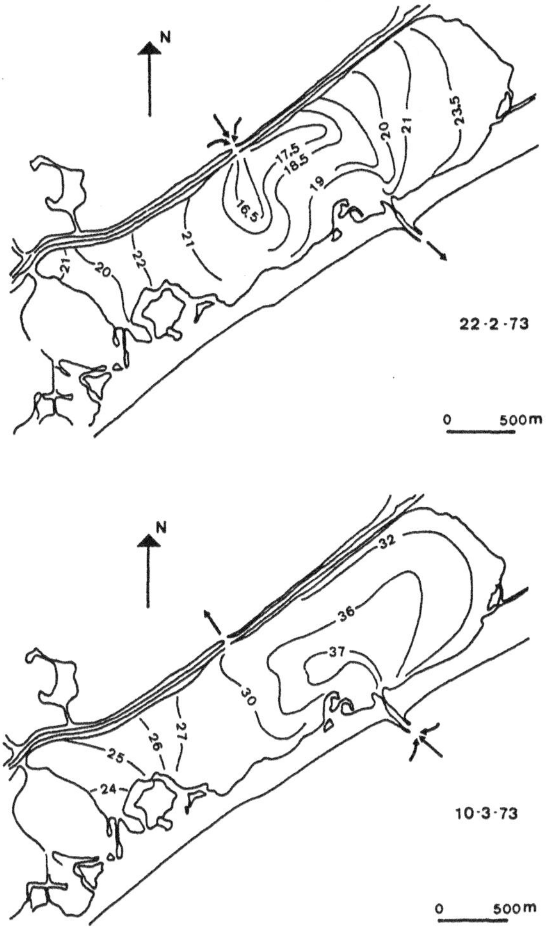


*Figure 15.* Salinity fields (‰) in the Prévost lagoon during two different climatic conditions: wind blowing from the Northeast (22.02.1973) and wind blowing from the Southwest (10.03.1973) (redrawn from Guélorget & Perthuisot, 1992).

*Table 5.* Annual input and stocks of nitrogen and phosphorus, expressed in tonnes, in the Étang du Prévost (anonymous, 1991).

| Origin | Nitrogen | Phosphorus |
|---|---|---|
| Domestic input | 60.2 | 11.1 |
| Agriculture | 5.5 | 0.5 |
| Industry | 1.4 | 0.2 |
| Sediment stock (10 cm) | 475 | 140 |
| Annual flux in sediment | 22 | 2.4 |
| *Ulva* stock | 20.6 | 1.8 |

mous, 1991). As a comparison, the input of nitrogen in the Bassin d'Arcachon is only 3 t $10^6$ $m^{-3}$.

Such large inputs of nutrients lead to the development of high biomass of primary producers in the

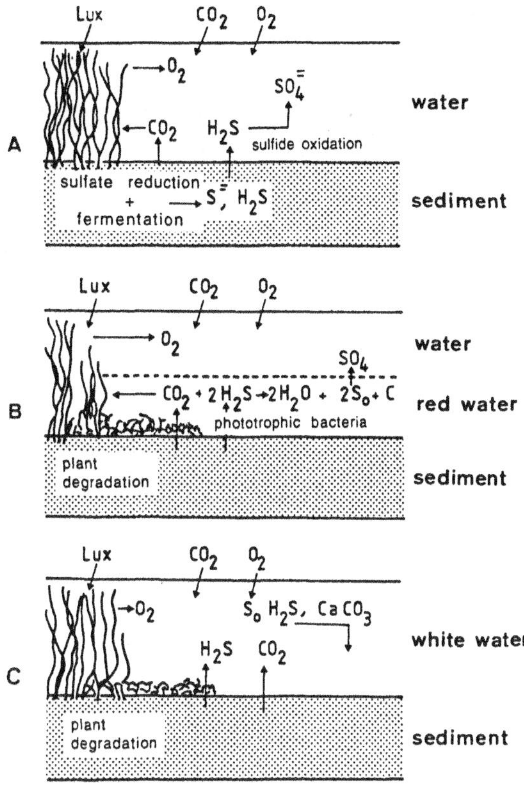


*Figure 16.* Schematic description of the biogeochemical processes occurring during a dystrophic crisis (redrawn from Baleux et al., 1979). A: Equilibrium state of the sulphur cycle, B: Dystrophic crisis with formation of 'red waters', C: Dystrophic crisis with formation of 'white water'.

Étang du Prévost. Its flora is essentially composed of green algae (*Ulva* and *Enteromorpha* spp.). The biomass of these algae may reach 5 kg W W $m^{-2}$ and sometimes accumulates at densities up to 30 kg W W $m^{-2}$. In summer, this biomass is rapidly degraded by aerobic heterotrophic bacteria whose numbers and activities increase rapidly. During this period the water turns anoxic and becomes rich in sulphide, which leads to severe dystrophic crisis.

### *The dystrophic crisis*

The Étang du Prévost is regularly subject to dystrophic crises, locally called the 'malaïgues'. Such dramatic events appear every year during the warm summer months for a period of about 15 days. They have a dramatic consequence for aquaculture, particularly oyster culture that takes place in the lagoon. Macroscopically, the dystrophic crises are characterized by coloured waters (white, red or black waters). They originate from a perturbation of the sulphur cycle.

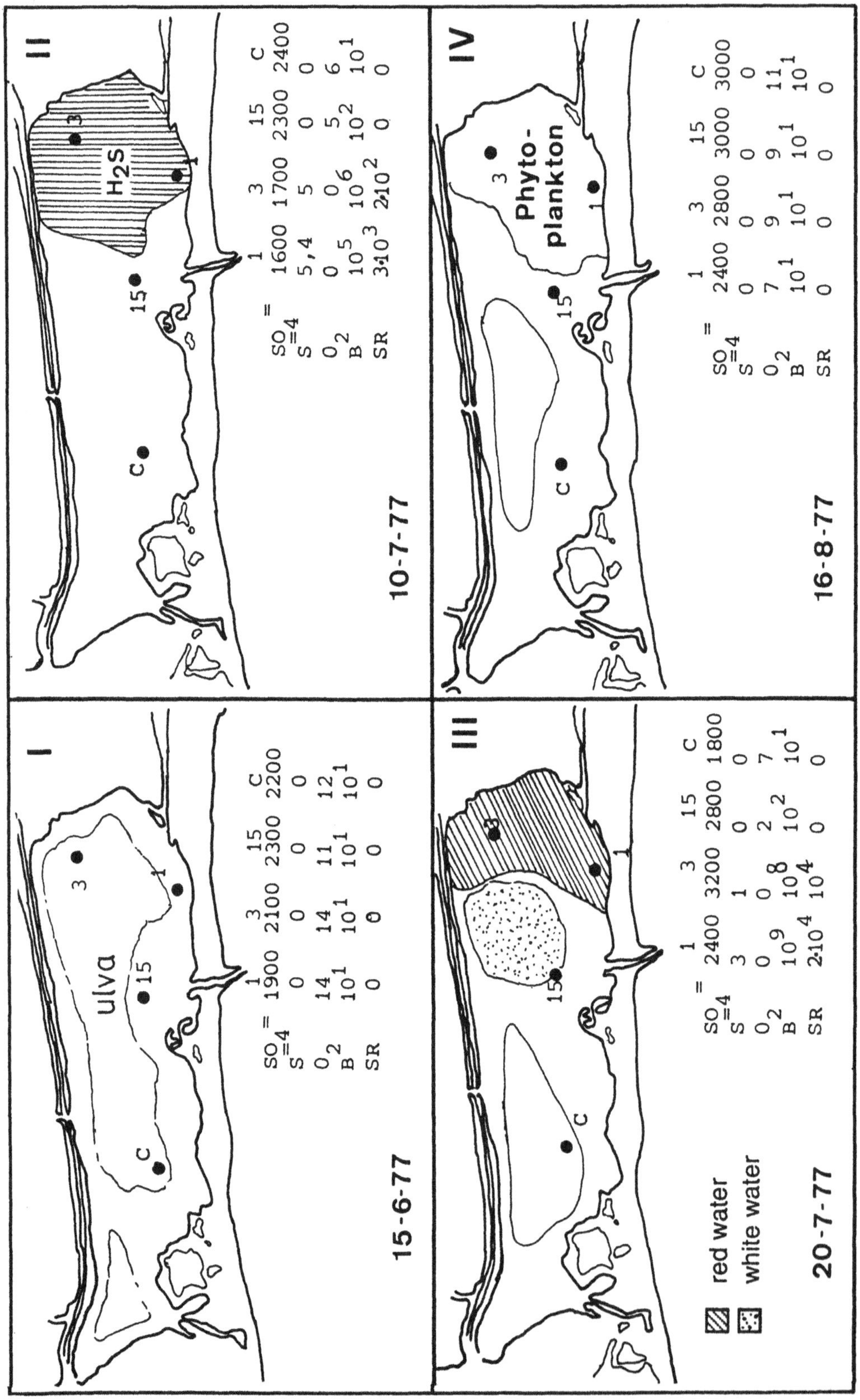


*Figure 17*. The Prévost lagoon during a dystrophic crisis (June–August 1977). I: before the dystrophic crisis, II: appearance of anoxic water with presence of sulfide, III: appearance of red and white waters, IV: after the dystrophic crisis. 1, 2, 3 and C refer to sampling stations. $SO_4^{=}$, $S^{=}$ and $O_2$ are expressed in mg $l^{-1}$. B = phototrophic bacteria, SR = sulfate reducers. Bacterial counts are expressed in numbers $ml^{-1}$.

In normal conditions, there is a steady-state at the benthic level between sulfate-reducing bacteria producing hydrogen sulfide and bacteria oxidizing this reduced sulphur compounds or elemental sulphur (Figure 16). The reactions of sulphate reduction lead to the production of hydrogen sulfide which is then rapidly oxidized to sulphate in the presence of oxygen. After an increase in organic load these oxido-reduction reactions accelerate. The $SO_4$/Cl ratio increases considerably indicating a solubilization of the sulphate. This ratio, which is typically around 11–12, can exceed 16 during a period of eutrophication. Under anoxic conditions white waters can be observed together with the production of $H_2S$. They usually occur when the light intensity is low. Concentrations of sulphur and carbonates are high which imparts a white colour to the water. The pH is generally low ($< 7.5$) due to the presence of considerable quantities of sulphur. This induces precipitation of carbonates in the form of white suspension (Figure 16).

Under certain conditions, red waters can be observed. They are due to the development of phototrophic sulphur oxidizing bacteria (e.g. *Thiocapsa*). The conditions for their development are: a sufficient light intensity, a pH value near 8 and a $H_2S$ concentration $< 1$ ppm. In anoxic conditions, in the light, only phototrophic bacteria are able to oxidize sulfide and to use it as an electron donor for reducing carbon dioxide.

Chronologically, white water is observed before red water. In favourable conditions, phototrophic bacteria metabolize the toxic compounds generated by anaerobic bacteria during fermentative processes and reduction of sulfate. In a reduced environment, rich in hydrogen sulfide, they are able to induce a new steady-state of the sulphur cycle by oxidizing the sulphur, under anaerobic conditions, using light energy. In the Étang du Prévost lagoon a succession of such events is commonly observed in summer (Amanieu et al., 1975; Caumette & Baleux, 1980). Coloured waters can invade most of the lagoon within a few days and disappear almost as quickly (Figure 17). In contrast to the Étang du Prévost, in the fish ponds of Arcachon only white waters have been observed. This is probably due to a lower light intensity, and the absence of complete anoxia in the water column (some *Ruppia* are always present).

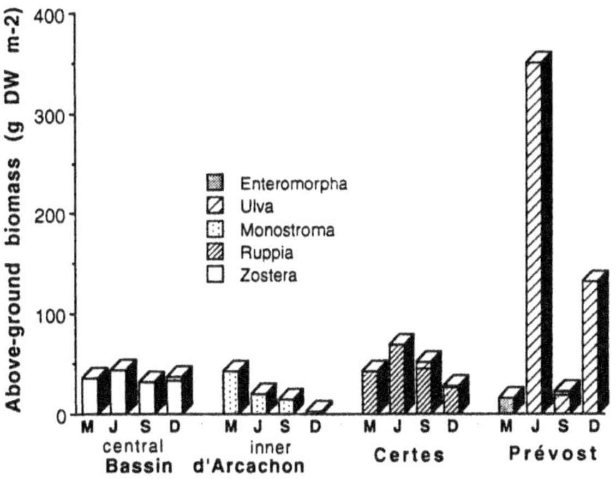


*Figure 18.* Biomass of living, above-ground parts of macrophytes in the Bassin d'Arcachon (central and inner parts), fish ponds of the Bassin d'Arcachon and Prévost lagoon, in March, June, September and December 1993 (modified from Auby et al., 1993 and Bachelet et al., 1994).

**Conclusions: the eutrophication gradient**

The three lagoon systems: Bassin d'Arcachon, fish réservoirs and étang du Prévost clearly differ by their degree of eutrophication. An evidence of eutrophication gradient is the increase in nutrient concentrations between the Bassin d'Arcachon and the Étang du Prévost (Table 6). As a consequence, the biomass of primary producers also increases.

This is shown by a recent study (Auby et al., 1993) where species composition and biomass of macrophytes have been studied in spring and summer 1993. One station (station A) was located in a seagrass bed (*Zostera noltii*) on a sandy mudflat in the central part of the Bassin d'Arcachon. Total living biomass (leaves + roots + rhizomes) of seagrass amounted to 141–167 g AFDW $m^{-2}$ (173–211 g DW $m^{-2}$), which is close to values reported in an earlier study (Auby, 1991) in the Bassin d'Arcachon (140–260 g DW $m^{-2}$). The above-ground biomass did not show any clear seasonal trend (Figure 18). In the inner part of the Bassin d'Arcachon (station B) living biomass was dominated by the green alga *Monostroma obscurum.* It declined from March to September (29–13 g AFDW $m^{-2}$, or 43–15 g DW $m^{-2}$) and never reached the high values found for other Ulvaceae in the Étang du Prévost (Figure 18). A high biomass of *Zostera* debris was measured in the sediments (92–186 g AFDW $m^{-2}$). The shallow fish ponds of Certes (station C1) were colonized by *Ruppia cirrhosa.* There was a clear seasonal trend with the highest biomass (63 g AFDW $m^{-2}$) occurring in

*Table 6.* Dissolved nitrate and ammonia in the water of the Bassin d'Arcachon (central and inner parts), fish réservoirs of Certes and Étang du Prévost. Values are given for the summer 1993 (Sloth et al., 1993) except for the fish réservoirs (summer 1985, Castel, unpublished).

| Lagoon system | $NO_3^-$ ($\mu$M) | $NH_4^+$ ($\mu$M) |
|---|---|---|
| Bassin d'Arcachon, central part (station A) | 0.32 – 2.0 | 2.0 – 2.4 |
| Bassin d'Arcachon, inner part (station B) | 0.32 – 1.3 | 1.5 – 2.5 |
| Fish réservoirs (station C1) | 0.7 – 1.01 | 4.6 – 8.3 |
| Étang du Prévost (station 11) | 1.2 – 4.9 | 5.7 – 10.5 |

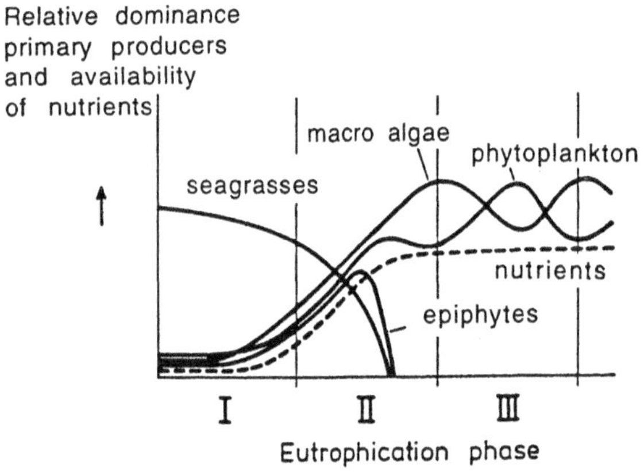


*Figure 19.* Tentative model depicting the relation between primary producers and nutrients, and the successive stages in the process of eutrophication (after Nienhuis, 1992).

June. Some *Monostroma* were also collected in September. In the Prévost lagoon (Station 11) the March samples contained a small amount of *Enteromorpha flexuosa* (Wulfen exRoth) J. Agardh and *E. intestinalis* (Linn.) Link (12 g AFDW $m^{-2}$). These algae disappeared in June when they were replaced by *Ulva* sp. (270 g AFDW $m^{-2}$) which filled the whole water column and induced anoxia in the benthos. In September algal biomass (*Enteromorpha* + *Ulva*) decreased to a low value (19 g AFDW $m^{-2}$). From these observations it appears that macrophytes did not show any seasonal trend in the Bassin d'Arcachon. Some evidence of eutrophication were visible in the Certes lagoons. The most obvious changes in macrophyte biomass occurred in the Prévost lagoon, especially in its inner part, with a massive development of green algae in June, followed by their complete disappearance in September.

A tentative model has been developed by Nienhuis (1992) describing the relation between the relative dominance of primary producers connected to the availability of nutrients and the successive phases in the process of increasing eutrophication (Figure 19). This model applies to the three types of lagoon environments described here. In 'healthy' lagoons seagrasses dominate. Nitrogen load and concentrations are low and the relative importance of phytoplankton in the shallow seagrass beds is insignificant. The Bassin d'Arcachon is an example of phase I. In brackish waters were eutrophication increases, revealed by higher nitrogen loads and nitrogen concentrations and generally lower, unstable salinities, seagrasses are outcompeted by macroalgae. Epiphyte growth on seagrass and algae increases considerably together with the relative dominance of phytoplankton. The fish ponds of the Bassin d'Arcachon are an example of phase II. In hypereutrophicated systems (phase III) nutrient concentrations are continuously high. Dense uncontrolled phytoplankton blooms alternate with mass growth of macroalgae and rooted plants have completely disappeared. Bottom sediments suffer from permanent anoxia. The Étang du Prévost is an example of phase III.

Such macroscopic observations should, now, be explained by the study of microbiological processes including meiofauna, protozoa, bacteria and all the components of the microbial loop.

## Acknowledgements

This paper provides the opportunity to J C and P C to express their gratitude to Prof. M. Amanieu and P. Lasserre who have promoted ecological research of lagoon systems in France. This is a contribution to the E.C. Environment programme (CLEAN contract EV5V-CT92-0080).

## References

Amanieu, M., B. Baleux, O. Guélorget & P. Michel, 1975. Etude biologique et hydrologique d'une crise dystrophique (malaïgue) dans l'étang du Prévost à Palavas, Hérault. Vie Milieu 25B: 175–204.

Anonymous, 1991. Efficacité de la réduction de la masse des nutriments dans la prévention des malaïgues. Application aux étangs Palavasiens. Summary Report, Région Languedoc-Roussillon, Agence de l'Eau Rhône-Méditerranée-Corse, 7 pp.

Auby, I., 1991. Contribution à l'étude des herbiers de *Zostera noltii* dans le Bassin d'Arcachon: dynamique, production et dégradation, macrofaune associée. Doct. Thesis, Univ. Bordeaux I, 162 pp.

Auby, I., 1993. Evolution de la richesse biologique du Bassin d'Arcachon. Scientific Report 91 5 527 019, IFREMER, Soc. Sci. Arcachon, 222 pp.

Auby, I., G. Bachelet & P. J. Labourg, 1993. Biomass and species composition of macrophytes in Arcachon Bay and Prévost lagoon, with a compilation of data on primary production in Arcachon Bay. In P. Caumette (ed.), CLEAN Progress report, part II, EU Environment Programme DG XII, Brussels: 65–72.

Auby, I., F. Manaud, D. Maurer & G. Trut, 1994. Etude de la prolifération des algues vertes dans le Bassin d'Arcachon. Scientific Report, IFREMER, Arcachon, 163 pp. + annex.

Bachelet, G., X. de Montaudouin, I. Auby & P. J. Labourg, 1994. A comparative study of the seasonal changes in macrophytes and macrozoobenthos assemblages in three coastal lagoons under varying degrees of eutrophication. In P. Caumette (Coord.) C.L.E.A.N. Progress Report 1994. EU Environment Programme DG XII, Brussels: 353–367.

Baleux, B., P. Caumette & M. Troussellier, 1979. Numération et approche qualitative des populations bactériennes des lagunes aménagées de Certes (Arcachon). II. Bactéries sulfo-oxydantes photosynthétiques. Publ. Sci. Tech. CNEXO, Brest, Actes colloques 7: 529–542.

Bouchet, J. M., 1968. Etude océanographique des chenaux du Bassin d'Arcachon. Doct. Thesis, Univ. Bordeaux, 200 pp.

Carman, K. R. & D. Thistle, 1985. Microbial food partitioning by three species of benthic copepods. Mar. Biol. 88: 143–148.

Carmouze, J. P. & P. Caumette, 1985. Les effets de la pollution organique sur les biomasses et activités du phytoplancton et des bactéries hétérotrophes dans la lagune Ebrié (Côte d'Ivoire). Rev. Hydrobiol. trop. 18: 183–212.

Castel, J., 1978. Plancton estival dans les étangs saumâtres du Bassin d'Arcachon. Bull. Off. natn. Pêch. Tunisie 2: 303–319.

Castel, J., 1989. Structure et évolution d'un écosystème lagunaire aménagé: les 'réservoirs à poissons' du Bassin d'Arcachon. In Connaissance et gestion de la frange littorale et du proche plateau continental. Conseil de l'Europe, Strasbourg: 417–428.

Castel, J., 1992. The meiofauna of coastal lagoons and their importance in the food web. Vie Milieu 42: 125–135.

Castel, J., P. J. Labourg, V. Escaravage, I. Auby & M. E. Garcia, 1989. Influence of seagrass beds and oyster parks on the abundance and biomass patterns of meio- and macrobenthos in tidal flats. Estuar. coast. Shelf Sci. 28: 71–85.

Caumette, P., 1986. Phototrophic sulfur bacteria and sulfate-reducing bacteria causing red waters in a shallow brackish lagoon. FEMS Microbiol. Ecol. 38: 113–124.

Caumette, P., 1989a. Les lagunes et les marais maritimes. In M. Bianchi, D. Marty, J. C. Bertand, P. Caumette & M. Gauthier (eds), Micro-organismes dans les écosystèmes océaniques. Masson éditions, Paris: 249–282.

Caumette, P., 1989b. Ecology and general physiology of phototrophic bacteria in benthic environments. In Y. Cohen & E. Rosenberg (eds), Microbial mats, Ecological physiology of benthic microbial communities. ASM Publications, Washington DC, USA: 283–304.

Caumette, P. & B. Baleux, 1980. Etude des eaux rouges dues à la prolifération des bactéries photosynthétiques sulfo-oxydantes dans l'étang du Prévost, lagune saumâtre méditerranéenne. Mar. Biol. 56: 183–194.

Caumette, P., M. Pagano & L. Saint-Jean, 1983. Répartition verticale des bactéries, du phytoplancton et du zooplancton dans une partie stratifiée d'une lagune tropicale (lagune Ebrié, Côte d'Ivoire). Hydrobiologia 106: 135–148.

Coull, B.C., 1973. Estuarine meiofauna: A review: trophic relationships and microbial interactions. In L. H. Stevenson & R. R. Colwell (eds), Estuarine Microbial Ecology. University of South Carolina Press, Columbia: 499–512.

Decho, A. W. & R. W. Castenholz, 1986. Spatial patterns and feeding of meiobenthic harpacticoid copepods in relation to resident microbial flora. Hydrobiologia 131: 87–96.

Erckenbrecher, C. & L. H. Stevenson, 1975. The influence of tidal flux on microbial biomass in salt marsh creeks. Limnol. Oceanogr. 20: 618–625.

Escaravage, V., 1990. Daily cycles of dissolved oxygen and nutrient content in a shallow fishpond: The impact of water renewal. Hydrobiologia 207: 131–136.

Fenchel, T. & B. B. Jørgensen, 1977. Detritus food chain in aquatic ecosystems: the role of bacteria. Adv. Microb. Ecol. 1: 1–55.

Ferguson, S. J., 1988. The redox reactions of the nitrogen and sulphur cycles. In J. A. Cole & S. J. Ferguson (eds), The nitrogen and sulphur cycles. Cambridge University Press: 1–29.

Gophen, M., B. Z. Cavari & T. Berman, 1974. Zooplankton feeding on differentially labelled algae and bacteria. Nature (London) 247: 393–394.

Guélorget, O. & J. P. Perthuisot, 1992. Paralic ecosystems. Biological organization and functioning. Vie Milieu 42: 215–251.

Herbert, R. A., 1975. Heterotrophic nitrogen fixation in shallow estuarine sediments. J. exp. mar. Biol. Ecol. 18: 215–225.

Herbert, R. A., 1982. Nitrate dissimilation in marine and estuarine sediments. In D. B. Nedwell & C. M. Brown (eds), Sediment microbiology. Academic Press, New York: 53–71.

Herbert, R. A. & D. B. Nedwell, 1990. Role of environmental factors in regulating nitrate respiration in intertidal sediments. In N. P. Revsbech & J. Sørensen (eds), Denitrification in Soil and Sediment. Plenum Press, New York: 77–90.

Jørgensen, B. B., 1982. Ecology of the bacteria of the sulphur cycle with special reference to anoxic-oxic interface environments. Philos. Trans. r. Soc. London, Ser. B 298: 543–561.

Jørgensen, B. B., 1983. The microbial sulphur cycle. In W. Krumbein (ed.), Microbial Geochemistry. Blackwell Scientific Publications, Oxford: 91–124.

Koike, I. & R. Hattori, 1978. Denitrification and ammonia formation in anaerobic coastal sediments. Appl. envir. Microbiol. 35: 278–282.

Labourg, P. J., 1975. Contribution à l'hydrologie des étangs saumâtres de la région d'Arcachon: description des phénomènes d'eaux blanches. Bull. Soc. Linn. Bordeaux 5: 3–8.

Labourg, P. J., 1979. Structure et évolution de la macrofaune invertébrée d'un écosystème lagunaire aménagé (réservoirs de Certes). Publ. Sci. Tech. CNEXO, Brest, Actes colloques 7: 575–590.

Labourg, P. J., 1985. Ecologie et utilisation des zones humides du Bassin d'Arcachon. In Actes du colloque sur les zones humides du littoral Aquitain, D.R.A.E. Aquitaine, Bordeaux: 5–28.

Lasserre, P., 1979a. Coastal lagoons. Sanctuary ecosystems, cradles of culture, targets for economic growth, Nature & Resources, Unesco 15: 2–21.

Lasserre, P., 1979b. Programme coordonné Ecotron sur le site d'Arcachon (Aquitaine, France): contrôle de la production biologique marine dans un écosystème lagunaire aménagé (réservoirs à poissons). Bilan, synthèse et perspective. Publ. Sci. Tech. CNEXO, Brest, Actes colloques 7: 463–502.

Lasserre, P. & H. Postma (eds), 1982. Proceedings of the international symposium on coastal lagoons, Bordeaux, France, 8–14 September 1981. Oceanol. Acta, No. sp., 461 pp.

Lasserre, P., J. Renaud-Mornant & J. Castel, 1976. Metabolic activities of meiofaunal communities in a semi-enclosed lagoon. Possibilities of trophic competition between meiofauna and mugilid fish. In G. Persoone & E. Jaspers (eds), Proc. 10th Europ. Symp. Mar. Biol., vol. 2. Population dynamics, Universa Press, Wetteren: 393–414.

Marty, D., G. Esnault, P. Caumette, E. Ranaivoson-Rambeloarisoa & J. C. Bertrand, 1990. Dénitrification, sulfato-réduction et méthanogenèse dans les sédiments superficiels d'un étang saumâtre méditerranéen. Oceanol. Acta 13: 199–210.

McFarlane, G. T. & R. A. Herbert, 1984. Dissimilatory nitrate reduction and nitrification in estuarine sediments. J. Gen. Microbiol. 130: 2301–2308.

Montagna, P. A., 1995. Rates of metazoan meiofaunal microbivory: a review. Vie Milieu 45: 1–9.

Nienhuis, P. H., 1992. Ecology of coastal lagoons in the Netherlands (Veerse Meer and Grevelingen). Vie Milieu 42: 59–72.

Nishio, T., I. Koike & A. Hattori, 1982. Denitrification, nitrate reduction and oxygen consumption in coastal and estuarine sediments. Appl. envir. Microbiol. 43: 648–653.

Nishio, T., I. Koike & A. Hattori, 1983. Estimates of denitrification and nitrification in coastal and estuarine sediments. Appl. envir. Microbiol. 45: 444–450.

Rieper, M., 1982. Feeding preferences of marine harpacticoid copepods for various species of bacteria. Mar. Ecol. Prog. Ser. 7: 303–307.

Robert, R., N. Guillocheau & Y. Collos, 1987. Hydrological parameters during an annual cycle in the Arcachon Basin. Mar. Biol. 95: 631–640.

Sloth, N. P., N. Risgaard-Petersen, S. Rysgaard & S. P. Pelegri, 1993. Nitrification, denitrification and nitrate ammonification in sediments of two coastal lagoons in Southern France. In P. Caumette (Coord.) C.L.E.A.N. Progress Report 1994. EU Environment Programme DG XII, Brussels: 159–185.

Soriano-Sierra, E., 1988. Contribution à l'écologie des étangs saumâtres du Bassin d'Arcachon: structure, dynamique et productivité d'une phytocénose dominée par *Ruppia cirrhosa* (Petag.). D.E.A., Univ. Bordeaux I, 18 pp. + annex.

Souza Santos, L. P., J. Castel & P. J. P. Santos, 1996. The role of phototrophic bacteria as food for meiobenthic harpacticoid copepods inhabiting eutrophic coastal lagoons. Hydrobiologia (this volume).

Torreton, J. P., 1991. Importance des bactéries hétérotrophes aérobies dans une lagune eutrophe tropicale (Lagune Ebrié, Côte d'Ivoire). Biomasse, production, exportations. Thèse Doctorat. Faculté des Sciences de Luminy, 246 pp.

Van Gemerden, H., 1983. Physiological ecology of purple and green bacteria. Ann. Inst. Pasteur Microbiol. 134: 73–92.

Visscher, P. T., P. Quist & H. van Gemerden, 1991. Methylated sulfur compounds in microbial mats: *in situ* concentrations and metabolism by a colorless sulfur bacterium. Appl. Envir. Microbiol. 57: 1758–1763.

Zobell, C. E., 1946. Marine Microbiology. Chronica Botanica Co. Waltham, Mass., 240 pp.

# PART I

## Eutrophication effects on population dynamics, biodiversity and trophic relationships in coastal lagoons

*Hydrobiologia* **329**: 3–17, 1996.
*P. Caumette, J. Castel & R. Herbert (eds), Coastal Lagoon Eutrophication and ANaerobic Processes (C.L.E.AN.).*



# Heterotrophic bacteria, activity and bacterial diversity in two coastal lagoons as detected by culture and 16S rRNA genes PCR amplification and partial sequencing

Susana Benlloch, Francisco Rodríguez-Valera*, Silvia G. Acinas &
Antonio J. Martínez-Murcia
*Departamento de Genética y Microbiologia, Universidad de Alicante, Campus de San Juan, Apartado 374, 03080 Alicante, Spain*
* *Author for correspondence. Fax: 34-6-5941787; E-mail: FRVALERA@UA.ES*

*Key words:* Bacterial diversity, coastal lagoons, 16S rRNA, eutrophication

## Abstract

Activity and numbers of heterotrophic bacteria have indicated that, as expected, Prévost Lagoon is more eutrophic than Arcachon Bay. Amplification and sequence analysis of the 16S rRNA genes from DNA samples extracted directly from the environment allow the determination of phylogenetic relationships among members of microbial communities in natural ecosystems without the need for cultivation. Analysis of partial 16S rRNA gene sequences obtained from Stations A and 11 revealed that, in both environments, a relatively large number of clones related to *Cytophaga/Flexibacter/Bacteroides* as well as to *α-Proteobacteria* were found. One hundred percent similarity with the sequences of the data bases were not found for any of the more than a hundred clones studied. In fact for most clones maximum similarity was below 95% for the nucleotide series sequenced. Similarity was not higher with any of the sequences found for the 14 isolates (pure cultures) obtained from the same samples. Redundancy, i.e. number of identical sequences, was higher in the samples from Arcachon. In addition, sequences related to representatives of ten major phylogenetic branches of Eubacteria were obtained from Prévost Lagoon, however only five branches were represented by the data from Arcachon. These findings indicate a higher bacterial diversity in Prévost Lagoon.

## Introduction

Coastal lagoons are important aquatic environments that, due to their location in the continental-ocean interphase, are particularly sensitive to anthropogenic influence and pollution. They have a tendency to become eutrophic. Relatively closed shallow basins tend to accumulate nutrients and are extremely vulnerable to nutrient input from the surrounding land. When heavy pollution occurs productivity by planktonic and/or macro-algae can lead to dystrophic crises in which the water column becomes anoxic, rich in $H_2S$ thus leading to catastrophic ecological and aesthetic consequences. Moreover, many coastal lagoons are exploited for recreational or fishing/aquaculture purposes and their ecological status has important economic consequences (Caumette, 1986).

Bacteria are considered to be the major decomposers of organic matter in aquatic ecosystems. It is widely accepted that about 50% of the primary production is processed by bacteria each day (Ducklow & Carlson, 1992), the rapid growth rates of aquatic bacteria have suggested that the production of heterotrophic bacterial biomass represents an important link between detritus, dissolved organic matter and higher trophic levels in pelagic ecosystems. The most common procedure for estimating bacterial production in aquatic systems is the thymidine incorporation into DNA used to determine activity, growth rates and carbon production of aerobic bacteria (Fuhrman & Azam, 1980) although it presents problems that must be resolved.

One such problem is the need to use a conversion factor to transform thymidine incorporation into bacterial production (Fuhrman & Azam, 1980).

The bacterial diversity present is obviously a key factor in actually carrying out the fast degradation of the organic matter accumulated by primary production. The diversity of marine prokaryotes is not well known, most of them are not yet cultured (Jannasch & Jones, 1959), showing the inability of standard microbiological techniques to cultivate most microorganisms. As a powerful alternative, a molecular approach based on the analysis of rRNA sequences has been used to determine the species composition of microbial communities (Pace et al., 1986). The sequences of rRNAs directly extracted from naturally occurring organisms are determined using molecular techniques and compared with known rRNA sequences, using molecular phylogeny techniques. The differences in sequence are the basis of the phylogenetic analysis to identify organisms in the environment by hybridization with organism-specific probes and may help to determine the community structure (Amann et al., 1995).

Previous 16S rRNA studies in the oligotrophic waters of the Pacific and Atlantic Oceans (Fuhrman et al., 1993; Giovannoni et al., 1990; Schmidt et al., 1991) have disclosed phylogenetic lines not described by classical microbiological methods. No sequence directly recovered from an environmental sample had 100% similarity with a previously cultured and described microorganism. In the Sargasso Sea, Giovannoni et al. (1990) determined that their clones formed two different groups represented by clone SAR7 and SAR11. SAR7 grouped with phototrophic oxygenic organisms (*Cyanobacteria*, chloroplasts and prochlorophytes) and SAR11 grouped with the *α-Proteobacteria* group. In the North Pacific Ocean, Schmidt et al. (1991) recovered clones related to *Cyanobacteria* and to a new phylogenetic group described by Giovannoni et al. (1990) in the Sargasso Sea. Other clones were close to *Proteobacteria*, but two clones (ALO30 and ALO33) did not belong to any phylogenetic lines previously described (Schmidt et al., 1991). No archaeal clones were recovered in this sample. In the Pacific Ocean again, Fuhrman et al. (1993) took samples at 100 and 500 m to observe the difference due to the varying depth of water. In both sample depths they found clones with 100% similarity, not with cultured microorganisms, but with clones belonging to the phylogenetic line SAR11 described by Giovannoni et al. (1990). Another group of clones from the 100 m depth sample was related to *Cyanobacteria*, *Synechococcus* and group SAR7 (Giovannoni et al., 1990). These authors also described a novel group, the so-called Marine GroupA, unrelated to any subdivisions described by Woese (1987). They also found clones related to Gram-positives, *α*- and *γ-Proteobacteria* and *Cytophaga/Bacteroides/Flexibacter*.

DeLong (1992) described, in the Pacific Ocean, two novel groups related to the domain *Archaea*, group I within the group *Crenarchaeota* and group II within the group *Euryarchaeota* and related to methanogenic microorganisms. In cold Antarctic waters, DeLong et al. (1993) found several clones related to group I, and a group of sequences appeared close to group II. In the study by Fuhrman et al. (1993) in the Pacific Ocean, the fourth group of clones appeared to be within the archaeal group described by DeLong (1992). Studies have been made to estimate the diversity of free-living organisms and aggregate-attached organisms (DeLong et al., 1993). Clones of free-living microorganisms appeared to be relatd to the SAR11 group (Giovannoni et al., 1990), *γ-Proteobacteria* and the *Cytophaga/Flexibacter/Bacteroides* group. Those of aggregate-attached microorganisms were phylogenetically close to *Cyanobacteria, Cytophaga/Flexibacter, Planctomyces* and *γ-Proteobacteria.* The most abundant phylogenetic types detected in macroaggregate-associated bacterial populations fell within the *Cytophaga/Flexibacter* group (DeLong et al., 1993).

In the present work two coastal lagoons that represent two model systems have been studied. The Arcachon lagoon on the French Atlantic coast is a typical tidal system in which the water in the lagoon is flushed by strong currents keeping a relatively low nutrient concentration and a strong connection with the neighbouring Atlantic waters. The Prévost Lagoon, on the other hand, is located on the Mediterranean coast. With a relatively narrow communication with the Mediterranean Sea that is virtually free of tides, the connection between this lagoon and the marine habitat is comparatively restricted. The Prévost lagoon is much more eutrophic than that of Arcachon and affected to a far greater degree by adjacent urban developments, with a history of dystrophic crises (Caumette, 1986).

We have studied the activity and numbers of heterotrophic bacteria by classical techniques (thymidine incorporation and glucose utilization/incorporation and plating on marine nutrient agar). Sediment samples were also obtained for viable counts and isolation. Besides, water samples from two sampling sites (A and 11) were processed to obtain DNA from the bacterioplankton, and then libraries of 16S rDNAs directly amplified from both environments were constructed, to study phylogenetic diversity. The 16S rDNA from some isolates obtained from the two environments were also partially sequenced and analyzed.

## Materials and methods

*Sampling sites and periodicity*

Sampling was always performed at high tide to obtain water samples and at low tide to get sediment cores. Station A was reached by boat and water samples were collected in sterilized 1 l glass bottles. Stations A and B (in Arcachon Bay), Stations $C_1$ and $C_2$ (Certes Domain) and Stations X and 11 (Prévost Lagoon) were selected. In station 11 samples were taken about 30 m from the shore with a water depth of about 1 m. Samples were always obtained from superficial water by opening the flask just below the surface. Four sampling campaigns were carried out in May and September 1993 and May and September 1994. Sediment samples were obtained in 5 ml syringes and the two upper ml separated from the lower three and processed independently. Sampling sites are described in Castel et al. (1995).

*Heterotrophic bacteria enumeration and isolation*

Total heterotrophic aerobes were counted by plating on nutrient agar supplemented with 2% w/v artificial sea water salts and incubated at 20 and 30 °C for 24 or 48 hours. For sediment samples dilutions in sterile artificial sea water were performed. Colonies were isolated at random and subcultured twice. 1 ml of bacterial cultures were harvested in sterile 2% marine salt water by centrifuging at 10000 rpm for 2 min and pellets were stored at −20 °C.

Some samples were plated onto selective media for enumeration of specific physiological groups. Sulphite reducing clostridia were counted from every sample by the most probable number (MPN) method using TSC selective medium (Merck) with cycloserine to inhibit the vegetative forms (APHA-AWWA-WPCF, 1975).

*Faecal coliforms*

These pollution indicators were determined by the standard membrane filtration method, using the selective medium mFC (Difco) (APHA-AWWA-WPCF, 1975).

*Classification of isolates*

Some colonies were classified to the genus level performing biochemical tests by the standard methods (Bradshaw, 1992): oxidase, catalase, O/F, sensitivity to vibriostatic agent O/129 (all the media were supplemented with 2% w/v artificial sea water salts). Morphology, Gram reaction and motility were determined using a microscope.

*Biomass collection*

Different volumes of water samples, ranging from 1–3 l of Station 11 to 10–15 l of Station A were obtained in plastic containers previously treated with 0.1 N HCl to eliminate contaminant DNA. They were processed to obtain bacterial biomass and DNA. Debris and eukaryotic cells were removed by using glass microfibre filters (Whatman GF/D). Microorganisms were collected by positive pressure on a 9 cm ⌀ 0.22 $\mu$m pore-size filter (Durapore; Millipore Corp., Bedford, MA). Water was pumped till filter clogging decreased flow rate to a few ml per min. Filters were stored at − 80 °C until DNA extraction.

*Determination of microbial heterotrophic activities*

Tritiated thymidine incorporation: Water samples were incubated at *in situ* temperature in the dark, on a rotary shaker for 40 min with a final concentration of 10 nM of tritiated thymidine (Amersham Life Science), labelled in the methyl position with a specific activity of 70–85 Ci/mmol. A blank was obtained by adding formaldehyde solution (final concentration 2%). Five ml from each sample were taken in propylene tubes pre-chilled on ice for 1 min and an equal volume of cold TCA 10% was added. After 5 min the mixture was filtered and rinsed three times with 5 ml of ice-cold 5% TCA. The filtration was carried out in an ice bath. The filters were dried at 40 to 50 °C for 15 min, chilled, placed in scintillation vials and radio-assayed by liquid scintillation counting (Fuhrman & Azam, 1980).

Glucose metabolism: Water samples were incubated at *in situ* temperature in the dark for 1 hour with 3 $\mu$Ci of D-U$^{14}$C Glucose (250 mCi/mmol), obtained from Amersham Life Science, in a bottle plugged with Whatman No. 1 paper. Another sample (blank) was fixed by adding formaldehyde solution for 1 hour. After incubation 0.10 ml of $H_2SO_4$ was added to the sample. The Whatman No. 1 paper was soaked with 0.2 ml of $\beta$-phenethylamine. In the dark, under the same conditions, the sample was incubated again for 1 hour. The sample was filtered through 0.2 $\mu$m pore size (Celotate, Millipore Corp., Bedford, MA.) filters then rinsed with 25 ml of sterile distilled water. The filters were dried at 40 to 50 °C for 15 min and counted. The Whatman No. 1 paper was placed in scintillation vials with 4 ml of scintillation liquid and counted (Wright & Hobbie, 1966).

Thymidine incorporation estimates were measured in pmoles of thymidine per unit volume and time to allow comparisons between studies. In order to transform thymidine incorporation into bacterial production an empirical conversion factor, $2 \times 10^{18}$ cells $mol^{-1}$ (Fuhrman & Azam, 1980), was used. The ratio between dividing cells and bacterial plate counts gave the specific growth rate ($\mu$). We assumed a value of 20 fg C $cell^{-1}$ (Moriarty, 1990), to represent the productivity estimate in carbon units. We also estimated the doubling time from the growth rate ($\ln 2/\mu$) (Moriarty, 1990).

*DNA extraction and purification*

Thawed filters were cut with a sterile razor blade into small pieces and vortexed in 2 ml of STE buffer (10 mM Tris-HCl [pH 8], 1 mM EDTA, 100 mM NaCl) in 50 ml conical based polypropylene centrifuge tubes. A 0.1 volume of 10% w/v sodium dodecyl sulphate (SDS) was added and the tubes placed in a boiling water bath for 2–5 min (Fuhrman et al., 1988). After spinning at 5000 rpm for 10 min, the supernatant fluids were transferred into 1.5 ml Eppendorf tubes and the DNA was purified with the GeneClean II kit (BIO 101, Inc.). DNA from bacterial cultures was extracted as previously described (Martínez-Murcia & Rodríguez-Valera, 1994). Quantification and purity (referred to protein content) of the DNA obtained were estimated and 5 $\mu$l aliquots of DNA dilutions were electrophoresed (Martínez-Murcia & Rodríguez-Valera, 1994).

*PCR Amplification of 16S rRNA genes*

Reaction mixtures contained 50 mM KCl, 10 mM Tris-HCL [pH 9], 1.5 mM MgCl2, 0.1% Triton X-100, 200 $\mu$M each of deoxyribonucleotide triphosphate (dATP, dCTP, dGTP, dTTP; Pharmacia LKB Biotechnology), 2 U of Taq I DNA polymerase (Promega), 0.2 $\mu$M of each oligonucleotide primer and 100 ng of template DNA in a total volume of 100 $\mu$l. The sequence of the forward primers was 5′-AGAGTTTGATCATGGCTCAG-3′ (Lane et al., 1985) at nucleotide positions 8-27 of the *E. coli* numbering (Brosius et al., 1981). Reverse primer was 5′-GGTTACCTTGTTACGACTT-3′ (Lane et al., 1985) at positions 1509–1491 (*E. coli* numbering). Reaction mixtures were overlaid with mineral oil (Light White Oil; Sigma) prior to a thermal cycling regime (35 cycles on a Perkin Elmer Cetus 480) of 94 °C for 1 min, 50 °C for 1 min, and 72 °C for 2 min. After the final cycle was completed, the amplicons were allowed to extend at 72 °C for 10 min. Negative controls were included with no addition of template DNA. Five $\mu$l of the PCR products were analyszed on 1% (w/v) agarose gels.

*Purification and cloning of the PCR products*

PCR products were purified using GeneClean II kit and cloned into a T-tailed-vector prepared as follows: Bluescript plasmid (Stratagene, La Jolla, CA) was digested with *Eco*RV and incubated with Taq I DNA polymerase (1 U/$\mu$g plasmid/20 $\mu$l volume) using standard buffer conditions in the presence of 2 mM dTTP for two hours at 72 °C (Marchuk et al., 1990) and ligated at 16 °C overnight. XLBlue *Escherichia coli* strain was transformed with a standard protocol (Maniatis et al., 1982). Plasmidic DNA was extracted using the Wizard Minipreps DNA Purification System (Promega).

*Primers and sequencing reactions*

Primers with 5′-TTACCGCGGCTGCTGGCACG-3′ sequence (Martinez-Murcia, 1993) at positions 533–514 (*E. coli* numbering) were chosen to sequence ca. 200 bp of a relatively variable region of each selected clone. Primers were synthesized by the phosphoramidite method (Beaucage & Caruthers, 1981) on a Model 392 RNA/DNA Applied Biosystems synthesizer. Clones randomly selected were sequenced by the dideoxy-chain termination method (Sanger et al., 1977) with a Sequenase Version 2.0 Kit (United States Biochemicals) following the manufacturer's protocol.

*Analysis of the sequence data*

The determined sequences were compared with 16S rRNA sequences obtained from the GenBank and the EMBL Data Library by using programs of the Wiscon-

*Table 1.* Viable bacteria count (cels/ml) in water samples from stations in Arcachon Bay (A and B), Certes Domain ($C_1$ and $C_2$) and Prévost lagoon (X and 11). ND = Non-determined data.

| Station | Date | Water column | | Superficial sediment | | Deep sediment | |
|---|---|---|---|---|---|---|---|
| A | May 93 | 60 | $5.0 \times 10^2$ | $1.0 \times 10^6$ | $1.0 \times 10^6$ | ND | $1.0 \times 10^5$ |
| | Sept 93 | $8.5 \times 10^2$ | $9.0 \times 10^2$ | $1.3 \times 10^6$ | $1.3 \times 10^6$ | $2.8 \times 10^4$ | $4.4 \times 10^4$ |
| | May 94 | 30 | $2.0 \times 10^2$ | $2.0 \times 10^4$ | $6.0 \times 10^4$ | $1.6 \times 10^4$ | $2.2 \times 10^4$ |
| | Sept 94 | $2.8 \times 10^2$ | $2.2 \times 10^2$ | $1.0 \times 10^4$ | $2.0 \times 10^4$ | $2.0 \times 10^3$ | $2.0 \times 10^4$ |
| B | May 93 | ND | ND | $3.2 \times 10^7$ | $2.5 \times 10^7$ | $2.0 \times 10^6$ | $2.0 \times 10^6$ |
| | Sept 93 | ND | ND | $4.5 \times 10^5$ | $3.6 \times 10^5$ | $1.2 \times 10^5$ | $1.5 \times 10^5$ |
| | May 94 | $2.3 \times 10^3$ | $3.2 \times 10^3$ | $3.2 \times 10^4$ | $5.9 \times 10^4$ | $2.0 \times 10^4$ | $5.0 \times 10^4$ |
| | Sept 94 | $3.5 \times 10^4$ | $2.9 \times 10^4$ | $1.5 \times 10^4$ | $1.0 \times 10^4$ | $8.0 \times 10^4$ | $9.2 \times 10^4$ |
| $C_1$ | May 93 | 70 | $1.7 \times 10^2$ | $6.0 \times 10^4$ | $1.5 \times 10^5$ | $4.0 \times 10^4$ | $6.0 \times 10^4$ |
| | Sept 93 | $1.8 \times 10^3$ | $2.0 \times 10^3$ | $1.7 \times 10^5$ | $2.5 \times 10^5$ | $1.2 \times 10^5$ | $1.7 \times 10^5$ |
| | May 94 | $9.0 \times 10^2$ | $1.4 \times 10^3$ | $1.9 \times 10^4$ | $2.5 \times 10^4$ | $7.4 \times 10^4$ | $8.8 \times 10^5$ |
| | Sept 94 | $1.6 \times 10^3$ | $1.6 \times 10^3$ | $1.7 \times 10^4$ | $1.2 \times 10^4$ | $2.0 \times 10^4$ | $2.0 \times 10^4$ |
| $C_2$ | May 93 | $3.7 \times 10^2$ | $1.0 \times 10^3$ | $2.0 \times 10^4$ | $3.0 \times 10^4$ | $5.0 \times 10^3$ | $2.0 \times 10^4$ |
| | Sept 93 | $1.8 \times 10^3$ | $4.2 \times 10^3$ | $9.0 \times 10^5$ | $1.0 \times 10^6$ | $1.4 \times 10^5$ | $2.5 \times 10^5$ |
| | May 94 | $1.8 \times 10^3$ | $1.8 \times 10^3$ | $6.0 \times 10^4$ | $7.6 \times 10^4$ | $9.0 \times 10^4$ | $6.7 \times 10^4$ |
| | Sept 94 | $1.3 \times 10^3$ | $1.9 \times 10^3$ | $9.0 \times 10^4$ | $7.0 \times 10^4$ | $6.0 \times 10^3$ | $4.0 \times 10^3$ |
| X | May 93 | $1.5 \times 10^3$ | $1.0 \times 10^3$ | ND | ND | ND | ND |
| | Sept 93 | $5.0 \times 10^2$ | $1.4 \times 10^3$ | $1.0 \times 10^5$ | $3.7 \times 10^5$ | $2.0 \times 10^5$ | $4.0 \times 10^5$ |
| | May 94 | $1.2 \times 10^2$ | $9.6 \times 10^2$ | $6.0 \times 10^4$ | $7.2 \times 10^4$ | $1.2 \times 10^4$ | $1.0 \times 10^4$ |
| | Sept 94 | $4.4 \times 10^2$ | $1.3 \times 10^3$ | ND | ND | ND | ND |
| 11 | May 93 | ND | ND | $7.5 \times 10^3$ | $2.5 \times 10^4$ | $2.3 \times 10^4$ | $3.0 \times 10^4$ |
| | Sept 93 | $4.5 \times 10^3$ | $7.0 \times 10^3$ | $5.5 \times 10^4$ | $7.0 \times 10^4$ | $6.8 \times 10^4$ | $9.9 \times 10^4$ |
| | May 94 | $2.9 \times 10^2$ | $2.2 \times 10^2$ | $3.5 \times 10^4$ | $6.0 \times 10^4$ | $2.0 \times 10^3$ | $1.4 \times 10^4$ |
| | Sept 94 | $1.0 \times 10^3$ | $7.4 \times 10^3$ | $2.6 \times 10^4$ | $2.7 \times 10^4$ | $1.3 \times 10^4$ | $2.0 \times 10^4$ |

sin Molecular Biology Package Version 8. Phylogenetic analysis were carried out with DNASTAR Package (Lasergene) in a Quadra 840 Macintosh Computer.

## Results

### *Heterotrophic bacterial numbers and activities in the two lagoons*

The study of the role of bacterioplankton starts with the estimation of microbial biomass and activities in water samples. The results of the viability counts (average of the three sampling campaigns) at the sampling sites, faecal coliforms and sulphite reducing clostridia are shown in Table 1 and Table 2.

About 100 isolates were obtained from different samples of Stations A and 11 during the September 1993 campaign. They were characterized to provide a primary classification at the genus level. As shown in Table 3, in both cases the predominant heterotrophic microflora in the water column were related to the genera *Vibrionaceae* as expected, and *Pseudomonaceae*. In the case of the Prévost Lagoon there also appeared to be a significant number of *Enterobacteriaceae* that could be allochthonous microorganisms due to previously described waste water pollution detected in this environment. In the sediment there was also a significant proportion of Gram Positive bacteria, both rods and cocci.

Thymidine incorporation estimates, assimilation and respiration of glucose values, growth rates and

*Table 2*. Faecal coliforms and clostridia numbers (cels/100 ml) in samples from stations in Arcachon Bay (A and B), Certes Domain ($C_1$ and $C_2$) and Prévost lagoon (X and 11). ND = Non-determined data.

| Station | Date | Faecal coliforms | | Clostridia | |
|---|---|---|---|---|---|
| | | Water column | Water column | Superficial sediment | Deep sediment |
| A | May 93 | ND | 0 | 60 | 80 |
| | Sept 93 | 2 | 0 | 170 | 20 |
| | May 94 | 0 | 2 | 1560 | 3316 |
| | Sept 94 | 2 | 4 | 43 | 250 |
| B | May 93 | ND | ND | 0 | 0 |
| | Sept 93 | ND | ND | 10 | 20 |
| | May 94 | ND | 4 | 4137 | 3320 |
| | Sept 94 | ND | 9 | 9 | 4 |
| $C_1$ | May 93 | ND | ND | 300 | 1100 |
| | Sept 93 | 1 | 0 | 200 | 760 |
| | May 94 | 39 | 780 | 13500 | 10800 |
| | Sept 94 | 0 | 620 | 80 | 65 |
| $C_2$ | May 93 | ND | ND | 150 | 1000 |
| | Sept 93 | 6 | 0 | 20 | 80 |
| | May 94 | 284 | 640 | 11300 | 7466 |
| | Sept 94 | 4 | 4 | 43 | 50 |
| X | May 93 | ND | 0 | 130 | ND |
| | Sept 93 | 4 | 10 | 110 | 110 |
| | May 94 | 0 | 30 | 1080 | 630 |
| | Sept 94 | 33 | 4 | ND | ND |
| 11 | May 93 | ND | 0 | 350 | ND |
| | Sept 93 | 20 | 50 | 130 | 130 |
| | May 94 | 0 | 60 | 1115 | 3036 |
| | Sept 94 | 14 | 9 | 610 | 620 |

doubling times of the heterotrophic bacteria are shown in Table 4. In Station A values of bacterial production ranged from 0.014 to 0.52 $\mu$g C $L^{-1}h^{-1}$, in Stations $C_1$ and $C_2$ (Certes Domain) from 1.70 to 4.65, and 0.27 to 2.5 $\mu$g C $L^{-1}h^{-1}$ respectively. Estimates in Prévost were in the range of 0.11 to 3.55 $\mu$g C $L^{-1}h^{-1}$ (Station X) and 0.80 to 8.0 $\mu$g C $L^{-1}h^{-1}$ (Station 11). The highest values at most sampling stations were found at the end of summer in 1993.

In order to compare estimates of bacterial production in terms of thymidine incorporation and assimilation of organic compounds (glucose) those values were referred to the numbers of bacterial plate counts as the heterotrophic activities are related with: (i) bacterial numbers and (ii) distinct activity of different bacterial populations. We observed the highest values of assimilation of glucose in early summer 1993–1994 in all the Stations. The highest heterotrophic activity estimated by glucose assimilation was in both the Stations in Certes Domain ($C_1$ and $C_2$). Maximal incorporation of both thymidine and glucose was measured in Station 11 (Prévost Lagoon) rather than Station A (Arcachon Bay), showing that the microbial community of Station 11 presented more heterotrophic activity per cell. In

*Table 3.* Taxonomic status of the isolates from Stations A (Arcachon Bay) and 11 (Prévost lagoon) during the September 1993 campaign. The total number of isolates from water, superficial sediment and deep sediment from both stations and the percentage of each taxonomic group are shown.

| Station A | Water | Superficial sediment | Deep sediment |
|---|---|---|---|
| Total number | 24 | 10 | 23 |
| *Vibrio, Plesiomonas, Aeromonas* | 67% | 60% | 57% |
| *Pseudomonas* | 25% | 30% | 26% |
| *Enterobacteriaceae* | – | – | – |
| *Acinetobacter* | 8% | – | – |
| *Bacillus* | – | – | 13% |
| *Lactobacillus* | – | – | 4% |
| *Planococcus* | – | 10% | – |

| Station 11 | Water | Superficial sediment | Deep sediment |
|---|---|---|---|
| Total number | 15 | 17 | 24 |
| *Vibrio, Plesiomonas, Aeromonas* | 40% | 41% | 54% |
| *Pseudomonas* | 40% | 29% | 25% |
| *Enterobacteriaceae* | 20% | 6% | 12.5% |
| *Acinetobacter* | – | – | – |
| *Bacillus* | – | – | 8.5% |
| *Lactobacillus* | – | 18% | – |
| *Streptococcus* | – | 6% | – |

summary, all data point to a major difference between the heterotrophic bacterial biomass of each of the two lagoons. As had been expected from the higher level of eutrophication (Caumette, 1986), it was much greater in the Prévost lagoon.

*Analysis of bacterial biodiversity by direct amplification and sequencing of the small subunit rRNA genes*

Amplification products of the 16S rRNA genes, using DNA obtained from Stations A and 11 (samples obtained in September 1993) as templates, were cloned in *E. coli.* About 50 clones from each environment were randomly selected and *ca.* 210 bp determined of the 16S rDNA sequences (ranging from positions 283 to 499; *E. coli* numbering) (Brosius et al., 1981). The sequences were compared with all available 16S rDNA sequences of the GenBank and EMBL Data Library. Pairwise sequence comparisons with selected numbers of several phylogenetic branches were performed, and two dendrograms of sequence similarity were constructed from two matrices of calculated percentages, each containing clones from Stations A and 11 respectively. Some of these results have been reported elsewhere (Benlloch et al., 1995).

No identical sequences were found on comparing the clones with those of the data banks. The highest similarity values found were those corresponding to clone 38a (Station A) with a representative of an undescribed marine eubacterial group (Fuhrman et al., 1993) (98.9%), and to clone 5011 (Station 11) with *Tropherima whippelli*, a High G+C content Gram-positive (99%). Overall similarity values with sequences of the data banks were higher for clones from Station A than from Station 11. In Station A clones appear to be related to different phylogenetic branches. A total of eighteen clones (Figure 1) were relatively close to the *Cytophaga/Flexibacter/Bacteroides* group with similarity values ranging from 88.4 to 94.6%. Five clones were related to the $\alpha$-*Proteobacteria* branch (93.6 to 96.8%). Four clones appeared to be phylogenetically close, albeit at a very low similarity (ranging from 72.6 to 85.6%), to a group of 16S rDNA fragments generated by PCR from a soil sample (Liesack & Stackebrandt, 1992). Two clones clustered with a marine group of bacteria (Fuhrman et al., 1993) and another two with a representative of SAR 11 group from a Sargasso Sea sample (Giovannoni et al., 1990), both groups described only by 16S rRNA sequencing.

In Station 11, twenty-two clones were related to the *Cytophaga/Flexibacter/Bacteroides* group with similar values ranging from 79.1 to 94.7% (Figure 2). Eight clones clustered with the $\alpha$-*Proteobacteria* (sequence similarities between 74.7 and 95.2%) and seven clones were related to High G+C Gram-positive Bacteria (85.6–99.6%). Two clones appeared to be phylogenetically close to $\gamma$-*Proteobacteria* with similarity values ranging from 89.7 to 90%. One clone related to an unknown *Actinomycete* (85.6%), another to *Cyanobacteria* (98.5%) and another to $\delta$-*Proteobacteria.* Although in Station 11, only one clone did not cluster with the known bacterial phylogenetic branches of Figure 2 (viz: Bacterium soil clone group, *Cytophaga/Flexibacter/Bacteroides* group, $\alpha$-$\beta$-$\delta$-*Proteobacteria*, Marine group, High G+C Gram-positive, *Cyanobacteria, Actinomycetes*), the sequence analysis showed a relationship with some members

*Table 4.* Tritiated thymidine incorporation and estimates of bacterial secondary production, growth rates, assimilation and respiration of $^{14}C$-Glucose due to bacteria in water samples from stations in Arcachon Bay (A and B), Certes Domain ($C_1$ and $C_2$) and Prévost lagoon (X and 11).

| Station | Date | Incorporation estimate mol tim $L^{-1}h^{-1}$ $\times 10^{-12}$ | Specific incorporation estimate pmol time $cel^{-1}h^{-1}$ | Bacterial production $\mu g\ C\ L^{-1}h^{-1}$ | Growth rate $\mu$ ($h^{-1}$) | Doubling time $t_d$ (h) | Assimilation mol Gluc $L^{-1}h^{-1}$ $\times 10^{-9}$ | Respiration mol Gluc $L^{-1}h^{-1}$ $\times 10^{-9}$ | Specific assimilation estimate mol Gluc $cel^{-1}h^{-1}$ $\times 10^{-9}$ | Specific respiration estimate mol Gluc $cel^{-1}h^{-1}$ $\times 10^{-9}$ |
|---|---|---|---|---|---|---|---|---|---|---|
| A | May 1993 | 5.20 | $1.04 \times 10^{-5}$ | 0.26 | 0.026 | 26.65 | 1.78 (84.7%) | 0.32 (15.3%) | $3.56 \times 10^{-6}$ | $0.64 \times 10^{-6}$ |
| | Sept 1993 | 10.40 | $0.86 \times 10^{-5}$ | 0.52 | 0.029 | 23.9 | 0.95 (86.2%) | 0.15 (13.8%) | $1.12 \times 10^{-6}$ | $0.178 \times 10^{-6}$ |
| | May 1994 | 1.51 | $0.75 \times 10^{-5}$ | 0.07 | 0.014 | 49.5 | 0.47 (88.1%) | 0.06 (11.9%) | $2.36 \times 10^{-6}$ | $0.318 \times 10^{-6}$ |
| | Sept 1994 | 9.16 | $3.27 \times 10^{-5}$ | 0.46 | 0.081 | 8.55 | 0.42 (91.3%) | 0.04 (8.7%) | $1.48 \times 10^{-6}$ | $0.14 \times 10^{-6}$ |
| $C_1$ | May 1993 | 69.0 | $40.6 \times 10^{-5}$ | 3.45 | 1.006 | 0.69 | 2.71 (73.3%) | 0.98 (26.7%) | $15.9 \times 10^{-6}$ | $5.8 \times 10^{-6}$ |
| | Sept 1993 | 93.0 | $4.65 \times 10^{-5}$ | 4.65 | 0.110 | 6.30 | 6.58 (71.1%) | 2.67 (28.9%) | $3.29 \times 10^{-6}$ | $1.33 \times 10^{-6}$ |
| | May 1994 | 34.0 | $2.42 \times 10^{-5}$ | 1.70 | 0.060 | 11.55 | 20.8 (85.1%) | 3.62 (14.9%) | $6.7 \times 10^{-6}$ | $2.6 \times 10^{-6}$ |
| | Sept 1994 | 47.0 | $2.93 \times 10^{-5}$ | 2.35 | 0.073 | 9.50 | 5.21 (99.9%) | 0.99 (1.9%) | $3.25 \times 10^{-6}$ | $0.62 \times 10^{-6}$ |
| $C_2$ | May 1993 | 50.3 | $5.03 \times 10^{-5}$ | 2.51 | 0.100 | 6.93 | 8.74 (79.1%) | 2.31 (20.9) | $8.74 \times 10^{-6}$ | $2.31 \times 10^{-6}$ |
| | Sept 1993 | 27.0 | $0.64 \times 10^{-5}$ | 1.35 | 0.005 | 128.4 | 9.55 (74.6%) | 3.24 (25.4%) | $2.27 \times 10^{-6}$ | $0.77 \times 10^{-6}$ |
| | May 1994 | ND | ND | ND | ND | ND | 28.1 (84.3%) | 5.24 (15.7%) | $15.6 \times 10^{-6}$ | $2.91 \times 10^{-6}$ |
| | Sept 1994 | 5.44 | $0.28 \times 10^{-5}$ | 0.27 | 0.007 | 97.6 | 0.77 (79.3%) | 0.20 (20.7%) | $0.41 \times 10^{-6}$ | $0.11 \times 10^{-6}$ |
| X | May 1993 | 71.0 | $4.7 \times 10^{-5}$ | 3.55 | 0.174 | 3.94 | 51.3 (93.1%) | 3.82 (6.9%) | $34.2 \times 10^{-6}$ | $2.54 \times 10^{-6}$ |
| | Sept 1993 | 23.0 | $1.64 \times 10^{-5}$ | 1.15 | 0.041 | 16.9 | 1.90 (76.9%) | 0.57 (23.1%) | $1.35 \times 10^{-6}$ | $0.41 \times 10^{-6}$ |
| | May 1994 | 2.20 | $0.23 \times 10^{-5}$ | 0.11 | 0.006 | 123.7 | 0.52 (90.4%) | 0.05 (9.6%) | $0.53 \times 10^{-6}$ | $0.06 \times 10^{-6}$ |
| | Sept 1994 | 4.40 | $0.34 \times 10^{-5}$ | 0.22 | 0.008 | 82.5 | 0.12 (80.3%) | 0.03 (19.7%) | $0.09 \times 10^{-6}$ | $0.02 \times 10^{-6}$ |
| 11 | May 1993 | 70.4 | ND | 3.52 | ND | ND | 18.3 (67.5%) | 8.81 (34.5%) | ND | ND |
| | Sept 1993 | 160 | $1.0 \times 10^{-5}$ | 8.0 | 0.056 | 12.37 | 25.4 (83.5%) | 5.01 (16.5%) | $3.60 \times 10^{-6}$ | $0.71 \times 10^{-6}$ |
| | May 1994 | 21.0 | $7.20 \times 10^{-5}$ | 0.80 | 0.182 | 3.80 | 2.98 (84.7%) | 0.54 (15.3%) | $10.27 \times 10^{-6}$ | $1.84 \times 10^{-6}$ |
| | Sept 1994 | 62.5 | $0.84 \times 10^{-5}$ | 3.13 | 0.021 | 33.0 | 15.1 (91.7%) | 1.37 (8.3%) | $2.0 \times 10^{-6}$ | $0.18 \times 10^{-6}$ |

within these groups. Clone 1611 showed the highest similarity value with *Lactobacillus rimae* (80.4%).

Some clones were found more than once. From the analysis of Station A clone 13a (93.5% with *Flexibacter columnaris*) and clone 10a (93.8% with *Rhodobacter adriaticus*) were found 5 times; clone 41a (85.6% with Bacterium soil clone) and clone 33a (93.9% *Cytophaga latercula*) four times; clone 48a (95.9% with *Clavibacter xyli*) and clone 60a (94.4% *Flexibacter maritimus*) three times; clones 37a (91.7% with *Cytophaga aquatilis*), 14a (93.4% with *Antarcticum vesiculatum*) and 26a (93.4% with *Flexibacter maritimus*) twice each. In Station 11 the following clones appeared twice: clone 1411 (91.6% with *Flavobacteria gondwanense*), clone 1111 (91.9% with *Cytophaga succinicans*), clone 3111 (88.9% with *Arthrobacter globiformis*), clone 4511 (79.5% with *Alcaligenes eutrophus*), clone 7011 (79.1% with *Cytophaga latercula*) and clone 911 (98.5% with *Prochlorococcus marinus*). Clone 2011 (95.2% with *Erytrobacter* sp.) appeared three times. Only one sequence (a 200 bp fragment; post. 283–499) was found to be identical in both Stations, showing 92.5% similarity with *Rhodobacter adriaticus*. An almost identical second sequence (only one nucleotide difference) was also found in Stations A and 11 and its closest relationship was with *Cytophaga marinoflava* (94.6%).

Fourteen randomly taken colonies (seven from each lagoon) isolated in rich medium from the same water samples used for direct DNA extraction were subjected to 16S rDNA analysis (partial 16S rDNA sequencing of the same variable region, 283–499; *E. coli*). All the isolate sequences were different from the sequences obtained by direct PCR amplification. Although again no perfect matching to sequences of the data bases was found here, in both environments most of the isolates had relatively high similarity to the typical heterotrophic aquatic bacteria (*Vibrio, Aeromonas, Alteromonas*). Of the seven isolates from Station A, three showed more than 95% similarity to sequences of known species of this group (95.8% *Alteromonas migrifaciens*, 96% *Vibrio marinus* and 94.9% *Alteromonas haloplanktis*). One showed 89.6% similarity to *Alteromonas haloplanktis*. Two showed over 94% similarity to *Bacillus* species. Finally one isolate showed a 93.8% similarity to the Gram-negative $\gamma$-*Proteobacteria*, *Moraxella phenylpyruvica*. Of the seven isolates from Station 11 three also showed relatively high similarity to be mentioned group of well known aquatic bacteria (96.6% to *Vibrio marinus* and 93.1% to *Alteromonas haloplanktis* and 92.7 to *Vibrio vulnificus*). One showed 90.7% similarity to *Aeromonas schubertii*. Two isolates showed about 90% similarity to *Bacillus cereus*. Finally one showed 90.7% similarity to *Yersinia ruckeri*.

Four strains, representatives of phototrophic bacteria isolated by using specific media and phenotypically characterized in another laboratory (Laboratoire d'Oceanographie Biologique, Bordeaux University) were also included in our study. Isolate 5811, identified as *Thiocapsa* sp. showed the highest similarity value with *Chromatium vinosum* (91%), *Rhodopseudomonas* sp. 2105 showed 95.7% similarity with *Rhodopseudomonas marina*, *Rhodobacter* sp. IO 2102 showed 100% similarity with *Rhodobacter sulfidophilus* and *Chromatium* sp. 2203 showed 87.2% with *Codakia orbicularis* gill symbiont (probably an artifact due to the relatively small number of phototrophic bacteria 16S rRNA sequences deposited in the data banks). Results from isolates representative of the Nitrogen Cycle Group (Dept. of Biological Sciences, Dundee University) were that isolate AAN 013 (ammonifier, Gram-positive, sporulated rod) showed an identical sequence to that of isolate AAN 036 (ammonifier, Gram-positive, sporulated rod), similarity value of 92.3% with *Clostridium bifermentans*. Isolate ANO 011 (denitrifier, Gram-positive cocci) revealed a similarity percentage of 100% with *Staphylococcus cohnii* (all three from Station A). Isolate PNO 019 (anaerobic, Gram-negative cocci) showed 92.7% similarity with *Carnobacterium funditum*, and isolate PAN 031 (amonifier, Gram-positive, sporulated rod), 98.1% with *Eubacterium tenue*, both from Station 11.

## Discussion

### *Heterotrophic bacterial numbers and activities in the sampling sites*

Out of the five sampling sites studied, station A is clearly different from the rest, that are all located in stagnant waters. Station A contains much less bacterial biomass and has less heterotrophic bacterial activity. This is to be expected given the effective tide flushing of the waters located at this site. Station A is very much influenced by the coastal Atlantic waters. In the water column (Table 1), the main difference occurs between Stations A and 11, Stations $C_1$, $C_2$ and X being somewhat intermediate. The numbers obtained were slightly higher at 30 °C in Stations $C_2$, X and 11, which are stagnant waters and probably reach higher tempera-

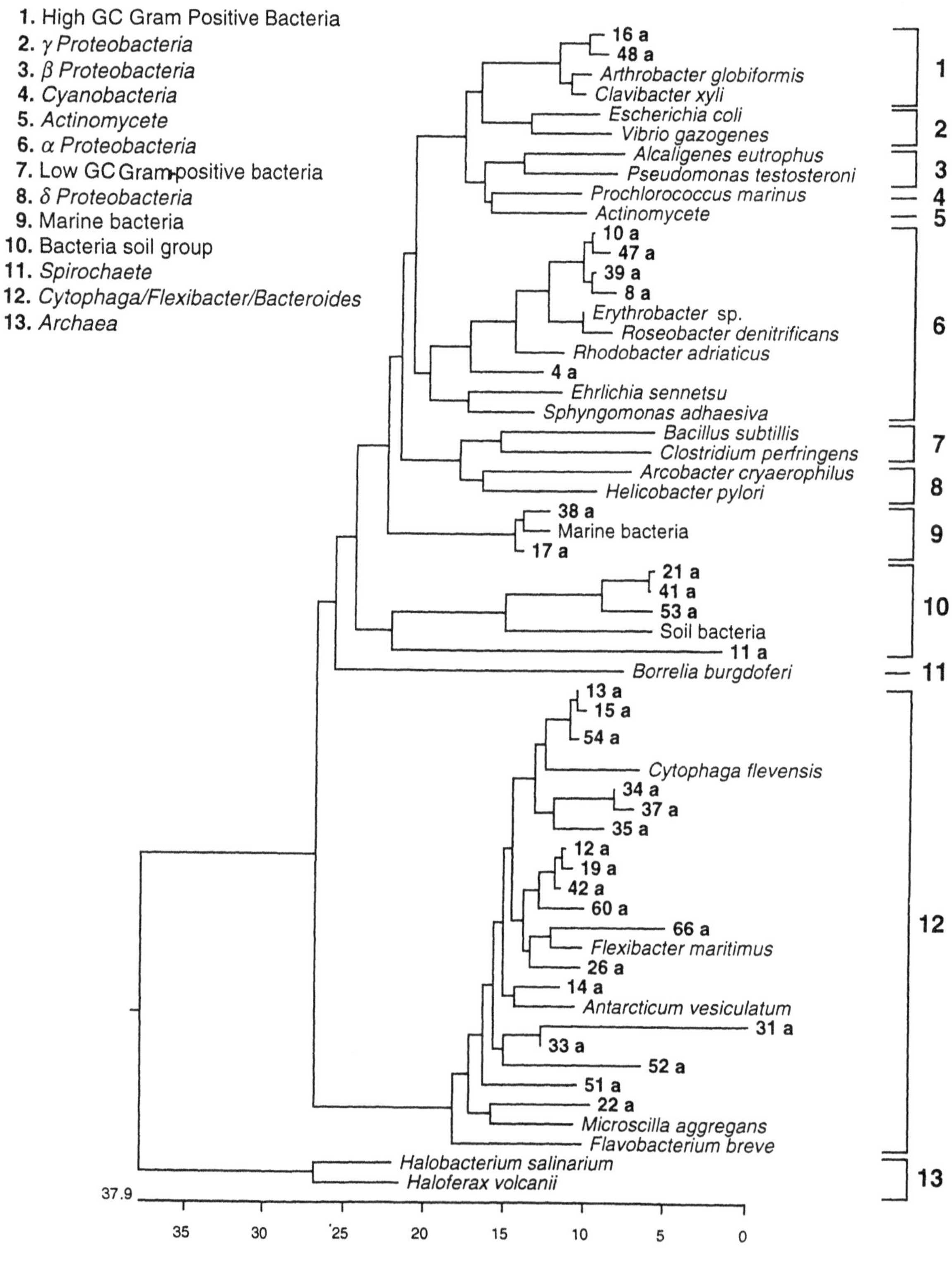


*Figure 1.* Similarity tree showing the relationships of partial 16S rDNA sequences from Arcachon Bay (Station A) and representative sequences of the data base. The scale bar indicates sequence divergence (%).

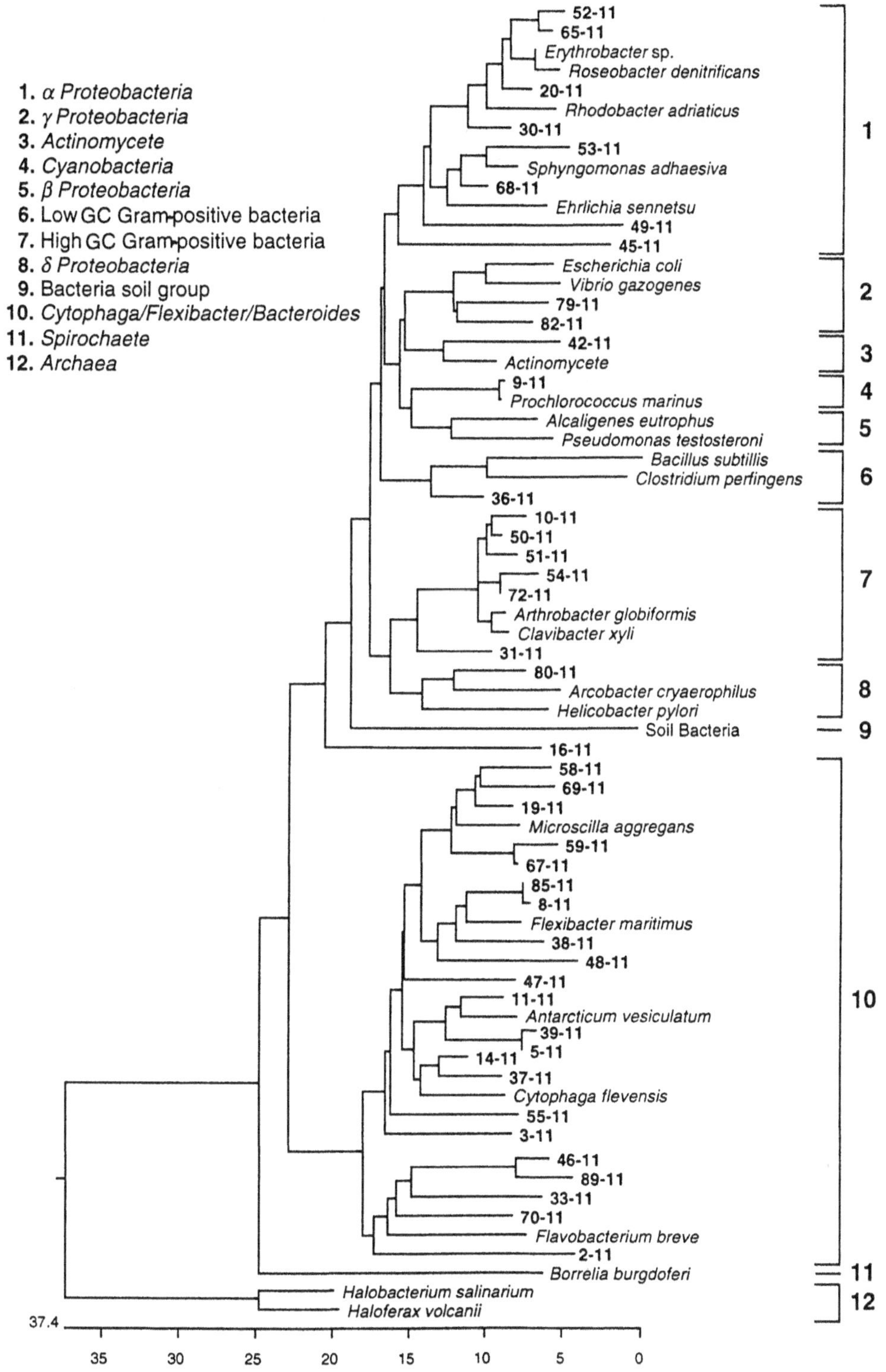


*Figure 2.* Similarity tree showing the relationships of partial 16S rDNA sequences from Prévost Lagoon (Station 11) and representative sequences of the data base. The scale indicates sequence divergence (%).

tures than Station A which is much more similar to open ocean waters known to contain important numbers of psychrophilic microbes. In any case the water in the Prévost Lagoon, particularly in the areas distant from the Mediterranean Sea Inlet (Station 11) is characterized by considerably higher viable counts, which are in agreement with the known higher productivity of this lagoon (Caumette, 1986). The presence of significant faecal contamination, as indicated by faecal coliforms (Table 2) and sulphite-reducing clostridia (Table 3) in Station 11 (14–20 FC and 50–60 clostridia/100 ml) is a clear indication that this lagoon is subject to considerable urban pollution from the surrounding population centres. In contrast, Arcachon Bay appears to be virtually free of significant faecal contamination. The presence of clostridia in the sediment is not significant since the spores can be accumulated in this environment for very long periods, nor is it an appropriate pollution indicator in this environment. Even so, the increase in clostridia counts at the end of the summer in Arcachon Bay is significant, probably as an indication of the cumulative pollution suffered throughout the summer season.

The sediments varied considerably from the above pattern. In the top two centimetres of the sediment (Table 1) the numbers of heterotrophs seem to be the equivalent of the water numbers, i.e. Station A showed the highest counts, followed by $C_2$, B and $C_1$, while the Prévost Lagoon stations showed lower numbers in the superficial sediment. This apparent discrepancy can be explained if we assume that while in Prévost Lagoon the larger (or a very significant) part of the primary productivity takes place in the water column due to the low light penetration in these eutrophic waters while in the more oligotrophic waters of Arcachon Bay most of the primary productivity takes place in the surficial sediment, richer in nutrients and directly exposed to sunlight at low tide. In the deep sediment (Table 1) sample results revert to the water pattern showing higher numbers in the Prévost Lagoon, probably because this deep anaerobic sediment is not related to the superficial sediment or water productivity. In any case the deep sediment samples are probably dominated by strict anaerobes that are not detected by present culture techniques used.

Ducklow & Carlson (1992), in an extensive study of oceanic bacterial production, concluded that bacterial biomass is controlled by resources rather than by predation. Bacterial abundance and bacterial production are statistically correlated with phytoplankton biomass (as Chl). Positive correlations between bacteria and Chl suggest that bacteria should respond to increases in phytoplankton nutrients. If bacterial populations were controlled by forces acting via bacteriovore ingestion, bacterioplankton abundance should be inversely correlated with bacteriovore abundance or biomass, and there is no evidence of this.

Heterotrophic bacterial production values in open oceans range from 0.08 $\mu$g C $L^{-1}h^{-1}$ in the Warm Core Ring Gulf to 0.45 $\mu$g C $L^{-1}h^{-1}$ in the Coral Sea, Australia. In coastal ocean waters, from 2.2 $\mu$g C $L^{-1}h^{-1}$ at Scripps Pier to 91 $\mu$g C $L^{-1}h^{-1}$ in Bietri, Ivory Coast (Ducklow and Carlson, 1992). Values are higher in coastal waters than in open ocean waters. Thymidine incorporation in Arcachon Bay ranged from 0.014 to 0.52 $\mu$g C $L^{-1}h^{-1}$, with values similar to those for open oceans. In the Certes Domain the average value was 2.8 $\mu$g C $L^{-1}h^{-1}$ and in Prévost Lagoon 4.55 $\mu$g C $L^{-1}h^{-1}$. These data support our original hypothesis that both systems represent very different situations that can be considered as different models. One (Arcachon) with a low level of eutrophication and urban pollution, relatively low productivity and strong influence of the surrounding ocean waters; the other (Prévost) highly eutrophic, with significant levels of urban pollution and very different (isolated) from the neighbouring Mediterranean Sea waters.

### *Analysis of bacterial biodiversity by direct amplification and sequencing of the small subunit rRNA genes*

In oligotrophic ocean waters the studies on microbial diversity of naturally occurring organisms by molecular approaches have shown that clones were mostly phylogenetically related with *Cyanobacteria* (Fuhrman et al., 1993; Giovannoni et al., 1990) SAR11 group ($\alpha$-*Proteobacteria*) (Giovannoni et al., 1990), $\gamma$-*Proteobacteria* (Fuhrman et al., 1993), Marine Group A (Fuhrman et al., 1993) (a new phylogenetic branch) and novel groups of the *Archaea* domain (DeLong, 1992; Fuhrman et al., 1992; DeLong et al., 1994). Minor numbers were related with Gram-positives (Fuhrman et al., 1993), *Cytophaga/Flexibacter/Bacteroides* (Fuhrman et al., 1993) and *Planctomyces* (Fuhrman et al., 1993). Other molecular approaches to compare coastal waters with open ocean waters have been used. For example, an oligonucleotide probe derived from the SAR11 (Giovannoni et al., 1990) sequence retrieved from the Sargasso Sea was hybridized against total RNA extracted from picoplankton samples obtained from both origins (Giovannoni et al., 1990). The results showed that the relative importance of these groups of *Proteobacteria* was lower in several coastal samples, suggesting that the organisms contributing these sequences may be more abundant in relatively oligotrophic open ocean waters. Another approach compared different samples by hybridization of total DNA obtained from different sampling sites (Lee & Fuhrman, 1991), the results of this work also suggested that the bacterial community of coastal waters may be different from that of open ocean water.

The bacterial biodiversity revealed by our 16S rDNA study in the two coastal lagoons showed some differences from any of those previous studies carried out in open ocean waters. Two clones from Station A (8a and 39a) and one from Station 11 (6511) were closely related to the widely distributed SAR11 group (94.2%, 96.8% and 88% respectively). But these are very few representatives in comparison with the larger number of clones found in ocean waters (Fuhrman et al., 1993). In both stations, we also found clones related with the Marine Group A described by Fuhrman et al. (1993): clone 38a and 17a (98.9% and 96.1%) and clones 1911, 211, 6911 and 5511 (86.7%, 70%, 89.8% and 82.0% respectively). So the novel phylogenetic lines described as being predominant in oligotrophic ocean waters are present in coastal lagoons although the proportion of clones belonging to these groups retrieved was much lower, suggesting that they could be less abundant in coastal lagoons. Another group that was relatively little represented in our study is the Cyanobacteria that was found as an important component of the picoplankton of open ocean waters (Giovannoni et al., 1990). Only a clone in Station 11 was similar to a Cyanobacteria. The above does not necessarily mean that this group of prokaryotes is not present in the environments studied. The large cellular size of most members of the well known groups of Cyanobacteria excludes them from our study since they would be retained by the glass fibre filter which nominal pore size is 2.7 $\mu$m.

On the other hand, in the coastal lagoons studied here we have observed a great number of representatives of the *Cytophaga/Flexibacter/Bacteroides* group. There is evidence that members of the *Cytophaga/Flexibacter/Bacteroides* group are present in ocean waters but in macroaggregate-associated bacterial populations (DeLong et al., 1993). These microorganisms are Gram-negative bacteria, gliding, aerobic, organotrophs. Many of them are able to degrade biomacromolecules like proteins, chitin, pectin, agar, starch or cellulose and probably play a major role in the turnover of matter in nature. Their main habitats are soils, decaying plant material and animal dung (Reichenbach, 1991). It seems that at the end of summer, after proliferation of macroalgae (*Ulva* sp., *Enteromorpha* sp.) decomposition of the macroalgae material would stimulate development of Cytophagales. They could be even more important in the aggregate-associated biota that is probably retained by the glass-fibre filter.

In Station A, four clones were phylogenetically related to a novel eubacterial soil group described by Liesack and Stackebrandt (1992). This phylogenetic line was not represented in oligotrophic ocean waters, it could reflect the continental influence in coastal lagoons. We also found clones related with members of Gram-positives (low and high G+C content). Fuhrman (1993) found 9 clones belonging to a new phylogenetic group apparently related to the high G+C Gram-positive Bacteria and represented by clone BDA1-5, a single isolate from the Pacific was identical to this one and the rest varied only at 1 or 2 positions. However, the Gram-positive type sequences that we found are not related to those found by Fuhrman (1993). They could also be of terrestrial origin. Remarkable also is the relatively low representation found for $\gamma$-*Proteobacteria* (only 2 clones in Station 11 and none in Station A). This group has been found in open ocean waters by the 16S rDNA technique (Fuhrman et al., 1993) but they were also in a minority, which is in contrast with previous studies about oligotrophic ocean waters. However, most of the heterotrophic bacteria that are easily isolated from the ocean belong to this phylogenetic cluster. Furthermore, our own isolates from the lagoons studied, although none showed a perfect match with any sequence deposited in data banks, all corresponded to organisms belonging to the $\gamma$-*Proteobacteria*. These results support the view that some groups of the $\gamma$-*Proteobacteria* (*Vibrio* for example) are particularly well fitted for growth in laboratory cultures, and although abundant in marine habitats, they are probably in a minority or at least not predominant in this kind of environment.

Regarding both coastal aquatic systems, the study of their diversity reflects some shared similarities. The major group represented is *Cytophaga/Flexibacter/Bacteroides* in both Stations and $\alpha$-*Proteobacteria*. However, they also differ in some aspects. An analysis in detail revealed that only one identical sequence was found in both environments (with a 92.5% similarity to *Rhodobacter adriaticus*). The Prévost Lagoon showed higher diversity by using the 16S rRNA gene amplification technique. The number of different sequences retrieved was larger: 43 different sequences in the 52 clones studied, while from Arcachon out of the same number of clones only 31 different sequences were found i.e. redundancy was higher. Furthermore, in Arcachon we found representatives of only five of the major phylogenetic branches of eubacteria (Woese, 1987) while in the Prévost lagoon we found representatives of ten different phylogenetic branches including one *Actinomycete*, one *Cyanobacteria* and a relatively large number of low G+C Gram-positives. These results suggest that bacterial biodiversity is higher in the eutrophic Mediterranean lagoon that in the relatively oligotrophic and colder Atlantic one and that there are also some differences regarding the groups represented in both environments.

## Acknowledgements

This work was supported by grants PB 93/0930 of the DGICYT, BIO 93/0750 of the CICYT, and the C.L.E.A.N. project contract No. EV5V-CT92-0080 of the European Commission. A.J.M.M. was recipient of a post-doctoral contract of the Spanish Ministry of Education and Science. Thanks are offered to Prof. Waltre Stühmer for technical support to carry out the phylogenetic analysis. Secretarial assistance by K. Hernández is gratefully acknowledged.

## References

Amann, R. I., W. Ludwig & K.-H. Schleifer, 1995. Phylogenetic identification and in situ detection of individual microbial cells without cultivation. Microbiol. Rev. 59: 143–169.

APHA-AWWA-WPCF, 1975. Standard methods for the examination of the water and wastewater. 14th ed., (Rand, M. C., A. E. Greenberg & M. J. Taras, Eds), American Public Health Association, Washington.

Beaucage, S. L. & M. H. Caruthers, 1981. Dideoxynucleotide phosphoramidites – a new class of key intermediates for deoxypolynucleotide synthesis. Tetrahedrom Lett. 22: 1859–1862.

Benlloch, S., F. Rodríguez-Valera & A. J. Martínez-Murcia, 1995. Bacterial diversity in two coastal lagoons deduced from 16S rDNA PCR amplification and partial sequencing. FEMS Microbiol. Ecol. 18: 267–280.

Bradshaw, L. J., 1992. Laboratory Microbiology. Fourth Edition. Saunders College Publishing.

Brosius, J. P., T. J. Dull, D. D. Sleeter & H. F. Noyer, 1981. Gene organization and primary structure of a ribosomal RNA operon from *E. coli*. J. Mol. Biol. 148: 107–127.

Castel, J., P. Caumette & R. Herbert, 1996. Eutrophication gradients in coastal lagoons. Hydrobiologia 329: xi–xxx.

Caumette, P., 1986. Phototrophic sulfur bacteria and sulfate/reducing bacteria causing red waters in a shallow brackish coastal lagoon (Prévost Lagoon, France). FEMS Microbiol. Ecol. 38: 113–124.

DeLong, E. F., 1992. Archaea in coastal marine environments. Proc. Natl. Acad. Sci. USA 89: 5685–5689.

DeLong, E. F., D. G. Franks & A. L. Alldredge, 1993. Phylogenetic diversity of aggregate-attached vs free-living marine bacterial assemblages. Limnol. Oceanogr. 38: 924–934.

DeLong, E. F., K. Y. Wu, B. B. Prézelin & R. V. M. Jovine, 1994. High abundance of Archaea in Antarctic marine picoplankton. Nature 371: 695–697.

Ducklow, H. W. & C. Carlson, 1992. Oceanic Bacterial Production. Adv. Microb. Ecol. 12: 113–181.

Fuhrman, J. A. & F. Azam, 1980. Bacterioplankton secondary production estimates for coastal waters of British Columbia, Antarctica and California. Appl. envir. Microbiol. 39: 1085–1095.

Fuhrman, J. A., D. E. Comeau, A. Hagstron & A. Chan, 1988. Extraction from natural planktonic microorganisms of DNA suitable for molecular biological studies. Appl. envir. Microbiol. 54: 1426–1429.

Fuhrman, J. A., K. McCallum & A. A. Davis, 1992. Novel major archaebacterial group from marine plankton. Nature 356: 148–149.

Fuhrman, J. A., K. McCallum & A. A. Davis, 1993. Phylogenetic diversity of subsurface marine microbial communities from the Atlantic and Pacific Oceans. Appl. envir. Microbiol. 59: 1294–1302.

Giovannoni, S. J., T. B. Britschgi, C. L. Moyer & K. G. Field, 1990. Genetic diversity in Sargasso Sea bacterioplankton. Nature 345: 60–63.

Jannasch, H. W. & G. E. Jones, 1959. Bacterial population in sea water as determined by different methods of enumeration. Limnol. Oceanogr. 4: 128–140.

Lane, D. J., B. Pace, G. J. Olsen, D. A. Stahl, M. L. Sogin & N. R. Pace, 1985. Rapid determination of 16S rDNA sequences for phylogenetic analysis. Proc. natn. Acad. Sci. USA 82: 6955–6959.

Lee, S. & J. A. Fuhrman, 1991. Spatial & temporal variation of Natural Bacterioplankton assemblages studied by total genomic DBA cross-hybridization. Limnol Oceanogr. 36: 1277–1287.

Liesack, W. & E. Stackebrandt, 1992. Occurrence of novel groups of the Domain Bacteria as revealed by analysis of genetic material isolated from an Australian Terrestrial Environment. J. Bacteriol. 174: 5072–5078.

Maniatis, T., E. F. Fritsch & J. Sambrook, 1982. Molecular cloning: a laboratory manual. Cold Spring Harbor Laboratory Press, Cold Spring Harbor, N.Y.

Marchuk, D., M. Drumm, A. Saulino & F. S. Collins, 1990. Construction of T-vectors, a rapid and general system for direct cloning of unmodified PCR products. Nucl. Acids. Res. 19: 1154.

Martinez-Murcia, A. J., 1993. Doctoral Thesis. Department of Microbiology, University of Reading, Reading, U.K.

Martinez-Murcia, A. J. & F. Rodriguez-Valera, 1994. The use of artibrarily primed PCR (AP-PCR) to develop taxa specific DNA probes of known sequence. FEMS Microbiol. Lett. 124: 265–270.

Moriarty, D. J. W., 1990. Techniques for estimating bacterial growth rates and production of biomass in aquatic environments. Methods in Microbiology. Vol. 22. Pag. 211–234.

Pace, N. R., D. A. Stahl, D. J. Lane & G. J. Olsen, 1986. The analysis of natural microbial populations by ribosomal RNA sequences. Adv. Microb. Ecol. 9: 1–55.

Reichenbach, H., 1991. The Order Cytophagales. In: The Prokaryotes, Second Edition (Balows, A., H. G. Trüper, M. Dworkin, W.

Harder & K.-H. Schleifer, Eds), pp. 3631–3675, Springer Verlag, New York.

Sanger, F., S. Nicklen & A. R. Coulson, 1977. DNA sequencing with chain-terminating inhibitors. Proc. natn. Acad. Sci. USA 74: 5463–5467.

Schmidt, T. M., E. F. DeLong & N. R. Pace, 1991. Analysis of a marine picoplankton community by 16S rRNA gene cloning and sequencing. J. Bacteriol. 173: 4371–4378.

Woese, C. R., 1987. Bacterial Evolution. Microbiol. Rev. 51: 221–271.

Wright, R. T. & J. E. Hobbie, 1966. Use of glucose and acetate by bacteria and algae in aquatic ecosystems. Ecology 47: 447–464.

Hydrobiologia **329**: 19–31, 1996.
P. Caumette, J. Castel & R. Herbert (eds), Coastal Lagoon Eutrophication and ANaerobic Processes (C.L.E.AN.).




# Description of prokaryotic biodiversity along the salinity gradient of a multipond solar saltern by direct PCR amplification of 16S rDNA

Susana Benlloch, Silvia G. Acinas, A. J. Martínez-Murcia & Francisco Rodríguez-Valera*
*Departamento de Genética y Microbiología, Universidad de Alicante, Campus de San Juan, Apartado 374, 03080 Alicante, Spain*
* *Author for correspondence. Fax: 34-6-5941787; E-mail: FRVALERA@UA.ES*

*Key words:* Biodiversity, 16S rRNA, halophilic *Bacteria*, halophilic *Archaea*, hypersaline environments

## Abstract

New methods based on PCR amplification of 16S rRNA genes from DNA samples extracted directly from the environment allow the description of microbial diversity in natural ecosystems without the need for cultivation. We have applied this technique to an extreme environment presumed to have very low diversity: the crystallizer ponds of a marine saltern with salinity over NaCl saturation. The molecular methodology has shown that indeed very low diversity can be found here. Prokaryotes belonging to the *Bacteria* domain are a minor component and only members of a closely related cluster of sequences were found, all relatives of the *α-Proteobacteria* (ca. 83% to *Rhodopseudomonas marina*). Halophilic *Archaea* were as expected the largest component of biomass in this environment. All the clones sequenced corresponded again to a highly homologous cluster (probably members of the same genus). However, all the sequences diverged considerably from the ones of the described genera of halophilic *Archaea*, in fact the data are consistent with the idea that the 16S rRNA genes directly amplified from the saltern correspond to members of an undescribed genus. This is remarkable since many collection strains sequenced come specifically from this saltern. Furthermore, 16S rDNA obtained from archaeal cultures isolated from the same sample had no homology to the sequences obtained by PCR amplification, instead they appear to be members of the well known genus *Haloarcula*. However, this concurs with the findings of other authors who obtained different organisms by culture from those detected by the sequences retrieved directly by PCR. A possible explanation is that culturability, in standard media, is the exception rather than the rule. To study the biodiversity gradient present along the salinity gradient found in a multi-pond solar saltern we have also applied a novel molecular strategy. This method is based on the restriction digestion of a population of 16S rDNA sequences directly amplified from an environmental sample. Digested fragments separated by polyacrylamide gel electrophoresis generate characteristic profile data for estimation of diversity and overall similarities between the organisms of different environments. The methodology has been applied to a set of five ponds covering the salinity gradient from about twice that of seawater (6.4%) to NaCl precipitation (30.8%). Bacterial (eubacterial) diversity estimated from the complexity of the banding pattern obtained by restriction of the amplicons from the different ponds decreased with increasing salinity while for *Archaea* (archaebacteria) the reverse was true i.e. the higher the salinity the higher the number of bands. The similarities in taxonomic composition of the prokaryotic populations present in those ponds were evaluated from the number of restriction bands shared by the different samples. The relationships found among the different environments were independent of the enzyme used for digestion and were consistent with previous descriptions obtained by the study of isolates from the different environments. This technique appears to be promising as a rapid method for microbial biodiversity fingerprinting useful to compare several environments and detect major shifts in species composition of the microbial population.

## Introduction

To assess the true degree of microbial diversity, particularly prokaryotic, in natural environments is a key question for microbiologists. The classical techniques that involve culture and isolation of the microorganisms present have the essential disadvantage of the so-called non-culturable (or non-cultivated) species. For example in marine picoplankton only 1 in $10^5$ to 1 in $10^3$ of microscopically observed cells develop colonies on marine agar plates (Giovannoni et al., 1990b), while it is known that about half are metabolically active. DNA reassociation techniques have shown that only 1 in about 200 genomes present in a Norwegian soil was recovered by culture (Torsvik et al., 1990). This major drawback has recently been bypassed – at least to some extent – by the use of molecular techniques (Pace et al., 1986). The most widely used and, up to the moment, most fruitful approach has been the PCR amplification (or cloning) of 16S rDNA directly from the sample (DeLong, 1992; DeLong et al., 1994; Fuhrman et al., 1993; Giovannoni et al., 1990a; Giovannoni et al., 1990b; Liesack & Stackebrandt et al., 1992; Schmidt et al., 1991; Stackebrandt et al., 1993; Torsvik et al., 1990). The sequences of the rDNA genes retrieved from the environment give an estimate of the microbial diversity present. The comparison of the sequences with data banks allows the phylogenetic placement of the microorganism. If the amplified DNA were to show enough homology to a cultivated taxonomic group it would even be possible to predict the physiology of the microorganism detected and concomitantly its putative ecological role. However, the number of samples and sequences per sample is limited by the sequencing capabilities of the laboratory and therefore most studies that have been carried out investigate only about 100 sequences at the most. To be able to evaluate the biodiversity present in a sample rapidly, and also to compare several samples, for example when time or spatial gradients are studied, a more expedient methodology is required. Restriction fragment length polymorphism of 16S rDNA has been shown to be highly informative (Avaniss-Aghajani et al., 1994; Moyer et al., 1994; Ralph et al., 1993) and could therefore be used to obtain a fingerprint of microbial biodiversity in the environment under study with a relatively small investment of time and cost, so allowing several samples to be analyzed simultaneously.

The cloning and PCR amplification of 16S rDNA technique has already been applied to a number of marine, soil and thermal (extreme) natural environments (Barns et al., 1994; DeLong, 1992; DeLong et al., 1994; Fuhrman et al., 1993; Giovannoni et al., 1990a, 1990b; Liesack & Stackebrandt et al., 1992; Schmidt et al., 1991; Stackebrandt et al., 1993; Torsvik et al., 1990; Ward et al., 1990; Weller et al., 1991). This has lead to the discovery of a remarkable unknown diversity of both *Eubacteria* and *Archaea*. A new eubacterial group distantly related to *α-Proteobacteria* represented by clone SAR 11 first described by Giovanonni et al. (1990a) is widely distributed, being found in the Atlantic and Pacific Oceans (Fuhrman et al., 1993). A new archaeal phylogenetic branch diverging deeply from the thermophilic branch or *Crenarchaeota* has also been found in ocean waters including deep ocean samples (500 m) and permanently cold Antarctic waters (DeLong et al., 1994; Fuhrman et al., 1992). The low temperature of the environments from which these sequences were recovered, as well as the low G + C of the 16S rDNA lead the authors to suggest that the phenotype of the corresponding organism would not be thermophilic at all and therefore very different from other *Crenarchaeota*. A novel line of *Bacteria* was also discovered in soil with less than 75% homology to any known organism (Liesack & Stackebrandt et al., 1992). The fact that no perfect matching to any known sequence has been found for any of the 16S rDNAs directly retrieved from the environment, not even in soil, an environment lengthily studied by classical microbial ecology approaches, is also remarkable. There are also a number of papers reporting the results of the technique applied to thermal environments. Since extreme environments are characterized by low biological diversity, results could be simpler to interpret. In fact the number of different sequences found was in general lower. For example, from a pink filament community in Octopus Spring (84–88 °C, located in Yellowstone National Park) of 35 clones studied 26 were identical (Reysenbach et al., 1994). In the cyanobacterial mat also at Octopus Spring (waters at 55 °C), 5 identical sequences were found in 14 clones studied (Weller et al., 1991). By restriction analysis of 16S rRNA genes the clones obtained in a deep-sea hydrothermal vent, after application of statistical tests, showed that only about 10 OTU (Operational Taxonomic Units) were present in the environment (Moyer et al., 1994). In a study carried out with the sediment in another thermal environment, Jim's Black Pool (74 °C), again in Yellowstone, of 98 clones obtained with archaeal primers, three sequences were repeated 23, 8 and 7 times respectively while the remainder were represented 1–4 times in the collection (Barns et al., 1994). However, in this work a considerable diversity of archaeal 16S rRNA sequences was found and although some clones had high sequence similarity to cultivated thermophiles (about 97%) others showed great evolutionary distance from any known organism. Nevertheless, a perfect matching between clones and 16S rRNA sequences of cultivated microorganisms has never been described, even when isolated from specifically the same environment and location.

Hypersaline environments represent a very interesting model, particularly the multipond saltern in which a gradient of salinities develops spatially, different ponds representing different salinity values that range from sea water (ca. 3.5% w/v) to NaCl precipitation (over 35% w/v) (Javor, 1983a; Rodríguez-Valera et al., 1985). Along this gradient biodiversity decreases as the environment becomes more hypersaline i.e. more extreme. In the crystallizer ponds, where NaCl is harvested, only members of the *Archaea* domain are recovered by culture and they belong to only 3 or 4 species of haloarchaea (formerly halobacteria) (Oren, 1990; Rodríguez-Valera et al., 1985). Microscopically, samples of the crystallizer water also show very little diversity, long slender rods predominating, often gas vacuolated similar to the cells of some species of the genus *Halobacterium* and some square bacteria like those described by Walsby (Javor, 1983b; Oren, 1990; Rodríguez-Valera et al., 1985; Walsby, 1980).

In the present work direct 16S rDNA amplification and sequencing were applied to the DNA obtained from a crystallizer pond. Bacterial and archaeal populations were also characterized by the analysis of the 16S rRNA genes directly amplified from the environment and digested with frequent cutting restriction endonucleases. The technique has indeed allowed us to characterize bacterial biodiversity in five selected ponds with different salinities (6.4–30.8%) and to compare the similarity of prokaryotic populations present in those ponds.

## Materials and methods

### *Salinity*

The total salt concentrations were determined by the ash content of 10 ml of water dried at 110 °C for 24 h and afterwards at 400 °C for 4 h.

### *Media and strains*

To isolate pure cultures from the crystallizer water, 100 $\mu$l were inoculated onto plates containing 25% (W/V) salt water agar supplemented with 0.5% yeast extract (Rodríguez-Valera et al., 1981). Colonies were isolated at random and subcultured twice; 1 ml of bacterial cultures was harvested in sterile 25% salt water by centrifuging at 10000 rpm for 2 min and pellets were stored at −20 °C. Culture collection strains, *Haloferax mediterranei* ATCC 33500$^T$, *Haloferax volcanii* ATCC 29605$^T$, *Haloferax gibbonsii* ATCC 33959$^T$, *Haloferax denitrificans* ATCC 35960$^T$, *Haloferax 'larsenii'* ATCC 49116, *Haloarcula hispanica* ATCC 33960$^T$, *Haloarcula californiae* ATCC 33799$^T$, *Haloarcula marismortui* ATCC 43049$^T$, *Halobacterium salinarium* ATCC 33171$^T$, *Halobacterium lacusprofundii* ATCC 49239$^T$, *Halococcus morrhuae* CCM 537$^T$, *Halococcus saccharolyticus* ATCC 49257$^T$, were cultivated on 25% (w/v) salt water agar supplemented (Rodríguez-Valera et al., 1981) with 0.5% yeast extract, 1% starch and 1% glucose at 37 °C for approximately one week. *Escherichia coli* ATCC 25922, *Pasteurella multocida* ATCC 43137$^T$, *Listeria monocytogenes* ATCC 15313$^T$, *Proteus mirabilis* ATCC 29906$^T$, *Salmonella* sp. ATCC 35664 and *Aeromonas salmonicida* NCIMB 1102$^T$ were cultivated as recommended by the American Type Culture Collection.

### *Biomass collection*

Approximately 5 l of superficial water (10 cm below the surface) were collected from ponds with different salinity of a solar saltern located in Santa Pola (Alicante, Spain). Debris and eukaryotic cells were removed by using glass microfiber filters (Whatman GF/D). Microorganisms were collected by positive pressure filtration on a 9 cm, 0.22 $\mu$m pore-size filter (Durapore; Millipore Corp., Bedford, MA). Water was pumped until filter clogging decreased the flow rate to a few ml min$^{-1}$. Filters were stored at −80 °C until DNA extraction.

### *DNA extraction and purification*

Thawed filters were cut with a sterile razor blade into small pieces and vortexed in 5 ml of 1% w/v SDS, TE buffer (100 mM Tris-HCl; pH 8; 10 mM EDTA), in 50 ml conical based polypropylene centrifuge tubes. Acid-washed glass beads (Sigma) were added and vortexed until resuspension (1–2 min). The sample was incubated at 37 °C for 1 h after addition of 50 $\mu$l of lysozyme (Boehringer) in TE buffer (1% w/v). After boiling for 10 min and vortexing for 2 min, 1 ml of lysis buffer (4% w/v SDS; 100 mM EDTA; 50 mM Tris-HCl; pH 8), and 30 $\mu$l proteinase K (14.7 mg ml$^{-1}$ stock solution) was added. The samples were vortexed for 1 min, incubated at 56 °C for 1 h and boiled for 5 min. As no complete lysis was observed at this step, a thermal shock (5 min at −80 °C and 5 min at boiling) was applied. The DNA was extracted once with a volume of phenol-Tris (pH 8) and a volume of chloroform-$H_2O$; 0.1 volume of 2M NaAc (pH 4.8) and 3 volumes of absolute ethanol were added and kept overnight at −20 °C. After centrifuging at 12 000 rpm for 10 min, the pellet was dried and DNA purified using the GeneClean II Kit. DNA from bacterial cultures was extracted as previously described (Martínez-Murcia & Rodríguez-Valera, 1994). Quantification and purity (referred to protein content) of the DNA obtained were estimated and 5 μl aliquots of DNA dilutions were analyzed (Martínez-Murcia & Rodríguez-Valera, 1994).

*PCR amplification of 16S rDNA*

Reaction mixtures contained 50 mM KCl, 10 mM Tris-HCl (pH 9), 1.5 mM $MgCl_2$, 0.1% Triton X-100, 200 μM each of deoxyribonucleotide triphosphate (dATP, dCTP, dGTP, dTTP; Pharmacia LKB Biotechnology), 2U of *Taq* I DNA polymerase (Promega), 0.2 μM each of oligonucleotide primer and 100 ng of template DNA in a total volume of 50 μl. The sequences of the forward primers were 5′-TTCCGGTTGATCCTGCCGGA-3′ (DeLong, 1992) ranging from nucleotide positions 2 to 21 of the *Halobacterium salinarium* (Larsen & Grant, 1989) numbering system (Mankin et al., 1985) for the Archaeal Domain and 5′-AGAGTTTGATCATGGCTCAG-3′ (Lane et al., 1985) at nucleotide positions 8 to 27 of the *E. coli* numbering (Brosius et al., 1981) for the Eubacterial Domain. Reverse primer for both Domains was 5′-GGTTACCTTGTTACGACTT-3′ (Lane et al., 1985) at positions 1509–1491 (*E. coli* numbering). Reaction mixtures were overlaid with mineral oil (Light White Oil; Sigma) prior to a thermal cycling regime (35 cycles on a Perkin Elmer Cetus 480) of 94 °C for 1 min, 55 °C for 1 min, and 72 °C for 2 min. After the final cycle was completed, the amplicons were allowed to extend at 72 °C for 10 min. Negative controls were included with no addition of template DNA. Five μl of the PCR products were analyzed on 1% (w/v) agarose gels to check for purity. PCR products were precipitated, dried and resuspended in 25 μl of sterile distilled water.

*Purification and cloning of the PCR products*

PCR products were purified using GeneClean II kit and cloned into a T-tailed-vector prepared as follows: Bluescript plasmid (Stratagene, La Jolla, CA) was digested with EcoRV and incubated with *Taq* I DNA polymerase (1 U/μg plasmid/20 μl volume) using standard buffer conditions in the presence of 2 mM dTTP for two hours at 72 °C (Marchuk et al., 1990) and ligated at 16 °C overnight. XLBlue *Escherichia coli* strain was transformed with a standard protocol (Maniatis et al., 1982). Plasmidic DNA was extracted using the Wizard Minipreps DNA Purification System (Promega).

*Primers and sequencing reactions*

Primers with 5′-TTACCGCGGCTGCTGGCACG-3′ sequence (Martínez-Murcia, 1993) at positions 533–514 (*E. coli* numbering) for the Bacterial Domain and 5′-CCCCGTAGGGCCCGG-3′(Barns et al., 1994) at positions 330–316 (*H. salinarium* numbering) for the Archaeal Domain were chosen to sequence ca. 200 bp of a relatively variable region of each selected clone. Sequencing primers 5′-GGATTTCACTCCTACCCC-3′ and 5′-CCATTGTAGCCCGCGTGT-3′ at positions 646–629 and 1203–1186 respectively (*H. salinarium* numbering) were designed in the present study from a 16S rRNA sequence alignment of members of *Halobacteriaceae* (Kamekura & Seno, 1993). These together with the primers 5′-ACCGGCGTTGACTCCAATT-3′ (DeLong, 1992) and 5′-GACGGGCGGTGTGTGCA-3′ (Barns et al., 1994) at positions 929–912 and 1368–1352 respectively (*H. salinarium* numbering) and the Bluescript sequencing primers, were used when determination of the complete archaeal 16S rDNA sequences were required. Primers were synthesized by the phosphoramidite method (Beaucage & Caruthers 1981) on a Model 392 RNA/DNA Applied Biosystems synthesizer. Clones randomly selected were sequenced by the dideoxy-chain termination method (Sanger et al., 1977) with a Sequenase Version 2.0 Kit (United States Biochemicals) following the manufacturer's protocol.

*Southern hybridization*

Different amounts of DNA extracted from the crystallizer were dot blotted onto a Hybond-N-Nylon membrane (Amersham). Total 16S rDNA amplified with either eubacterial or archaeal primers were digoxigenin labelled by random primed synthesis according to the procedure recommended by Boerhinger Mannhein (DIG-System, Boerhinger Mannheim) and then separately hybridized to the total DNA blots. Hybridization was carried out overnight at high stringency conditions (68–70 °C). Filters were washed twice for 5 min at room temperature with 2 × SSC, 0.1% SDS and twice for 15 min at 68 °C with 0.1 × SSC and 0.1% SDS. Chemoluminescent detection was performed using Lumiphos™ as a substrate for the alkaline phosphate linked to the anti-digoxigenin antibody (Boerhinger Mannheim) following the recommendations of the manufacturer. After two washes for 15 min in 0.2 N NaOH, 0.1% w/v SDS at 37 °C, and equilibration in 2 × SSC, blots were stored at −20 °C and reused up to six times. Probe solutions were also maintained at −20 °C, and reused following denaturation (10–15 min at 90–100 °C).

*Restriction endonuclease digestions*

Enzymatic digestions were performed by incubating 25 μl of the amplified products with 5U of each endonuclease and the corresponding enzyme buffer. Digestions were continued overnight at 37 °C for *Alu*I, *Hinf*I, and *Mbo*I. Digested products were precipitated with ethanol, dried and resuspended in 5 μl of dye loading buffer.

*Electrophoresis*

The digested products were heated for 2 min at 96 °C and chilled on ice before loading on 55 cm wedge-shaped (0.2–0.6 mm) 6% (w/v) polyacrylamide denaturing (7 M urea)-TBE gels. Electrophoreses were performed in TBE buffer at 55 °C using a LKB Macrophor 2010 sequencing unit operated at 50 W per gel. The bromophenol blue was allowed to migrate to the bottom; the gel was washed in 10% v/v glacial acetic acid for 20 min, and silver stained by using the Silver Sequence TM Kit (Promega) following the manufacturer's protocol. The gel was photographed using a 14″ × 17″ EDF film (Promega).

*Analysis of the sequence data*

The determined sequences were compared with 16S rRNA sequences obtained from the GenBank and the EMBL Data Library by using programs of the Wisconsin Molecular Biology Package Version 8. Complete 16S rDNA sequences of archaeal clones HAC1 and HAC14 were manually aligned to those from other halophilic *Archaea* (Kamekura & Seno 1993). A matrix of calculated similarity percentages was analyzed by UPGMA. Pairwise comparisons of the band patterns were manually performed, and a matrix (presence-absence for each character) constructed. The data was computed using the NTSYS program version in IBM PC.

*Nucleotide sequence accession numbers*

Amplified products were cloned and the sequences determined in the present study were deposited in the EMBL data base with accession numbers: X84084, X84330, X84331, X84322, X84323, X84324.

## Results

The multi-pond solar saltern selected for this study has been studied previously by other techniques and the gradient of physicochemical conditions as well as the microbial populations that develop along this gradient have been described (Rodríguez-Valera et al., 1985; Rodríguez-Valera et al., 1981; Rodríguez-Valera, 1986). The water of the five ponds sampled had salinities representing the increase gradient from about twice the seawater salinity (Preparation Pond PR3, 6.4%) located in the so-called carbonate domain (Barbe et al., 1990) to saturation for sodium chloride, halite domain (Crystallizer Pond CR30, 30.8%), the concentrator ponds have intermediate salinities (gypsum domain) (as shown in Table 1). Bacterial and archaeal primers gave good amplification from all the samples except the two ponds of lower salinity PR3 and PR6 that gave no amplification when archaeal primers were used, indicating a low presence or non-existence of archaeal representatives at low salinity. Archaeal groups have been found in marine samples of different origins by 16S rDNA direct PCR amplification and sequencing but the samples were always obtained from open ocean, relatively oligotrophic waters (DeLong, 1992; DeLong et al., 1994; Fuhrman et al., 1992). Apparently the much more eutrophic, as well as more saline waters of these two ponds are devoid of a significant proportion of this phylogenetic group.

Although the 16S rRNA genes were amplified from the crystallizer with both eubacterial and archaeal primers, the relative abundance of both domains was very different as shown by hybridization of the products from both amplifications to the total DNA extracted and blotted on membranes (see Figure 1). The amplified archaeal rDNA gave much stronger signal, the weak hybridization shown by the bacterial rDNA indicates a relatively small representation, particularly considering that a part could be due to hybridization of highly conserved 16S rDNA regions to the archaeal

*Table 1*. Number of digested fragments of total 16S rDNA PCR-amplified from environmental samples using bacterial and archaeal primer sets. Clone HACl was obtained from the Crystallizer Pond 30 in a previous study.

| Primers | Samples | Salinity % | *Hinf*I | *Mbo*I + *Hinf*I | *Alu*I + *Hinf*I |
|---|---|---|---|---|---|
| | Crystallizer Pond 30 | 30′8 | 37 | 73 | 42 |
| | Concentrator Pond 111 | 21′6 | 59 | 74 | 44 |
| ANTI-S | Concentrator Pond 108 | 13′3 | 56 | 98 | 63 |
| (Bacteria) | Preparation Pond 6 | 9′2 | 75 | 96 | 56 |
| | Preparation Pond 3 | 6′4 | 64 | 107 | 60 |
| | Crystallizer Pond 30 | 30′8 | 28 | 37 | 58 |
| 21F-S | Concentrator Pond 111 | 21′6 | 20 | 34 | 52 |
| (Archaea) | Concentrator Pond 108 | 13′3 | 18 | 30 | 48 |
| | Clone HACl | 30′8 | 11 | 30 | 24 |

*Table 2*. Similarity values for 16S rRNA gene sequences of halophilic *Archaea*.

| | 1 | 2 | 3 | 4 | 5 | 6 | 7 | 8 | 9 | 10 | 11 | 12 | 13 | 14 |
|---|---|---|---|---|---|---|---|---|---|---|---|---|---|---|
| 1. CLONE HACl | 100 | 99.0 | 91.1 | 90.4 | 87.3 | 87.7 | 87.1 | 87.1 | 86.5 | 86.3 | 87.6 | 86.7 | 87.5 | 86.7 |
| 2. CLONE HACl4 | | 100 | 90.4 | 90.0 | 87.1 | 87.4 | 87.2 | 87.0 | 90.7 | 85.4 | 87.4 | 86.7 | 87.1 | 86.5 |
| 3. *H. volcanii* | | | 100 | 98.5 | 87.8 | 88.6 | 90.1 | 91.2 | 88.8 | 89.6 | 88.4 | 88.1 | 88.9 | 89.7 |
| 4. *H. mediterranei* | | | | 100 | 89.3 | 88.3 | 88.2 | 89.5 | 88.6 | 87.3 | 87.8 | 87.5 | 90.1 | 88.2 |
| 5. *H. sinaiiensis* M | | | | | 100 | 97.7 | 91.4 | 88.8 | 87.0 | 86.7 | 87.0 | 87.2 | 86.1 | 86.1 |
| 6. *H. marismortui rrnb* | | | | | | 100 | 94.9 | 89.4 | 87.6 | 88.7 | 87.7 | 88.2 | 87.1 | 87.0 |
| 7. *H. marismortui rrna* | | | | | | | 100 | 90.1 | 90.0 | 90.0 | 90.1 | 88.8 | 87.6 | 87.4 |
| 8. *N. magadii* | | | | | | | | 100 | 90.1 | 89.4 | 90.5 | 87.6 | 93.0 | 88.0 |
| 9. *H. cutirubrum* | | | | | | | | | 100 | 88.3 | 99.8 | 87.8 | 90.0 | 88.5 |
| 10. *H. morrhuae* | | | | | | | | | | 100 | 88.4 | 86.3 | 90.1 | 88.6 |
| 11. *H. salinarium* | | | | | | | | | | | 100 | 87.4 | 90.1 | 88.6 |
| 12. *H. sodomense* | | | | | | | | | | | | 100 | 87.4 | 94.7 |
| 13. *N. occultus* | | | | | | | | | | | | | 100 | 87.9 |
| 14. *H. saccharovorum* | | | | | | | | | | | | | | 100 |

DNA. In any case considering the salt concentration in the pond (30,8%) not many *Bacteria* would be physiologically active (Oren, 1990; Rodríguez-Valera, 1986). Previous studies using differential inhibitors for halophilic *Archaea* showed that in the crystallizer ponds most (if not all) the active biomass was archaeal (Oren, 1990).

Five clones of the amplified eubacterial 16S rDNA from the crystallizer pond were randomly selected and sequence analysis of ca. 200 bp (ranging from 283 to 487; *E. coli* numbering), revealed they belong to a highly homogeneous phylogenetic group. Clones HBC1, HBC5 and HBC7 showed identical sequences (HBC for Halophilic Bacterial Clone). Clones HBC2 and HBC8 showed differences with HBC1 in only 2 and 14 nucleotide positions respectively. Sequence comparisons with all 16S rRNAs available in the data bases were carried out and the highest sequence similarity value was found to be that with *Rhodopseudomonas marina* (82.6% and 83.6%, clones HBC1 and HBC8 respectively). Summarizing, the bacterial DNA present in the crystallizer belonged mainly to a single group (possibly a single genus) distantly related to the *α-Proteobacteria*.

A total of 12 archaeal 16S rDNA clones from the crystallizer were randomly selected and sequences of ca. 200 bp (positions 100–300; *H. salinarium* numbering) were determined. Analysis of data showed the twelve clones have identical sequence at these positions. The above cloned-fragments were obtained from several amplification reactions and cloning experiments. Some of these selected archaeal clones were subjected to analysis of another variable region of the 16S rDNA sequence (positions 384–

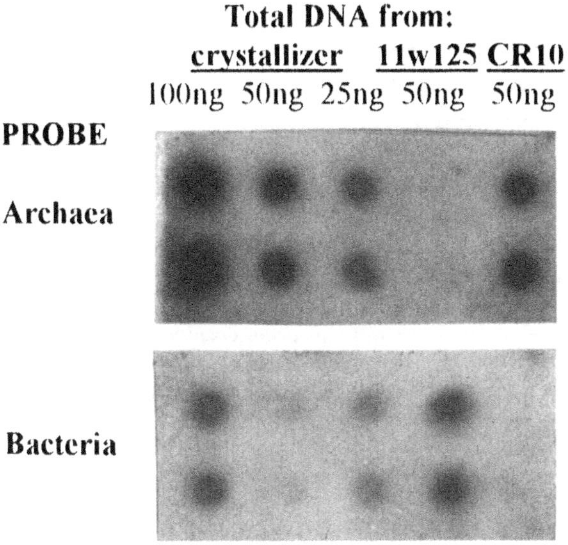


*Figure 1.* Southern blot showing hybridizations of different amounts of total DNA from a crystallizer pond using archaeal and eubacterial amplified 16S rDNA products as probes. DNA extracted from archaeal and eubacterial isolates (CR10 and 11W125 respectively) were used as controls (Benlloch et al., 1995).

600; *H. salinarium* numbering). Clones HAC4, HAC8, HAC11, HAC12, HAC14 and HAC16 had an identical sequence at this stretch indicating that they may belong to a single species (HAC for Halophilic Archaeal Clone). However, this sequence showed 12 and 11 nucleotide differences from those of HAC1 and HAC6 respectively. Moreover, the complete 16S rDNA insert of clones HAC1 and HAC14 was sequenced (1436 and 1438 nucleotides respectively ranging from positions 1 to 1458 of the *H. salinarium* 16S rRNA sequence (Mankin et al., 1985). The two sequences were highly similar, although 15 nucleotide differences were found, indicating that they correspond to related species of the same genus. Similarities from the sequence comparisons with those of other *Halobacteriaceae* are shown in Table 2 and Figure 2. As described by Kamekura and Seno (1993) it has been shown that in general the accepted taxonomy of haloarchaea fits well with the phylogenetic relationships derived from the known 16S rDNA sequences. Sequence similarities between genera range from 86 to 91% while species inside each genus are normally over 98% homology. An exception is the species *Halobacterium saccharovorum* which should be included in a new genus, as has already been suggested (Grant & Ross, 1986). As shown in Figure 2, *H. sodomense* does not belong to the genus *Halobacterium*, but is instead related to *H. saccharovorum*. The similarities between the clones HAC1 and HAC14 and other haloarchaea (as shown in Table 2) ranged from ca. 86% (*Halococcus morrhuae*) to 91% (viz. *Haloferax volcanii*, *Haloferax mediterranei*). The dendrogram of Figure 2 shows that both clones are, although relatively related, as distant from members of the genus *Haloferax* as the other genera are from one another. These results revealed that clones HAC1 and HAC14 clearly correspond to a group distinct from all known haloarchaea, at least at the genus level. Signature sequences for the different genera are found in the region between positions 151–200 (see Figure 3). In this region 16 positions differentiate *Haloarcula* from *Haloferax*, 22 discriminate *Halobacterium* from *Haloferax*, 11 separate *Haloarcula* from *Halococcus*, and 17 *Halococcus* from *Haloferax*. This region is also highly divergent in all the crystallizer clones with at least 20 positions different to any of the other genera. An inspection in detail of the complete 16S rDNA sequences of clones HAC1 and HAC14 revealed different nucleotide composition at 21 positions which are conserved in all haloarchaea examined. An example is illustrated in Figure 3; at position 196 all sequences exhibited thymine, but all selected clones showed cytosine. The considerably long branch in the dendrogram (see Figure 2), the genus-specific signature, and changes in relatively conserved positions of the sequenced molecule all strengthen the view that they belong to a new, undescribed genus of haloarchaea. In our study a high frequency of cloning 16S rDNA fragments with highly similar sequence (i.e. with the same sequence at 100–300 region) has been taken as a suspicion that members of this unknown genus are apparently numerous in the environment analyzed. This is particularly noteworthy considering that three of the species and culture collection strains of haloarchaea (*Haloferax mediterranei*, *Haloferax gibbonsii*, *Haloarcula hispanica*) have been isolated from the same saltern studied here. The results of the present study are reminiscent of that of Ward et al. (1990) where a well known community, the Octopus Spring cyanobacterial mat was studied by 16S rDNA direct amplification. Like our results, 16S rDNA clones were markedly different from the 16S rDNA sequences of the organisms commonly isolated from this spring and considered to be the main components of the population – *Chloroflexus auranticus* and *Synechococcus lividus* (highest similarity around 90%).

DNA of four strains of haloarchaea isolated from the same sample was extracted and a part of their 16S rDNA sequenced. All the strains isolated had the same sequence in the region 130–270 (*H. salinarium* num-

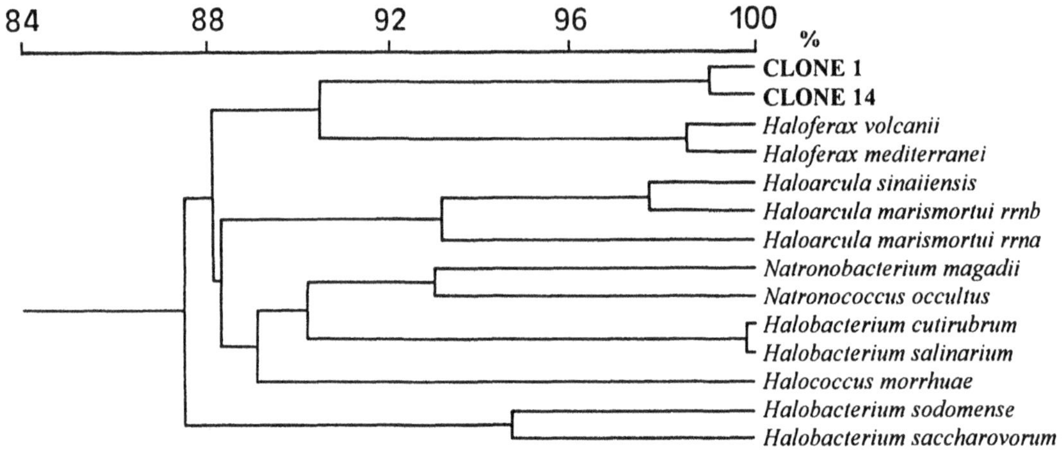


*Figure 2.* Dendrogram of 16S rDNA sequence similarities from clones HAC1 (Clone 1), HAC14 (Clone 14) and members of *Halobacteriaceae* (Benlloch et al., 1995).

| | 151 … 200 |
|---|---|
| *Natronococcus occultus* | TCCCCATACCGCTCGACGCCTGGAAC-TGCGCGAGCCCGAAACGCTCCGG |
| *Natronobacterium magadii* | CTGGAATACGACTCTCATCCTGGAGT-GGAGAGAGTCCGAAATGCTCCGG |
| *Halobacterium salinarium* | CCCCGATAACGCTTTGCTCCTGGAAG-GGGCAAAGCCGGAAACGCTCCGG |
| Y12 | CCCCGATAACGCTTTGCTCCTGGAAG-GGGCAAAGCCGGAAACGCTCCGG |
| *Halobacterium cutirubrum* | CCCCGATAACGCTTTGCTCCTGGAAG-GGGCAAAGCCGGAAACGCTCCGG |
| *Haloarcula sinaiiensis* M | AGCGGATATAACTCTCATGCTGGAGT-GCAGAGAGTTAGAACCGTTCCGG |
| *Haloarcula marismortui rrn*B | AGCGGATATAACTCTCATGCTGGAGT-GCAGAGAGTTAGAAACGTTCCGG |
| *Haloarcula sinaiiensis* m | AGCGGATACAACTCTCATGCTGGAGT-GCCGAGAGTTGGAAACGTTCCGG |
| *Haloarcula marismortui rrn*A | AGCGGATATAACTCTCATGTTGGAGT-GCCGAGAGTTAGAAACGTTCCGG |
| *Halococcus morrhuae*-16008 | CCCAGATACTGCTTTCATGTTGGAAT-ACNGAAAGTCGGAAACGGTCCGC |
| *Halococcus morrhuae* | CCCAGATACTGCTTTCATGTTGGAAT-ACAGAAAGTCGGAAACGGTCCGC |
| *Halobacterium sodomense* | AGTGTATCCCATACCACCACTGGAAT-GAGTGGTATGCCAAACGCTCCGG |
| *Haloferax gibbonsii* | AGTTCATACGGGAGTCATGCTGGAAT-GCCGACTCCCCGAAACGCTCAGG |
| *Haloferax denitrificans* | AGTTCATACGGGAGTCATGCTGGAAT-GCCGACTCCCCGAAACGCTCAGG |
| *Haloferax volcanii* | AGTTCATACGGGAGTCATGCTGGAAT-GCCGACTCCCCGAAACGCTCAGG |
| *Haloferax mediterranei* | AGTTCATACGGGAGTCATGCTTGAAT-GCCGACTCCCCGAAACGCTCCGG |
| Clone 1 HA | CCGTCATACCGATCTCAGGCTTGAATCGCAGAGATCACAAAACGCCCCGG |
| Clone 14 HA | CCGTCATACCGATCTCAGGCTTGAATCGCAGAGATCACAAAACGCCCCGG |

*Figure 3.* Alignment of a 16S rDNA region from members of *Halobacteriaceae* and cloned fragments. The boxed areas are nucleotide signatures that differentiate at the genus level. The start showed a nucleotide position where all selected HAC clones have changed a conserved nucleotide. Numbers indicate positions following *Hb. salinarium* numbering (19). Clone 1, identical sequence for all selected HAC clones. Y12, *Halobacterium* sp. (14) (Benlloch et al., 1996).

bering) and showed the highest homology (97%) to *Haloarcula marismortui* belonging very probably to a closely related species.

Enzymes *Alu*I, *Hinf*I and *Mbo*I were selected after a computer analysis of many restriction endonucleases that are known to cleave the 16S rDNA sequences frequently. The specific discriminatory power of the digestion enzymes was tested by computing against the data from all species of the genus *Aeromonas*, the 16S rDNAs of which show only 1 to 33 nucleotide differences (Martínez-Murcia et al., 1992). As a conclusion of this test, enzymes such as *Aci*I cutting at C.CGC, *Hae*III(GG.CC), *Fnu*4HI(GC.NGC), *Mnl*I(CCTC[N]$_7$) and *Scr*FI(CC.NGG) generated ca. 10 to 20 digested products of different sizes, but discrimination was not satisfactory. *Alu*I (AG.CT), *Hinf*I (G.ANTC) and *Mbo*I (.GATC) yielded specific patterns for all the *Aeromonas* species and were selected. Obviously the discriminatory power for the specific case of *Aeromonas* cannot be extrapolated to all *Bacteria* and *Archaea* but we preferred the use of this criterium rather than a purely arbitrary selection or one based on the sequences of other more divergent genera. To check the reliability of the technique 16S rDNA were amplified from cultures of *Escherichia coli* ATCC 25922, *Pasteurella multocida*, ATCC 43137$^T$, *Listeria monocytogenes* ATCC 15313$^T$, *Proteus mirabilis* ATCC 29906$^T$, *Salmonella* sp. ATCC 35664, *Aeromonas salmonicida* NCIMB 1102$^T$, then a mixture of all the amplified 16S rDNAs were digested with *Hinf*I. All the bands detected for every individual reference strain could be detected as well in the digested mixture (data not shown).

Bacterial and archaeal 16S rDNA fragments amplified from the saltern waters were separately digested using three different enzyme combinations: *Hinf*I, *Mbo*I + *Hinf*I and *Alu*I + *Hinf*I. The products were separated by polyacrylamide gel electrophoresis (see Figure 4), the numbers of detected bands for each sample are listed in Table 1. In the case of *Bacteria* it is clear that the total number of bands decreases at increasing salinities in all the enzyme/enzyme combinations used. For this phylogenetic group the environment becomes more extreme at increasing salinities (over 1M NaCl), therefore a general decrease in biodiversity is to be expected (Rodríguez-Valera et al., 1981). On the other hand, the archaeal 16S rDNA gave more restriction fragments at higher salinities. Halophilic *Archaea* have generally very high growth optima (over 20% NaCl) and most of the species do not grow below 10% NaCl (Larsen & Grant, 1989), therefore for this phylogenetic group the low salinity conditions can be considered an environmental extreme, and thus their decrease in diversity.

The patterns of each sample were compared by identifying fragments of identical size in the same digestion from different samples. In this way a similarity coefficient was obtained for every pair of samples and a similarity dendrogram generated. The dendrogram shows the similarity level regarding archaeal or bacterial populations in the systems studied. Analysis of bacterial patterns using *Hinf*I (see Figure 5A) revealed that 16S rDNA sequences from two concen-

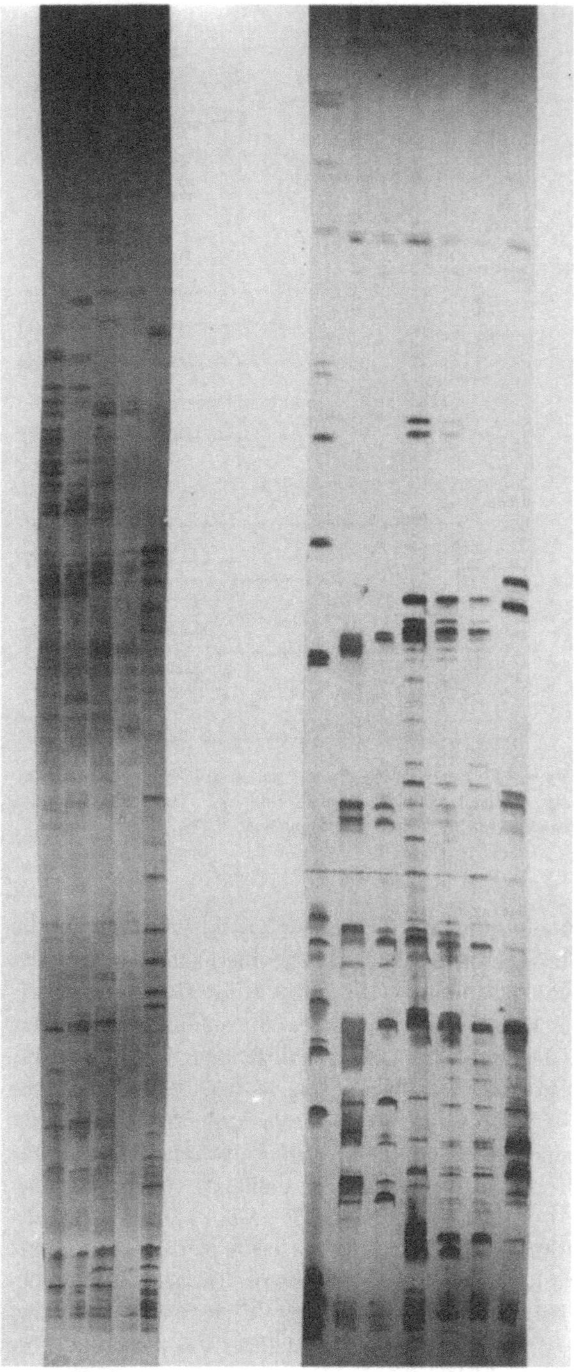

*Figure 4.* Polyacrylamide gel showing *Hinf*I + *Mbo*I digestions of (A) archaeal 16S rDNA amplified from: lane 1, Concentrator pond (CO)108; 2, CO111; 3, Crystallizer Pond (CR)30; 4, clone HAC1 (see reference 31); 5, isolate 9 from CR30; 6, isolate 10 from CR30; 7, molecular marker V (pBR 322 DNA.*Hae* III; Boehringer Minion); and (B) bacterial 16S rDNA from: lane 1, CR30; 2, CO111; 3, CO108; 4, Preparation pond (PR) 6; 5, PR3 (Martínez-Murcia et al., 1995).

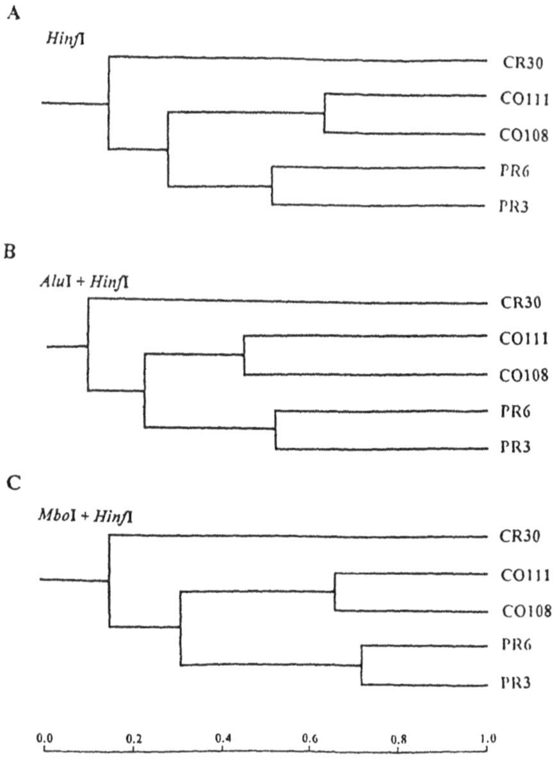


*Figure 5.* Dendrogram of bacterial 16S rDNA-RFLP similarities obtained by digestion with *HinfI* (A), *AluI+HinfI* (B), and MboI+HinfI (C) (Martínez-Murcia et al., 1995).

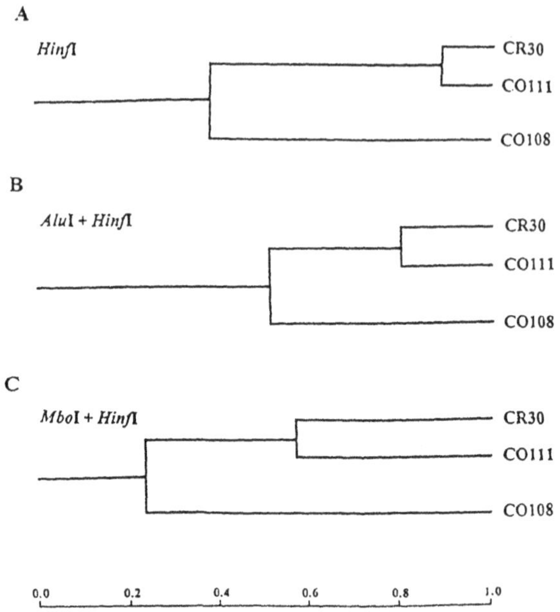


*Figure 6.* Dendrogram of archaeal 16S rDNA-RFLP similarities obtained by digestion with *HinfI* (A), *AluI+HinfI* (B), and *MboI+HinfI* (C) (Martínez-Murcia et al., 1995).

trator samples (CO111 and CO108) were more similar to each other than to those from other water samples. In the same way, 16S rDNA RFLP from the preparation ponds were relatively very similar to each other. The sample from the crystallizer pond (CR30) not only showed lower diversity but was also less related to the 16S rDNA restriction patterns from the concentrators and preparation ponds. The same relationships were found in the diagram of Figure 5B (using enzymes *AluI+HinfI*) and Figure 5C (*MboI+HinfI*). The relationships of archaeal 16S rDNA RFLP patterns were different from those of *Bacteria*. The sequences amplified from the concentrator CO111 were more related to those from the crystallizer CR30 than to those from CO108 (Figure 6A; *HinfI*). The same relationships were found when *AluI+HinfI* or *MboI+HinfI* were used (Figures 6B and 6C).

With regard to *Archaea* the situation is somewhat reversed. The crystallizer and pond CO111 (21.6% salinity) had similar archaeal populations. Again this range (20–30%) corresponds to the normal range of growth of halophilic *Archaea*. A different population seems to be present at CO108 (13.3%) that is close to the lower salinity tolerated by most halophilic *Archaea* (about 10%). Probably a restricted and physiologically specialized group of halophilic *Archaea* is present here. Our results illustrate the different salt tolerance that characterizes halophilic *Bacteria* and *Archaea*.

One possible application of the technique described here is to check the presence of an organisms characterized only by sequence, i.e. non-cultivated, in a given environment. The clone designated HAC1 was subjected to 16S rDNA restriction digestion, together with the amplicon of the crystallizer and with some halophilic archaeal collection strains of the genera *Haloferax, Haloarcula, Halobacterium, Halococcus*, and two isolates from the crystallizer CR30 as a reference (results not shown). The high intensity restriction bands of clone HAC1 were of the same size as the major bands of the crystallizer. However, profiles from clone HAC1, and reference strains were different from each other. The sample collected in which clone HAC1 was obtained (Benlloch et al., 1995) came from the same crystallizer pond but it was almost a year before the sample was analyzed by digestion of 16SrRNA genes. Therefore, our results support the hypothesis (Benlloch et al., 1995) that the group of clones characterized by 16S rDNA sequencing are predominant in the crystallizer and, so far, have not been isolated.

## Discussion

The extreme environments that have been studied by the PCR technique show, as might be expected, less diversity than sea water or soil, frequently having a higher redundancy in the sequenced 16S rDNA clones. However, as in high diversity environments high homologies to cultured organisms are very seldom found. The extremely hypersaline waters of the crystallizer studied here show indeed even less diversity by the direct rDNA amplification methodology than in a previous work by culture isolation (Rodríguez-Valera et al., 1985). However, the organisms retrieved by one technique are markedly different from those obtained by the other, even when both techniques are applied to the same sample. Our data can be interpreted if we consider that the culture methods strongly bias the results towards microorganisms that are a minority in the environment. If this hypothesis is correct and applicable to other environments it has a major impact on our understanding of microbial diversity.

The results reported to date from soil, ocean and thermal environments have shown a remarkably low level of homology to 16S rRNA sequences from cultured microorganisms present in the data bases. In fact, homologies over 92% are rarely found and very often values around 85% are the highest scores for a given sequence comparison, this is within the range of different phylum homology (Liesack & Stackebrandt, 1992). This has generally been interpreted as a reflection of the extremely high real bacterial diversity, orders of magnitude higher than that shown by culture and isolation, i.e. barely 3000 cultured bacterial species that are known and have 16S rRNA genes sequenced. If the real number of species is several orders of magnitude greater, the probability of retrieving any of the known bacteria is relatively small unless a very large number of sequences from the natural environment is screened. On the other hand, some data indicate less diversity. Most studies have detected considerable redundancy in the sequences retrieved from the environment, very often two representatives of each sequence are found and sometimes five or more repetitions have been recorded. Identical sequences have been obtained from samples from different oceans and by different authors (Fuhrman et al., 1993; Giovannoni et al., 1990a). Those findings seem to indicate a more moderate diversity.

A concise method for assessing microbiological diversity, based on the digestion of the 16S rDNA sequences directly amplified from a complex population of microorganisms, is also reported (Martínez-Murcia et al., 1996). This methodology allows rapid comparison of the microbial populations present in several samples since more than 30 samples can be run in a single gel. The technique may have several applications, for example, scanning serial samples in a gradient in order to identify major boundaries in terms of microbial biodiversity, or to estimate the contribution of some isolates or 16S rDNA clones to the total microbial population. The application of this method to a well understood environmental gradient, that found in the multi-pond saltern, illustrates its reliability since previous knowledge obtained from more classical microbial ecology approaches has been largely confirmed, independently of the enzyme or combination used.

Although the techniques based on direct detection of the 5S rRNA present in the sample are more direct (obviating the need for PCR amplification and restriction enzyme digest) (Hofle, 1992) they also have some disadvantages when compared with the methodology described here. First of all, the extraction and manipulation of RNA is time-consuming and the sensitivity to degradation also introduces the risk of not retrieving the less represented species. Also 5S rRNAs are very small and the size differences do not always reflect the variety of organisms present. We consider that the number of bands alone is an acceptable estimate of the total biodiversity present, particularly when different enzyme combinations are used. Of course, we cannot disprove that more than one sequence could be contained in a single band or that incidentally one single organism could give a disproportionately large or small number of bands leading to an inaccurate evaluation of the real diversity present. However, both events are sufficiently infrequent to allow statistically significant evaluations by means of the method described here. Another possible source of error may be incomplete digestion. To avoid this problem it is essential to adjust the ratio DNA/enzyme accurately and always digest a DNA control of which the restriction pattern is known; incubation times should always be very long to reduce this possibility.

It is remarkable that the topology of the dendrograms is similar in all cases irrespective of the enzyme/enzyme combination used. This is a strong argument in favour of the reliability of the technique. In the case of *Bacteria* the results of the pattern comparisons shown in the dendrograms revealed three groups of bacterial populations. One from samples below 10% salinity (PR3, PR6) that would correspond to the halotolerant and marine eubacteria described phenotypically. Recent studies (Mellado-Duran, 1994) have shown the halophilic *Bacteria* to be phylogenetically distant from their marine or non-halophilic taxonomic counterparts. This is again confirmed by our results that give a high similarity in the two ponds corresponding to 13.3 and 21.6% salinity respectively, environments that by pure culture were shown to be dominated by this group of *Bacteria*, forming a second similarity ecophysiological group. Then the *Bacteria* present in the NaCl saturated pond were a third group distantly related to the two previous ones. This one probably represents an extremely specialized niche for *Bacteria*. Previous reports obtained with differential inhibitors and measurement of activity in situ (Oren, 1990) as well as our own work carried out with Domain specific probes indicate that only a very small proportion of the crystallizer biomass corresponds to *Bacteria*. In fact, to our knowledge, very few described bacterial groups are able to proliferate in saturated salt (Kushner & Kamekura, 1988). The DNA of *Bacteria* present in the crystallizer could also be left over from previous concentration stages that are just carried over and are not actively growing here. In any case, our results indicate that, active or not, the eubacterial population present in the crystallizer is very different from that of the other environments studied.

The newly developed techniques of directly sequencing 16S rDNA genes retrieved from the environment allow a much more realistic view of the real microbial diversity present. However, the isolation of these new groups of non-cultured organisms is an essential step to follow if we want to understand the role that they play in their ecosystems. DNA probes derived from the sequences are very useful for that purpose (Amman et al., 1995). The study of restriction enzyme digests gives an additional tool for this type of search. The 16S rDNA restriction screening of a number of isolates from crystallizer CR30 to identify and isolate the type of *Archaea* represented by clone HACl is under way.

## Acknowledgments

This work was supported by grants PB 93/0930 of the DGICYT, BIO 93/0750 of the CICYT, and EV5V-CT92-0080 of the European Commission. A.J.M.M. was recipient of a post-doctoral contract of the Spanish Ministry of Education and Science. Secretarial assistance by K. Hernández is gratefully acknowledged.

## References

Amann, R. I., W. Ludwig & K-H. Schleifer, 1995. Phylogenetic identification and in situ detection of individual microbial cells without cultivation. Microbiol. Rev. 59: 143–169.

Avaniss-Aghajani, E., K. Jones, D. Chapman & C. Brunk, 1994. A molecular technique for identification of bacteria using small subunit ribosomal RNA sequences. BioTechniques, 17: 144–149.

Barbe, A., J. O. Grimalt, J. J. Pueyo & J. Albaiges, 1990. Characterization of model evaporitic environments through the study of lipid components. Org. Geochem. 8: 293–297.

Barns, S. M., R. E. Fundyga, M. W. Jeffries & N. R. Pace. 1994. Remarkable archaeal diversity detected in a Yellowstone National Park hot spring environment. Proc. Natl. Acad. Sci. USA. 91: 1609–1613.

Beaucage, S. L. & M. H. Caruthers, 1981. Dideoxynucleotide phosphoramidites – a new class of key intermediates for deoxypolynucleotide synthesis. Tetrahedrom Lett. 22: 1859–1862.

Benlloch, S., A. J. Martínez-Murcia & F. Rodríguez-Valera, 1996. Bacterial and archaeal diversity as shown by partial sequencing of 16S rRNA genes directly amplified from a hypersaline environment System. Applied Microbiol. (in press).

Brosius, J. P., T. J. Dull, D. D. Sleeter & H. F. Noyer, 1981. Gene organization and primary structure of a ribosomal RNA operon from *E. coli*. J. Mol. Biol. 148: 107–127.

DeLong, E. F., K. Y. Wu, B. B. Prézelin & R. V. M. Jovine; 1994. High abundance of *Archaea* in Antarctic marine picoplankton. Nature 371: 695–697.

DeLong, E. F., 1992. Archaea in coastal marine environments. Proc. Natl. Acad. Sci. USA. 89: 5685–5689.

Fuhrman, J. A., K. McCallum & A. A. Davis, 1992. Novel major archaebacterial group from marine plankton. Nature 356: 148–149.

Fuhrman, J. A., K. McCallum & A. A. Davis, 1993. Phylogenetic diversity of subsurface marine microbial communities from the Atlantic and Pacific Oceans. Appl. Envir. Microbiol. 59: 1294–1302.

Giovannoni, S. J., T. B. Britschgi, C. L. Moyer & K. G. Fieldt, 1990a. Genetic diversity in Sargasso Sea bacterioplankton. Nature 345: 60–63.

Giovannoni, S. J., E. F. DeLong, T. M. Schmidt & N. R. Pace, 1990b. Tangential flow filtration and preliminary phylogenetic analysis of marine picoplankton. Appl. Envir. Microbiol. 56: 2572–2575.

Grant, W. D. & H. N. M. Ross, 1986. The ecology and taxonomy of halobacteria. FEMS Microbiol. Rev. 39: 9–15.

Höfle, M. G., 1992. Bacterioplankton community structure and dynamics after large-scale release of nonindigenous bacteria as revealed by low-molecular-weight-RNA analysis. Appl. Envir. Microbiol. 58: 3387–3394.

Javor, B. J., 1983a. Planktonic standing crop and nutrients in a saltern ecosystem. Limnol. Oceanogr. 28: 153–159.

Javor, B. J., 1983b. Nutrients and ecology of the Western Salt and Exportadora de Sal saltern brines. 6th International Symposium on Salt. Vol. 1. Salt Institute.

Kamekura, M. & Y. Seno, 1993. Partial sequence of the gene for a serine protease from a halophilic archaeum *Haloferax mediterranei* R4, and nucleotide sequences of 16S rRNA encoding genes from several halophilic archaea. Experientia 49: 503–513.

Kushner, D. J. & M. Kamekura, 1988. Physiology of halophilic eubacteria. In: Rodríguez-Valera, F. (ed.), Halophilic Bacteria, Volume I. CRC Press, Inc. Boca Raton, Florida: 109–138.

Lane, D. J., B. Pace, G. J. Olsen, D. A. Stahl, M. L. Sogin & N. R. Pace, 1985. Rapid determination of 16S rRNA sequences for phylogenetic analysis. Proc. Natl. Acad. Sci. USA. 82: 6955–6959.

Larsen, H. & W. D. Grant, 1989. Genus I: *Halobacterium*. Elazari-Volcani 1957, $207^{AL}$, p. 2219. In J. T. Staley, M. P. Bryant, N. Pfennig & J. G. Holt (eds), Bergey's Manual of Systematic Bacteriology, Vol. 3. The Williams & Wilkins Co., Baltimore.

Liesack, W. & E. Stackebrandt, 1992. Occurrence of novel groups of the Domain Bacteria as revealed by analysis of genetic material isolated from an Australian Terrestrial Environment. J. Bacteriol. 174: 5072–5078.

Maniatis, T., E. F. Fritsch & J. Sambrook, 1982. Molecular cloning: a laboratory manual. Cold Spring Harbor Laboratory Press, Cold Spring Harbor, N.Y.

Mankin, A. S., V. K. Kagramanova, N. L. Teterina, P. M. Rubtsov, E. N. Belova, A. M. Kopylov, L. A. Baratova & A. A. Gogdanov, 1985. The nucleotide sequence of the gene coding for the 16S rRNA from the archaebacterium *Halobacterium halobium*. Gene 37: 181–189.

Marchuk, D., M. Drumm, A. Saulino & F. S. Collins, 1990. Construction of T-vectors, a rapid and general system for direct cloning of unmodified PCR products. Nucl. Acids. Res. 19: 1154.

Martínez-Murcia, A. J., S. Benlloch & M. D. Collins, 1992. Phylogenetic interrelationships of members of the genera *Aeromonas* and *Plesiomonas* as determined by 16S rDNA sequencing: lack of congruence with results of DNA-DNA hybridization. Int. J. Syst. Bact. 42: 412–421.

Martínez-Murcia, A. J., 1993. Doctoral Thesis. Department of Microbiology, University of Reading, Reading, U.K.

Martínez-Murcia, A. J. & F. Rodríguez-Valera, 1994. The use of arbitrarily primed PCR (APPCR) to develop taxa specific DNA probes of known sequence. FEMS Microbiol. Lett. 124: 265–270.

Martínez-Murcia, A. J., S. G. Acinas & F. Rodríguez-Valera, 1996. Evaluation of prokaryotic diversity by restrictase digestion of 16S rDNA directly amplified from hypersaline environments. FEMS Microbiol. Ecol. (in press).

Mellado Duran, M. E., 1994. Bacterias halofilas moderadas: filogenia y construccion de vectores de clonacion. PhD Thesis, University of Sevilla.

Moyer, C. L., F. C. Dobbs & D. M. Karl, 1994. Estimation of diversity and community structure through restriction fragment length polymorphism distribution analysis of bacterial 16S rRNA genes from a microbial mat at an active, hydrothermal vent system, Loihi Seamount, Hawaii. Appl. Envir. Microbiol. 60: 871–879.

Muyzer, G., E. C. De Waal & A. G. Uitterlinden, 1993. Profiling of complex microbial populations by denaturing gradient gel electrophoresis analysis of Polymerase Chain Reaction-Amplified genes coding for 16S rRNA. Appl. Envir. Microbiol. 59: 695–700.

Oren, A., 1990. Estimation of the contribution of halobacteria to the bacterial biomass and activity in solar salterns by the use of bile salts. FEMS Microbiol. Ecol. 73: 41–48.

Pace, N. R., D. A. Stahl, D. J. Lane & G. J. Olsen, 1986. The analysis of natural microbial populations by ribosomal RNA sequences. Adv. Microb. Ecol. 9: 1–55.

Ralph, D., M. McClelland, J. Welsh, G. Baranton & P. Perolat, 1993. *Leptospira* species categorized by arbitrarily primed polymerase chain reaction (PCR) and by mapped restriction polymorphisms in PCR-amplified rRNA genes. J. Bacteriol. 175: 973–981.

Reysenbach, A-L., G. S. Wickham & N. R. Pace, 1994. Phylogenetic analysis of the hyperthermophilic pink filament community in Octopus Spring, Yellowstone National Park. Appl. Envir. Microbiol. 60: 2113–2119.

Rodríguez-Valera, F., F. Ruiz-Berraquero & A. Ramos-Cormenzana, 1981. Characteristics of the heterotrophic bacterial populations in hypersaline environments of different salt concentrations. Microb. Ecol. 7: 235–243.

Rodríguez-Valera, F., 1986. The ecology and taxonomy of aerobic chemoorganotrophic halophilic eubacteria. FEMS Microbiol. Rev. 39: 17–22.

Rodríguez-Valera, F., A. Ventosa, G. Juez & J. F. Imhoff, 1985. Variation of environmental features and microbial populations with salt concentrations in a multi-pond saltern. Microb. Ecol. 11: 107–115.

Sanger, F., S. Nicklen & A. R. Coulson., 1977. DNA sequencing with chain-terminating inhibitors. Proc. Natl. Acad. Sci. USA. 74: 5463–5467.

Schmidt, T. M., E. F. DeLong & N. R. Pace, 1991. Analysis of a marine picoplankton community by 16S rRNA gene cloning and sequencing. J. Bacteriol. 173: 4371–4378.

Stackebrandt, E., W. Liesack & B. M. Goebel, 1993. Bacterial diversity in a soil sample from a subtropical Australian environment as determined by 16S rDNA analysis. The FASEB J. 7: 232–236.

Torsvik, V., S. Kãre, R. Sørheim & J. Goksøyr, 1990. Comparison of phenotypic diversity and DNA heterogeneity in a population of soil bacteria. Appl. Envir. Microbiol. 56: 776–781.

Walsby, A. E., 1980. A square bacterium. Nature 283: 69.

Ward, D. M., R. Weller & M. M. Bateson, 1990. 16S rRNA sequences reveal numerous uncultured microorganisms in a natural community. Nature 345: 63–65.

Weller, R., J. W. Weller & D. M. Ward, 1991. 16S rRNA sequences of uncultivated hot spring cyanobacterial mat inhabitants retrieved as randomly primed cDNA. Appl. Envir. Microbiol. 57: 1146–1151.

*Hydrobiologia* **329**: 33–43, 1996.
*P. Caumette, J. Castel & R. Herbert (eds), Coastal Lagoon Eutrophication and ANaerobic Processes (C.L.E.AN.).*



# Anoxygenic phototrophic bacteria in eutrophic coastal lagoons of the French Mediterranean and Atlantic Coasts (Prévost Lagoon, Arcachon Bay, Certes fishponds)

R. Guyoneaud[1], R. Matheron[2], R. Baulaigue[2], K. Podeur[2], A. Hirschler[2] & P. Caumette[1] *
[1] *Laboratoire d'Océanographie Biologique, Université de Bordeaux I, 2 rue du Professeur Jolyet, 33120 Arcachon, France (* author for correspondence)*
[2] *Laboratoire de Microbiologie, Université d'Aix-Marseille III, Faculté des Sciences et Techniques de Saint-Jérôme, Av. Escadrille Normandie-Niemen, 13397 Marseille Cedex 20, France*

*Key words:* Coastal lagoons, eutrophication, phototrophic bacteria, purple nonsulfur bacteria, purple sulfur bacteria, bacterial diversity

**Abstract**

Sediment samples collected from coastal lagoons on the French Mediterranean (Prévost Lagoon) and Atlantic coasts (Arcachon Bay and Certes fishponds) have been studied in order to determine the population densities and the species diversity of the different groups of anoxygenic phototrophic bacteria (purple sulfur bacteria, purple nonsulfur bacteria and green sulfur bacteria) present in these ecosystems. Several strains of each group were isolated in pure culture and characterized by their physiological properties. The occurrence of purple nonsulfur bacteria in organic rich sediments of the Arcachon Bay and the dominance of purple sulfur bacteria in the Prévost lagoon and Certes fishponds are discussed with respect to their community structure and abundance. The diversity differences of the phototrophic bacterial strains isolated from both environments are also discussed.

## Introduction

As a consequence of their location between land and sea, coastal lagoons are characterized by large fluctuations in physical and chemical conditions. Moreover, many shallow coastal lagoons receive freshwater inputs, rich in organic and mineral nutrients derived from urban, agricultural and/or industrial effluents and domestic sewages. Due to the high sulfate concentration in seawater, sulfate-reducing bacteria play a key role in the mineralization of organic matter in anoxic sediments. As a consequence of their metabolic activity they produce considerable quantities of hydrogen sulfide, which is a characteristic feature of these ecosystems. This sulfide is trapped in the sediments as ferrous sulfide (FeS) or pyrite ($FeS_2$) (Cahet, 1965; Jørgensen, 1977; Caumette, 1986).

In such ecosystems, the surface sediment layer, which is a few millimeters thick, is a transition zone between oxic and anoxic conditions. This zone is characterized by microgradients of oxygen, sulfide and light where anoxygenic photosynthesis can take place (Jørgensen et al., 1983; Pierson et al., 1987).

During the warm summer months, increased activity of sulfate-reducing bacteria leads to an overproduction of hydrogen sulfide which may extend through the entire water column in some eutrophic shallow lagoons situated on the mediterranean coast. Such dystrophic crises with periodic anoxia favor the bloom of phototrophic bacteria originating from benthic communities and the formation of red waters (Forti, 1933; Cerruti, 1938; Heldt, 1952; Devèze & Fauvel, 1966; Stirn, 1971; Caumette & Baleux, 1980). In contrast, in deeper coastal environments (Caumette et al., 1983; Guerrero et al., 1987) or in tide-influenced ecosystems (Van Gemerden et al., 1989a, b), such dystrophic crises rarely occur. Generally, part of the water column remains oxic, even under the most eutrophic conditions, and consequently phototrophic bacteria grow only in the surface layer of the sediments or in the bottom water below the oxic-anoxic interface.

In lagoonal ecosystems, phototrophic bacteria should be able to tolerate the fluctuating conditions of salinity, oxygen, sulfide and light. Therefore, the conditions of eutrophication or dystrophy will lead to the dominance of species of phototrophic bacteria that are well adapted to these fluctuating conditions. The objective of the present study was to compare the phototrophic bacterial communities of three coastal systems, the eutrophic and periodically dystrophic Etang du Prévost (Prevost Lagoon) on the French Mediterranean coast, the less eutrophic and tidal bay of the Bassin d'Arcachon on the French Atlantic coast, and the man-made fishponds at Certes which are separated from the tidal influences of the Arcachon Bay by a dike. The influence of environmental conditions on the selection of the dominant species in the phototrophic bacterial communities of both lagoons is discussed in this paper.

## Materials and methods

### *Sampling sites and procedures*

Two sampling stations in Arcachon Bay (A and B), one in Certes fishponds (C) and two in Prévost Lagoon (11 and X) were periodically investigated. The main characteristics of these sampling sites have been described by Castel et al. (1996).

Samples were collected at each sampling station in March, June and September 1993. The water of the non-tidal stations (C, 11 and X) was sampled in sterile screw-capped bottles, 10 cm below the air-water interface. Sediment samples were collected using plexiglass core samplers from zero to 20 cm depth. Samples were stored in cold bags at +4 °C and analysed in the laboratory within 12 hours of collection.

### *Chemical analyses*

Sulfide was determined in the sediments by the methylene blue method of Cline (1969) after distillation under nitrogen gas atmosphere according to the method of Jørgensen & Fenchel (1974). A Beckman model 54 spectrophotometer was used.

The organic matter content of the sediments was analysed using a total organic carbon analyser (LECO) by Dr Etcheber, Departement of Geology and Oceanography, University of Bordeaux I.

### *Bacteriological analyses*

Population densities of the different groups of phototrophic bacteria in the water and sediment samples were enumerated after dilution in anaerobic sterile lagoon water. The deep agar-shake method (deep agar dilution series) (Pfennig & Trüper, 1992) was used and pure strains were isolated from colonies obtained in the agar media.

### *Media for enumeration and isolation of anoxygenic phototrophic bacteria*

The same medium was used to count both purple and green anoxygenic phototrophic bacteria. The basal medium was prepared in a special 3-l flask (Pfennig & Trüper, 1992). Seawater medium was prepared according to Pfennig & Trüper (1992) and Guyoneaud et al. (1993). It contains per liter of filtered seawater (0.2 $\mu$m): $KH_2PO_4$ : 0.1 g, $NH_4Cl$ : 0.5 g, trace element solution (SL 12, modified after Eichler & Pfennig, 1986) : 1 ml, vitamin solution (V7, Pfennig & Trüper, 1992) : 1 ml, Na-acetate : 5 mM, $Na_2S_2O_3 \cdot 5H_2O$ : 2.5 mM, $NaHCO_3$ : 1.5 g, $Na_2S \cdot 9H_2O$ : 0.6 g; the medium was autoclaved and the final pH adjusted to 7.2. 1 ml of each sample or dilution was transferred into the deep agar medium according to the method of Pfennig & Trüper (1992). Series of 6 to 10 fold dilutions were prepared and incubated at 30 °C under an alternating light-dark cycle (16 hours 500 lux light – 8 hours dark) using tungsten lamps. Colony counts were made after two months incubation. Identification of the different phototrophic bacterial colonies was carried out on the basis of colony color, cell morphology, motility and presence/absence of sulfur globules. The agar-shake method was used for both the enumeration of viable bacterial populations and isolation of pure strains. The purity of the isolates was checked microscopically and by growth tests in AC medium (Difco Laboratories, Ann Arbor, Mich., USA) supplemented with NaCl (1.5% w/v), sodium thiosulfate (0.05% w/v) and glucose (0.05% w/v).

### *Methods for the identification of pure cultures of phototrophic bacteria*

*Morphology.* Microscopical observation and microphotographs of the individual isolates were carried out using a Zeiss photomicroscope according to the method of Pfennig & Wagener (1986).

*Physiological tests.* All physiological tests were performed in synthetic media according to Pfennig & Trüper (1992). Utilization of carbon sources and electron donors was tested in liquid medium supplemented with 0.02% (w/v) $Na_2S{\cdot}9H_2O$ and substrates final concentrations of either 2 or 5 mM. Tests for substrate utilization, NaCl requirement, and pH optima were performed in completely filled 25 ml screw-capped tubes. Growth was followed by measuring the optical density of the cultures at 650 nm using a Bausch and Lomb spectrophotometer (Spectronic 20) during 10 to 20 days of incubation.

Semi-aerobic growth was tested in uniformly inoculated deep agar tubes open to air (Kämpf & Pfennig, 1980). The vitamin requirements and capacity for assimilatory sulfate-reduction were tested in 60 ml screw-capped bottles by presence/absence of growth after 4 consecutive transfers. Capacity for dinitrogen fixation was checked in rubber stopper bottles half-filled with synthetic medium lacking bound nitrogen sources under a $N_2/CO_2$ (9/1 ratio) gas phase; four consecutive transfers were tested.

*Photopigments.* Absorption spectra of living cells were directly measured with a Kontron Uvikon (model 860) spectrophotometer after suspension of a cell pellet in a sucrose solution (Pfennig & Trüper, 1992).

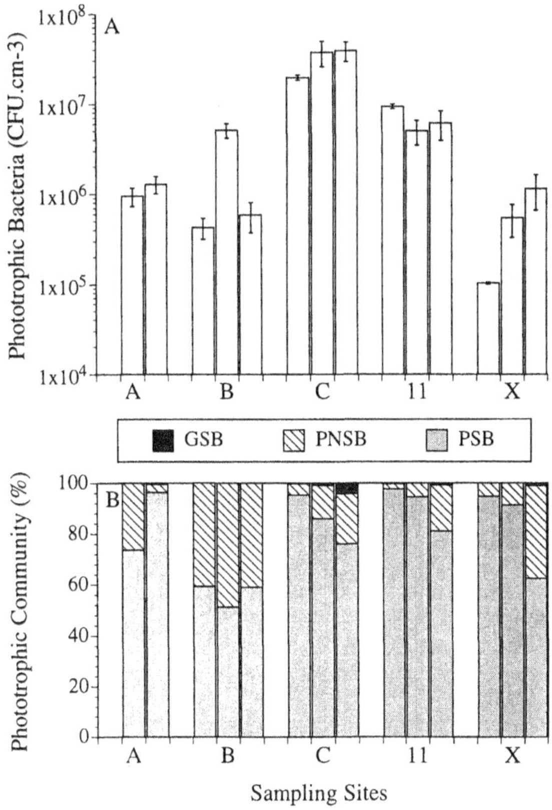


*Figure 1.* Numbers of phototrophic bacteria in CFU $cm^{-3}$ (A) and community composition (B) in percentages (PSB, purple sulfur bacteria; PNSB, purple nonsulfur bacteria; GSB, green sulfur bacteria) in the 0–2 cm horizon of the sediments at the sampling sites in March 1993 (left), June 1993 (center) and September 1993 (right). Error bars in panel A indicate standard error calculated from three samples.

## Results

### *Cell numbers of phototrophic bacteria and community structure*

The phototrophic bacterial population determined as colony forming units (CFU) in the 0 –2 cm layer of the sediment samples collected from each sampling site are presented in Figure 1A. Cell numbers increased from March to June; from June to September they were stable in most of the stations, with the exception of station B where populations decreased to numbers similar to those recorded in March. The highest numbers, ranging from $5 \times 10^6$ to $5 \times 10^7$ CFU $cm^{-3}$, were found in the Certes Fishponds (station C) and the inner stations of the Prévost Lagoon and Arcachon Bay (stations 11 and B). These stations are also characterized by hight amounts of acid-volatile sulfide (A. V. S.) and total organic carbon as shown in Figure 2. In contrast, at the two sites directly subjected to tidal influences (stations A and X), the numbers of phototrophic bacteria as well as A. V. S. and amounts of total organic carbon were much lower (Figures 1A and 2).

Analyses of colonies of phototrophic bacteria which grew in the agar shake tubes showed differences in the community structure between the sampling sites (Figure 1B). Purple sulfur bacteria of the *Chromatiaceae* were always dominant at the non-tidal stations (X, 11 and C), representing between 70 and 85% of the total phototrophic community. Nevertheless, at these sampling sites, the population of purple nonsulfur bacteria increased from March to September, increasing from 2–5% to 20–30%. In contrast, at station B, subjected to tidal influence in the Arcachon Bay, purple sulfur bacteria were less dominant and purple nonsulfur bacteria represented a significant component of the phototrophic bacterial community through the duration of this investigation. At station B purple nonsulfur bac-

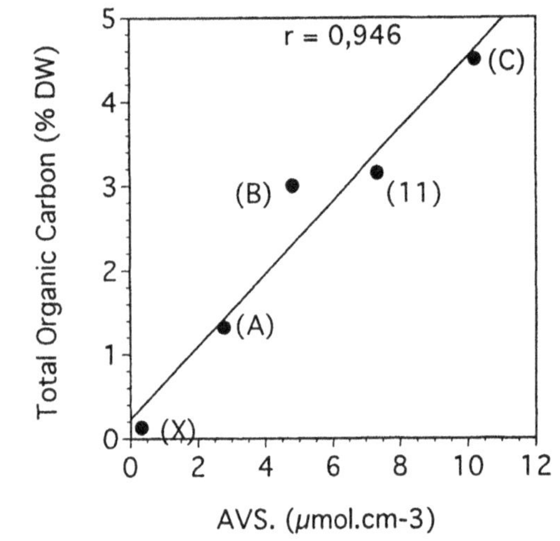


*Figure 2.* Total organic carbon and acid volatile sulfide in the 0–2 cm horizon of the sediment at the sampling stations in March 1993.

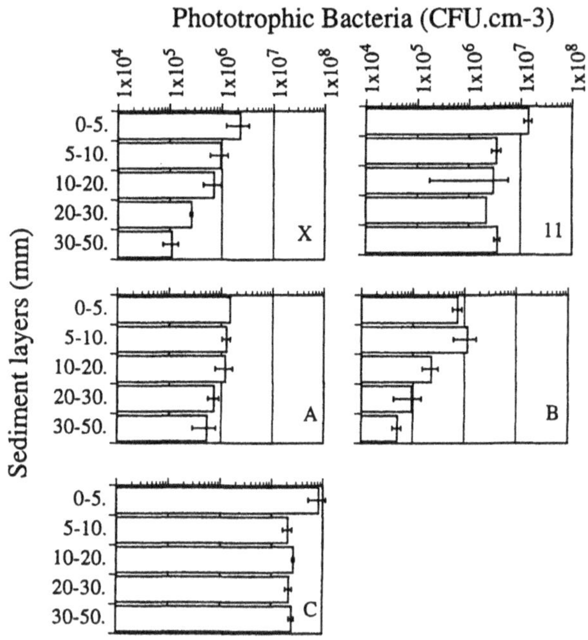


*Figure 3.* Vertical distribution of phototrophic bacteria in the top five cm of the sampling sites in September 1993. Error bars indicate standard error calculated from three samples.

teria composed up to 40% of the phototrophic bacterial community.

In September 1993 the phototrophic bacterial populations were determined from zero to five cm depth in the sediments (Figure 3). In the non-tidal stations, the highest numbers were found in the upper 5 mm of the sediment. Numbers decreased below this horizon however phototrophic bacteria were still present at 5 cm depth, ranging from $10^5$ (station X) to $10^7$ CFU $cm^{-3}$ (station C). At the tidal stations A and B, the highest numbers were found in the 5–10 mm depth horizon of the sediment.

*Diversity of phototrophic bacteria*

Analyses of the data on the diversity of the phototrophic bacteria in the samples collected in September 1993 show that there were significant differences between the sampling sites (Table 1). Among the *Chromatiaceae*, bacteria belonging to the genus *Thiocapsa* were dominant (40–45%) at station 11 in the Prévost Lagoon whereas in the Arcachon Bay and Certes Fishponds, the motile purple sulfur bacteria of the genera *Chromatium* and *Thiocystis* were dominant. Green sulfur bacteria were detected only in the non-tidal stations, and were a minor component of the phototrophic bacterial community (0,5 to 4%). Among them, bacteria of the genus *Prosthecochloris* were dominant. A significant feature was the presence of substential numbers of purple nonsulfur bacteria, mainly represented by the genera *Rhodobacter* and/or *Rhodovulum* and *Rhodopseudomonas* and/or *Rhodobium* at station B.

From the three different ecosystems several strains were isolated and identified according to their morphological and physiological properties, in order to assess their taxonomic position and their metabolic capacities. Data presented in Table 2 shows the major characteristics of some isolates. On the basis of their morphology and carotenoids composition, several strains could be classified with the genera *Chromatium (Chr.)*, *Thiocapsa (Tca.)*, *Thiocystis (Tcs.)*, *Rhodobacter (Rba.)* and/or *Rhodovulum (Rdv.)*, *Rhodopseudomonas (Rps.)* and/or *Rhodobium (Rbi.)*, *Rhodospirillum (Rsp.)* and *Prosthecochloris (Prs.)*.

Data from the physiological tests (Tables 3 and 4) enable several strains to be identified to the species level. Thus it was found that bacteria resembling *Chr. gracile* and *Tca. roseopersicina* as well as bacteria resembling the genera *Rhodobacter* and/or *Rhodovulum* were found in all the lagoons. Tidal station A in the Arcachon Bay was characterized by isolates belonging to the species *Chr. purpuratum* and *Rdv. sulfidophilum*, whereas at station B the last species was the dominant member of the purple nonsulfur bacteria isolated.

*Table 1*. Composition of phototrophic bacterial communities obtained from the different sampling sites in September. **Water**: Phototrophic bacterial numbers (CFU $ml^{-1}$) and percentages are calculated from a single enumeration, **sediments**: Community composition are calculated from three samples and from the average percentages of the different sediments depth (percentages ± SE).

| | A | B | C | 11 | X |
|---|---|---|---|---|---|
| | | | **Water** | | |
| CFU $ml^{-1}$ | | | $3.5 \times 10^3$ | $7.2 \times 10^4$ | $1.2 \times 10^2$ |
| Community (%) | | | | | |
| **PNSB** | | | **31.4** | **37.5** | **50.0** |
| **PSB** | No Water | No Water | **68.6** | **55.6** | **50.0** |
| *Chr./Tcs.* | | | 57.1 | 25.0 | 25.0 |
| *Chr.* (P) | | | 2.9 | 5.6 | 0 |
| *Tca.* | | | 8.6 | 25.0 | 25.0 |
| **GSB** | | | **N.D.** | 6.9 | **N.D.** |
| | | | **Sediments** | | |
| Community (%) | | | | | |
| **PNSB** | **4.1 ± 1.9** | **42.7 ± 2.9** | **21.4 ± 9.1** | **25.7 ± 13.1** | **37.6 ± 4.2** |
| **PSB** | **95.9 ± 1.9** | **57.3 ± 2.9** | **75.0 ± 10.9** | **72.5 ± 13.8** | **61.7 ± 3.9** |
| *Chr./Tcs.* | 63.9 ± 11.2 | 26.7 ± 3.7 | 39.5 ± 11.6 | 29.1 ± 6.9 | 29.8 ± 3.5 |
| *Chr.* (P) | 10.7 ± 7.2 | 12.4 ± 10.4 | 10.2 ± 0.9 | 8.5 ± 7.4 | 3.4 ± 2.3 |
| *Tca.* | 21.3 ± 12.9 | 18.2 ± 6.9 | 25.3 ± 3.2 | 34.9 ± 8.1 | 28.5 ± 3.2 |
| **GSB** | **N.D.** | **N.D.** | **3.6 ± 2.2** | **1.8 ± 2.6** | **0.7 ± 0.9** |

Community: **PNSB**, purple nonsulfur bacteria; **PSB**, purple sulfur bacteria (*Chr./Tcs.*, red *Chromatium* sp. and/or *Thiocystis* sp.; *Chr.* (P), purple *Chromatiaceae*; *Tca.*, *Thiocapsa* sp.); **GSB**, green sulfur bacteria; **N.D.**, not detected.

Station C in Certes Fishponds was characterized by a more diverse phototrophic bacterial community with the presence of isolates resembling *Chr. gracile, Chr. buderi, Tca. roseopersicina, Thiocystis* sp. (sp. nov.), *Rbi. marinum, Rhodovulum* sp., *Rhodospirillum* sp. (sp. nov.) and *Prs. aestuarii*. Green sulfur bacteria resembling the genera *Pelodictyon* and *Chlorobium* were also present but have not been isolated in pure culture.

In the Prévost Lagoon, *Tca. roseopersicina* and *Chr. gracile* were the dominant species of the phototrophic community. Further strains could be identified as species of the purple nonsulfur bacteria (*Rdv. sulfidofilum, Rbi. marinum, Rps. blastica*) and the green sulfur bacteria (*Prs. aestuarii*).

## Discussion

Sediments from the three coastal ecosystems contained substential populations of phototrophic bacteria, irrespective of sampling site. It is perhaps not surprising that such ecosystems are excellent habitats for phototrophic bacteria. Their coastal position promotes eutrophication and the sediments are rich in organic matter due to high primary production by algae and plants. These conditions lead to the anoxia of the surface layers of the sediments and consequently favor sulfate-reducing bacteria which generate large amounts of reduced sulfur compounds (Caumette, 1986). Due to shallow water, relatively high light intensities reach the sediments (Lassen et al., 1993)

*Table 2.* Main characteristics of some isolated phototrophic bacteria.

| Isolate code | Shape, cell division [(a)] | Size ($\mu$m) L. × l or D. [(b)] | Motile | S° in cells | Carotenoid [(c)] | Gen [(d)] |
|---|---|---|---|---|---|---|
| | | Prévost lagoon | | | | |
| EP 2201 | ovoid/rod-f | 1 – 1.5 × 2.5 – 3 | + | + | sp, ly, rh | Chr |
| EP 2204 | sphere-f | 0.8 – 1.5 | – | + | sp | Tca |
| EP 2205 | ovoid/rod-f | 1 – 2 × 2.5 – 4 | + | + | sp, ly, rh | Chr |
| EP 2208 | sphere-f | 1 – 2 | – | + | sp | Tca |
| EP 2209 | ovoid/rod-f | 0.9 – 1.6 × 2 – 4 | + | + | sp, ly, rh | Chr |
| EP 5811 [(e)] | sphere-f | 1.5 – 2.5 | – | + | ok | Tca |
| EP 2101 | rod-f | 0.4 – 0.6 × 0.7 – 1.5 | + | – | se, sn | Rdv. |
| EP 2104 | ovoid/rod-b | 0.4 – 0.7 × 1 – 1.5 | + | – | se, sn | Rps |
| EP 2105 | ovoid/rod-b | 0.5 – 0.7 × 1 – 2.5 | + | – | sp | Rbi. |
| EP 2402 | starlike-f | 0.7 – 1.3 | – | – | chl | Prs |
| | | Arcachon Bay | | | | |
| IO 2201 | ovoid/rod-f | 1.5 – 2 × 3 – 4.5 | + | + | sp, ly, rh | Chr |
| IO 2203 | rod-f | 0.9 – 1.3 × 2.5 – 5 | + | + | ok | Chr |
| IO 2204 | ovoid/rod-f | 1 – 1.5 × 2 – 5 | + | + | sp, ly, rh | Chr |
| IO 2102 | rod-f | 0.6 – 1 × 1.5 – 2.5 | + | – | se, sn | Rdv |
| IO 2103 | rod-f | 0.5 – 0.8 × 1.5 – 2.5 | + | – | se, sn | Rdv |
| BA 2101 | rod-f | 0.5 – 1 × 1 – 2 | + | – | se, sn | Rdv |
| BA 2102 | rod-f | 0.5 – 1 × 1 – 1.7 | + | – | se, sn | Rdv |
| | | Certes lagoon | | | | |
| CE 2201 | ovoid/rod-f | 0.9 – 1.6 × 2 – 4 | + | + | sp, ly, rh | Chr |
| CE 2203 | sphere-f | 1.5 – 2 | + | + | ly, sp, rh | Tcs |
| CE 2205 | ovoid/rod-f | 1 – 1.4 × 3 – 4.5 | + | + | sp, ly, rh | Chr |
| CE 2206 | ovoid/rod-f | 2.5 – 3.5 × 3.5 – 7.0 | + | + | $r^{al}$, $r^{ol}$ | Chr |
| CE 2207 | sphere-f | 1 – 2 | + | + | ly, sp, rh | Tcs |
| CE 2209 | sphere-f | 1.5 – 2.5 | – | + | sp | Tca |
| CE 2102 | rod-b | 0.5 – 0.9 × 1 – 2 | + | – | sp | Rbi. |
| CE 2104 | ovoid-f | 0.6 – 1.2 × 1.2 – 2 | + | – | se, sn | Rdv |
| CE 2105 | spiral-f | 0.6 – 1 × 2.5 – 5.5 | + | – | | Rsp |
| CE 2401 | starlike-f | 0.6 – 1 | – | – | chl | Prs |

(a) Cell division: f, binary fission; b, budding.
(b) Size: L, length; l, width; D, diameter.
(c) Carotenoids: sp, spirilloxanthin; ly, lycopene; rh, rhodopin; $r^{al}$, rhodopinal; $r^{ol}$, rhodopinol; se, spheroidene; sn, spheroidenone; chl, chlorobactene.
(d) Genus: Chr, *Chromatium*; Tca, *Thiocapsa*; Tcs, *Thiocystis*; Rdv, *Rhodovulum*; Rps, *Rhodopseudomonas*; Rsp, *Rhodospirillum*; Prs, *Prosthecochloris*; Rbi, *Rhodobium*.
(e) Data from Caumette et al., 1985.

and allow the phototrophic development of anoxygenic phototrophic bacteria. Thus, the availability of organic matter, anoxic conditions, reduced sulfur compounds and light will control the numbers and activities of the phototrophic bacteria (Caumette, 1992).

When the results obtained in this study at the five different sampling sites are compared, some important features are evident. The inner stations of these ecosystems (stations 11, B and C) appear to have more favorable conditions for the development and maintenance of phototrophic bacteria. These stations are

*Table 3.* Physiological characteristics of some phototrophic sulfur bacteria isolated from Etang du Prévost, Bassin d'Arcachon and Certes fishponds

| Site | Prévost. EP | | Arcachon. IO | | Certes. CE | | | |
|---|---|---|---|---|---|---|---|---|
| Strain N° | 2201 | 2208 | 2203 | 2204 | 2201 | 2203 | 2206 | 2209 |
| Genus | *Chr.* | *Tca.* | *Chr.* | *Chr.* | *Chr.* | *Tcs.* | *Chr.* | *Tca.* |
| Requirements | | | | | | | | |
| vitamins | – | – | – | – | – | – | – | – |
| NaCl (% w/v) | 0.5 | 0 | 1 | 0.5 | 0.5 | 0.5 | ND | ND |
| $N_2$-fixation | + | + | + | + | + | + | + | + |
| $SO_4$-red | + | + | – | + | + | – | – | + |
| Oxygen | m | m | an | m | m | m | an | m |
| Optimum pH | 7.1 | 7.2 | 7.2 | 7.2 | 7.0 | 7.15 | 7.15 | 7.2 |
| Electron donors | | | | | | | | |
| Sulfide | + | + | + | + | + | + | (+) | + |
| Sulfite | – | + | – | + | – | – | + | (+) |
| Thiosulfate | + | + | + | + | + | + | – | + |
| Sulfur | + | + | + | + | + | + | + | + |
| Formate | – | – | – | – | – | + | – | – |
| Acetate | + | + | + | + | + | + | + | + |
| Propionate | + | + | + | + | + | + | – | + |
| Butyrate | + | – | + | + | + | – | – | – |
| Valerate | + | – | + | + | + | – | – | – |
| Caprylate | – | – | – | – | – | – | – | + |
| Lactate | – | + | + | + | + | + | – | + |
| Glycolate | – | + | – | – | – | + | – | + |
| Pyruvate | + | + | + | + | + | + | + | + |
| Malate | + | + | + | + | + | (+) | (+) | + |
| Fumarate | + | + | + | + | + | + | (+) | + |
| Succinate | + | + | + | + | + | + | (+) | + |
| Citrate | – | – | – | – | – | – | – | – |
| Glucose | – | – | – | – | – | – | – | – |
| Fructose | – | + | – | – | – | + | – | + |
| Methanol | – | – | – | – | – | – | – | – |
| Ethanol | – | – | – | – | – | + | – | – |
| Propanol | – | – | – | – | – | + | – | – |
| Glycerol | – | + | – | – | – | – | – | + |
| Yeast extract | + | + | + | + | + | + | + | + |
| Casamino-acids | (+) | – | + | – | – | – | – | (+) |

Genus: *Chr, Chromatium*; *Tca, Thiocapsa*; *Tcs, Thiocystis*.
Oxygen: m, microaerophilic growth with sulfide (2.5 mM) and sulfide (0.8 mM) plus acetate (10 mM) in the dark; an, strictly anaerobic.
Electron donors: –: growth is similar to control; (+): growth is between 1 and 2 times more than control; +: growth is 2 times or more than control.

*Table 4.* Physiological characteristics of phototrophic nonsulfur bacteria isolated from Etang du Prévost, Bassin d'Arcachon and Certes fishponds.

| Site | Prévost. EP | | | Arcachon. IO | | Certes. CE | | |
|---|---|---|---|---|---|---|---|---|
| Strain N° | 2101 | 2104 | 2105 | 2102 | 2103 | 2102 | 2104 | 2105 |
| Genus | *Rdv.* | *Rdw.* | *Rbi.* | *Rdv.* | *Rdv.* | *Rbi.* | *Rdv.* | *Rsp.* |
| Requirements | | | | | | | | |
| NaCl (% w/v) | 0.5 | 0.5 | 0.5 | 0.5 | 0.5 | ND | ND | ND |
| $N_2$-fixation | + | + | + | + | + | + | + | + |
| $SO_4$-red | + | + | + | + | + | + | + | + |
| Oxygen | ae | m | m | ae | ae | ae | ae | ae |
| Optimum pH | 6.6 | 6.8 | 6.7 | 6.8 | 6.7 | 6.8 | 6.8 | 6.8 |
| Electron donors | | | | | | | | |
| Sulfide | + | + | + | + | + | + | + | – |
| Thiosulfate | – | + | + | + | + | – | + | – |
| Sulfur | – | + | – | – | – | – | – | – |
| Formate | + | + | (+) | + | + | (+) | + | + |
| Acetate | + | + | + | + | + | + | + | + |
| Propionate | + | + | + | + | + | + | + | + |
| Butyrate | + | + | + | + | + | + | + | + |
| Valerate | + | + | + | – | + | + | + | + |
| Caprylate | + | + | + | – | + | + | + | – |
| Pelargonate | – | + | + | ND | + | + | + | – |
| Lactate | + | + | + | + | + | + | + | + |
| Glycolate | + | + | + | + | + | (+) | + | – |
| Tartrate | – | – | – | – | – | – | – | – |
| Benzoate | – | – | – | – | – | – | – | – |
| Pyruvate | + | + | + | + | + | + | + | + |
| Malate | + | + | + | + | + | + | + | + |
| Fumarate | + | + | + | + | + | + | + | + |
| Succinate | + | + | + | + | + | + | + | + |
| Citrate | + | + | + | – | – | + | (+) | – |
| Glucose | + | + | + | + | + | + | + | – |
| Fructose | + | + | + | – | + | + | + | + |
| Gluconate | + | + | + | + | + | + | + | + |
| Methanol | ND | (+) | – | – | – | – | – | – |
| Ethanol | + | + | + | – | – | + | + | – |
| Glycerol | – | + | – | + | + | – | + | + |

Genus: *Rps, Rhodopseudomonas*; *Rbi, Rhodobium*; *Rsp, Rhodospirillum*; *Rdv, Rhodovulum.*
Oxygen: m, microaerophilic growth with sulfide (0.8 mM) plus acetate (10 mM) in the dark; ae, aerobic growth.
Electron donors: –: growth is similar to control; (+): growth is between 1 and 2 times more than control; +: growth is 2 times or more than control.

characterized by high levels of organic matter and have high acid volatile sulfide concentrations, indicating the strong activity of sulfate-reducing bacteria (Bourgues et al., 1994). In contrast, the two tidal stations (stations A and B) were less eutrophic and characterized by lower populations of phototrophic bacteria.

Whilst similarities in the total population of phototrophic bacteria were observed in all three lagoon systems, major differences were recorded in the community structure. For example, green sulfur bacteria were only found in the sediments at the sampling stations C and 11. These bacteria do not tolerate oxygen as the purple bacteria do, and therefore have never been found in the eutrophic but tidal station B in the Arcachon Bay. Phototrophic purple bacteria that are commonly found in coastal marine sediments (Pfennig, 1989) were also widely distributed at all the sampling stations investigated in this study. Other phototrophic bacteria such as members of the *Chloroflexaceae* or the *Heliobacteriaceae* have not been detected in these lagoon systems, even when enrichment cultures were applied.

Among the purple bacteria, the purple nonsulfur bacteria were well represented at all the sampling sites, especially at station B. These bacteria are widely distributed in organic rich aquatic environments such as sewage contaminated waters (Siefert et al., 1978; Kobayashi, 1977) or lagoons subjected to excessive pollution (Devèze & Fauvel, 1966; Caumette, 1987). At station B the sediments receive large quantities of organic matter. The total biomass of the macrophytes *Monostroma obscurum* and *Zostera* sp. at this site may reach 260 g $m^{-2}$ of dry mass (Auby et al., 1993). Decomposition of this organic matter favor high sulfate-reduction rates at this station (Finster et al., 1994) and the concomitant release of low molecular weight organic compounds which are metabolized by purple nonsulfur bacteria. The *Rhodovulum* and *Rhodobium* strains isolated from the Arcachon Bay were found to be able to utilise a wide range of simple organic compounds (organic acids, sugars, sugar-alcohols, amino acids) but also sulfide. These isolates are less sensitive to oxygen than purple sulfur bacteria, and therefore dominated the phototrophic bacterial community at station B because oxygen diffused down into the sediments at low tide periods.

At sampling station A, a tidal site dominated by the macrophyte *Zostera noltii Hornem.* that forms beds without decaying algae (Auby et al., 1993), purple nonsulfur bacteria only composed a minor component of the phototrophic bacterial community. At this site, purple sulfur bacteria dominated the community. Among them, several strains resembling *Chromatium purpuratum* were isolated. This is a typical marine and strictly anaerobic species isolated from marine sponges (Imhoff & Trüper, 1980) and marine sediments (Matheron, 1972) (Table 3). This bacterium also formed purple layers on sandy sediments adjacent to station A.

In the Certes Fishponds, the inner station C was characterized by high organic matter and acid-volatile sulfide levels as was station 11 in the Prévost Lagoon. The main difference between these two sampling sites was their macrophyte composition (Auby et al., 1993) and the abundance of meiobenthos (Buffan & Castel, 1993). At station C, *Ruppia cirrhosa* (Pet.) Grande was the dominant component of the macrophyte biomass and maintained a relatively stable standing crop throughout the year. At station 11, as well as station X, the principal primary producer was the alga *Ulva* sp. which could achieve biomass levels of 370 g $m^{-2}$ of dry mass. Station 11 was characterized by a sharp decrease in meiobenthos abundance during the summer, whereas in the Certes Fishponds meiobenthos biomass increased significantly in summer due to the presence of transient meiofauna populations (polychaete larvae) which could reach densities greater than 3000 individuals $cm^{-2}$ (Buffan & Castel, 1993). This meiofauna may play a significant role in the bioturbation of the sediments, leading to the resuspension of reduced sulfur compounds trapped in the deeper sediment layers. These ecological considerations may account for the observed differences in community structure of phototrophic bacteria at the two sites.

In the Certes Fishponds, numbers of phototrophic bacteria were higher than those at the Prévost Lagoon and the community structure was more diverse, particularly among the dominant purple sulfur bacteria which included *Chromatium gracile, Chromatium buderi, Thiocystis* sp. and *Thiocapsa roseopersicina.* In contrast, the community of purple sulfur bacteria in the Prévost Lagoon was strongly dominated by bacteria resembling *Thiocapsa roseopersicina*, as was previously reported by Caumette (1986). *Thiocapsa roseopersicina* is a purple sulfur bacterium well adapted to the fluctuating conditions of oxygen and sulfide (De Wit & van Gemerden, 1987, 1990; Schaub & Van Gemerden, 1994). Such fluctuating conditions occur periodically in the Prévost Lagoon. Macroalgae are rapidly degraded with a concomitant rapid consumption of dissolved oxygen. Further degradation is carried out by sulfate-reducing bacteria that form hydrogen sulfide in the entire water column, leading to dystrophic crises (Amanieu et al., 1975). During these crises, *Thiocapsa roseopersicina* is the dominant phototrophic bacterium that develops in the water column, producing red water and oxidizing the toxic sulfide (Caumette, 1986). In contrast, at station B which is exposed to air periodically, as well as at the non-tidal station C, the high levels of iron in the sediments play a significant role in trapping the hydrogen sulfide generated by sulfate reducing bacteria (Stal et al., 1994).

## Acknowledgements

This study was financed by a joint EU project on Coastal Lagoon Eutrophication and Anaerobic processes (CLEAN), Contract number EV5V – CT92 – 0080.

## References


Amanieu, M., B. Baleux, O. Guélorget & P. Michel, 1975. Etude biologique et hydrobiologique d'une crise dystrophique (Malaïgue) dans l'étang du Prévost à Palavas (Hérault). Vie Milieu, série B 25: 175–204.

Auby, I., G. Bachelet & P. J. Labourg, 1993. Biomass and species composition of macrophytes in Arcachon Bay and Prévost Lagoon, with a compilation of data on primary production in Arcachon Bay. In Caumette, P. (coord.), Proceedings of the CLEAN Meeting, part II, Alicante, Spain: 65–72. European Community Environmental Programme, DG XII, Brussels.

Bourguès, S., O. Pringault & R. de Wit, 1994. Spatial and temporal variations of oxygen uptake, sulfide production and population of sulfur cycle bacteria at the CLEAN stations. In Caumette, P. (coord.), Proceedings of the CLEAN Meeting, part II, Ferrara, Italy: 109–132. European Community Environmental Programme, DG XII, Brussels.

Buffan, E. & J. Castel, 1993. Distribution of benthic microbionta in Arcachon Bay and Prévost Lagoon. In Caumette, P. (coord.), Proceedings of the CLEAN Meeting, part II, Alicante, Spain: 73–89. European Community Environmental Programme, DG XII, Brussels.

Cahet, G., 1965. Contribution à l'étude des eaux et des sédiments de Bages-Sigean (Aude). III. Réduction des composés soufrés. Vie Milieu 16: 917–981.

Castel, J., P. Caumette & R. A. Herbert, 1996. Eutrophication gradients in coastal lagoons. Hydrobiologia 329: xi–xxx.

Caumette, P., 1986. Phototrophic sulfur bacteria and sulfate-reducing bacteria causing red waters in a shallow brackish coastal lagoon (Prévost lagoon, France). FEMS Microbiol. Ecol. 38: 113–124.

Caumette, P., 1987. Rôle des bactéries phototrophes et des bactéries sulfato-réductrices dans les milieux lagunaires. O.R.S.T.O.M. (ed.), Paris, 304 pp.

Caumette, P., 1992. Bacterial communities in coastal lagoons. An overview. Vie milieu 42: 111–123.

Caumette, P. & B. Baleux, 1980. Etude des eaux rouges dues à la prolifération des bactéries photosynthétiques sulfo-oxydantes dans l'étang du Prévost, lagune saumâtre méditerranéenne. Mar. Biol. 56: 183–194.

Caumette, P., M. Pagano & L. Saint-Jean, 1983. Répartition verticale du phytoplancton des bactéries et du zooplancton dans un milieu stratifié en baie de Biétri (Lagune Ebrié, Côte d'Ivoire). Hydrologia 106: 135–148.

Caumette, P., K. Schmidt, H. Biebl & N. Pfennig, 1985. Characterization of a *Thiocapsa* strain containing okenone as major carotenoid. System. Appl. Microbiol. 6: 132–136.

Cerruti, A., 1938. Le condizioni oceanographiche et biologiche del mar piccolo di Taranto durante l'Agosto del 1938. Boll. Pesca. Piscicolt. Idrobiol. 14: 711–751.

Cline, J. D., 1969. Spectrophotometric determination of hydrogen sulfide in natural waters. Limnol. Oceanogr. 14: 454–458.

Devèze, L. & Y. Fauvel, 1966. Un phénomène bactérien d'eaux rouges dans l'étang d'Ingril (Hérault). Rev. Travaux de l'Institut des Pêches Maritimes 30: 365–374.

de Wit, R. & H. van Gemerden, 1987. Chemolithotrophic growth of the phototrophic purple sulfur bacterium *Thiocapsa roseopersicina*. FEMS Microbiol. Ecol. 45: 117–126.

de Wit, R. & H. van Gemerden, 1990. Growth and metabolism of the purple sulfur bacterium *Thiocapsa roseopersicina* under combined light/dark and oxic/anoxic regiments. Arch. Microbiol. 154: 459–464.

Eichler, B. & N. Pfennig, 1986. Characterization of a new platelet-forming purple sulfur bacterium, *Amoebobacter pedioformis* sp. nov. Arch. Microbiol. 146: 295–300.

Finster, K., A. Gammelgaard & L. S. Hansen, 1994. Sulfate reduction rates, inorganic sulfur pools and the accumulation of volatile fatty acids in the Bassin of Arcachon and the Prévost lagoon, southern France: a comparative study. In Caumette, P. (coord.), Proceedings of the CLEAN Meeting, part II, Ferrara, Italy: 223–252. European Community Environmental Programme, DG XII, Brussels.

Forti, A., 1933. Il fenomeno del lago di sangue nello stagno di Pergusa in Sicilia. Nuovo Giorno Bot. Ital. 40: 76.

Guerrero, R., C. Pedrós-Alió, I. Esteve & J. Mas, 1987. Communities of phototrophic sulfur bacteria in lakes of the Spanish Mediterranean region. Acta Academiae Aboensis 47: 125–151.

Guyoneaud, R., A. Hirschler, R. Matheron & P. Caumette, 1993. Phototrophic bacterial communities in Arcachon bay and Prévost lagoon. In Caumette, P. (coord.), Proceedings of the CLEAN Meeting, part II, Alicante, Spain: 341–355. European Community Environmental Programme, DG XII, Brussels.

Heldt, J. H., 1952. Les eaux rouges. Soc. Sci. Nat. Tunisie 5: 103–106.

Imhoff, J. H. & H. G. Trüper, 1980. *Chromatium purpuratum*, sp. nov., a new species of the Chromatiaceae. Zbl. Bakt., I. Abt. Orig. C1: 61–69.

Jørgensen, B. B., 1977. The sulfur cycle of a coastal marine sediment (Limfjorden, Denmark). Limnol. Oceanogr. 22: 814–832.

Jørgensen, B. B. & T. Fenchel, 1974. The sulfur cycle of a marine sediment model system. Mar. Biol. 24: 189–201.

Jørgensen, B. B., N. P. Revsbech & Y. Cohen, 1983. Photosynthesis and structure of benthic microbial mats: Microelectrode and SEM studies of four cyanobacterial communities. Limnol. Oceanogr. 28: 1075–1093.

Kämpf, C. & N. Pfennig, 1980. Capacity of *Chromatiaceae* for chemotrophic growth. Specific respiration rates of *Thiocystis violacea* and *Chromatium vinosum*. Arch. Microbiol. 127: 125–135.

Kobayashi, M., 1977. Utilization and disposal of wastes by photosynthetic bacteria. In H. G. Schlegel & J. Barnea (eds), Microbial energy conversion. Pergamon Press, Oxford: 443–453.

Lassen, C., R. de Wit, J. Lorentzen & N. P. Revsbech, 1993. Distribution of light, $O_2$ and oxygenic photosynthesis. In Caumette, P. (coord.), Proceedings of the CLEAN Meeting, part II, Alicante, Spain: 23–40. European Community Environmental Programme, DG XII, Brussels.

Matheron, R. & R. Baulaigue, 1972. Bactéries photosynthétiques sulfureuses marines. Assimilation des substances organiques et minérales, et influence de la teneur en chlorure de sodium du milieu de culture sur leur développement. Arch. Mikrobiol. 86: 291–304.

Pfennig, N., 1989. Ecology of phototrophic purple and green sulfur bacteria. In H. G. Schlegel & B. Bowien (eds), Autotrophic Bacteria. Science Technique Publisher, Madison and Springer Verlag, Berlin: 97–116.

Pfennig, N. & S. Wagener, 1986. An improved method of preparing wet mounts for the photomicrography of microorganisms. J. Microbiol. Meth. 4: 303–306.

Pfennig, N. & H. G. Trüper, 1992. The family *Chromatiaceae*. In A. Balows, H. G. Trüper, M. Dworkin, W. Harder & K. H. Schleifer (eds), The Prokaryotes. Springer-Verlag, New-York: 3200–3221.

Pierson, B., A. Oesterle & G.L. Murphy, 1987. Pigments, light penetration, and photosynthetic activity in the multi-layered microbial mats of Great Sippewisset salt marsh, Massachussets. FEMS Microbiol. Ecol. 45: 365–376.

Schaub, B. E. M. & H. van Gemerden, 1994. Simultaneous phototrophic and chemotrophic growth in the purple sulfur bacterium *Thiocapsa roseopersicina* M1. FEMS Microbiol. Ecol. 13: 185–196.

Siefert, E., R. L. Irgens & N. Pfennig, 1978. Phototrophic purple and green bacteria in a sewage treatment plant. Appl. Envir. Microbiol. 35: 38–44.

Stal, L., F. Kruying, S. B. Behrens & M. Villbrandt, 1994. Spatial and seasonal variations of iron, phosphate and sulfide in the sediments of two eutrophic coastal lagoons. In Caumette, P. (coord.), Proceedings of the CLEAN Meeting, part II, Ferrara, Italy: 167–186. European Community Environmental Programme, DG XII, Brussels.

Stirn, J., 1971. Ecological consequences of marine pollution. Rev. Int. Oceanogr. medit. 24: 13–46.

van Gemerden, H., R. de Wit, C. S. Tughan & R. A. Herbert, 1989a. Development of mass blooms of *Thiocapsa roseopersicina* on sheltered beaches on the Orkney Islands. FEMS Microbiol. Ecol. 62: 111–118.

van Gemerden, H., C. S. Tughan, R. de Wit & R. A. Herbert, 1989b. Laminated microbial ecosystems on sheltered beaches in Scapa Flow, Orkney Islands. FEMS Microbiol. Ecol. 62: 87–102.

*Hydrobiologia* **329:** 45–55, 1996.
*P. Caumette, J. Castel & R. Herbert (eds), Coastal Lagoon Eutrophication and ANaerobic Processes (C.L.E.AN.).*



# Zooplankton variability related to environmental changes in a eutrophic coastal lagoon in the Po Delta

Sandra Sei, Giampaolo Rossetti, Ferdinando Villa & Ireneo Ferrari
*Dipartimento di Scienze Ambientali, Università di Parma, 43100 Parma, Italy*

*Key words:* Po Delta, coastal lagoons, dystrophic crisis, zooplankton, copepods

## Abstract

Zooplankton variability in a lagoon of the Po Delta, the Sacca di Goro, was studied in relation to changes in its hydrodynamic and hydrological features. From 1987 to 1992 the lagoon was affected each summer by severe anoxia due to the decomposition of dense and widespread beds of *Ulva rigida*. In August 1992 a canal was opened through the sand bank closing off the lagoon from the sea in order to improve the water circulation. This hydraulic intervention led to significant environmental changes in the lagoon: during the two subsequent years the summer dystrophic crises were brief and less severe, due to a sharp decrease in the *Ulva* cover. At the same time a clear increase in phytoplankton biomass values was observed. Following the opening of the channel, the most remarkable change in zooplankton was a significant density increase of calanoids, particularly of *Acartia tonsa*.

## Introduction

In recent years a great number of investigations has broadened our knowledge of the structure, variability and ecological role of zooplankton in coastal lagoons and estuarine environments (Menendez & Comin, 1986; Lam Hoai & Amanieu, 1989; Castel, 1993; Soetaert & van Rijswijk, 1993; Bakker, 1994; Bakker & van Rijswijk, 1994; Laprise & Dodson, 1994). The main results of research on zooplankton in Italian lagoons have been summarized by Carrada et al. (1987) and Ceccherelli et al. (1994). Since the seventies a particularly strong research effort has concentrated on the zooplankton in lagoons of the Po Delta: the spatial distribution pattern of the biocoenosis, the interactions connecting it to the other compartments and the exchange of plankton biomass between lagoon and sea have been investigated (Ferrari & Chieregato, 1981; Ferrari et al., 1985, Ambrogi et al., 1989).

Special attention has been paid to the study of plankton communities in the Sacca di Goro, a subtidal lagoon located in the southern part of the Po Delta (Ferrari et al., 1993) (Figure 1). From 1987 to 1992 this lagoon was affected by summer anoxic events, due to decomposition of *Ulva rigida* C. Agardh beds (Viaroli et al., 1992; 1993a). These dystrophic episodes were particularly severe in the sheltered eastern zone (Valle di Gorino), which was closed off from the sea by a large sand bank. At the beginning of August 1992, after a protracted and particularly severe anoxic event which resulted in a high mortality of farmed populations of mussels and clams, a channel, 300–400 m wide and more than 2 m deep, was cut through the sand bank (Figure 1): this hydraulic intervention led to a significant increase of sea water inflow and of water renewal in the Valle di Gorino. In the subsequent years a sharp reduction of the *Ulva* cover was observed and the summer dystrophic events were quite mild and brief (Viaroli et al., 1993b).

In this paper variability of zooplankton (particularly of the copepod component) in the Sacca di Goro has been investigated in relation to the environmental changes which occurred following the cutting of the channel. Analyses were performed on zooplankton samples collected monthly from 1987 to 1989 (Ferrari et al., 1993) and then for three subsequent years, from 1992 to 1994.

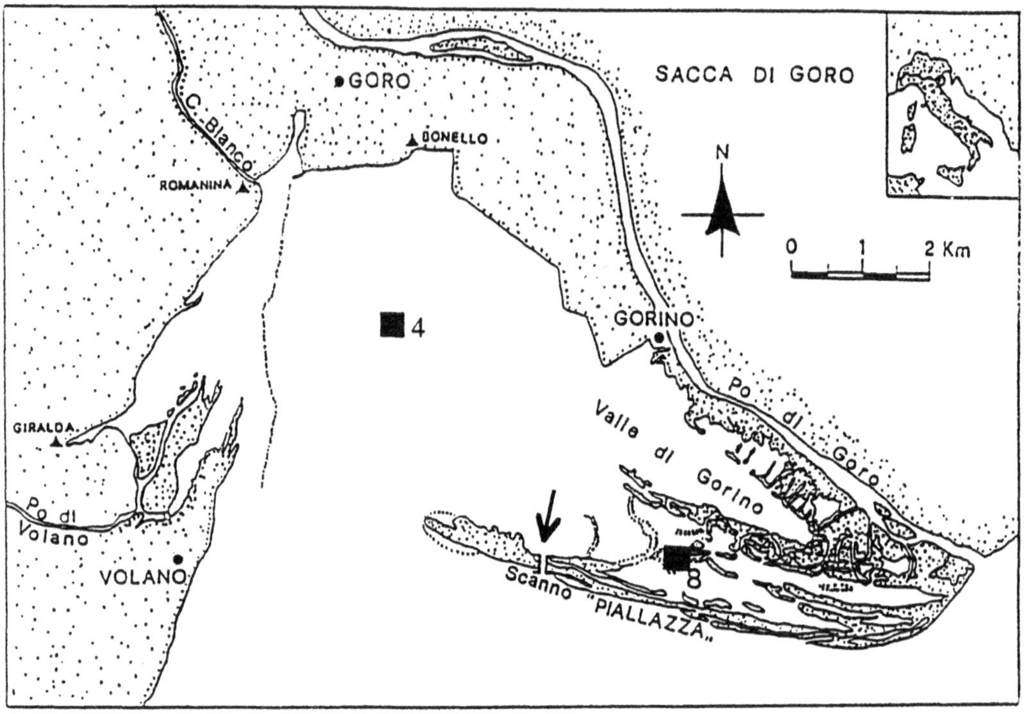


*Figure 1.* Sacca di Goro: sampling stations are indicated; the arrow shows the position of the sand bank cut.

## Materials and methods

Zooplankton samples were collected at two stations, St. 4 in the centre of the lagoon and St. 8 in the Valle di Gorino (Figure 1), at monthly intervals from November 1987 to October 1989 and then from March to December 1992, from April to November 1993 and from February to December 1994. Tidal phase variations were not the same for the different sampling dates.

Sampling was carried out with a 15 l Patalas trap: at least 150 l of water were taken from the entire water column (maximum depth was about 2 m) for each sample and then filtered through a 50 $\mu$m mesh net. A buffered formaldehyde solution (4%) was used as fixative.

Simultaneously with zooplankton sampling, water temperature, salinity, dissolved oxygen (Winkler method, APHA, 1975) and phytoplankton chlorophyll-*a* content (trichromatic method, APHA, 1975) were measured at both the surface and near bottom layers of the water column (for 1994 data are available only from March to August).

Zooplankton counting was carried out on subsamples taken by a Hansen-Stempel pipette. Taxonomic identification was, when possible, made to the species level for the copepods and cladocerans. For data processing all identified taxonomic units were subdivided into 18 major groups: *Noctiluca*, tintinnids, *Synchaeta*, other rotifer species, polychaete larvae, gastropod larvae, bivalve larvae, cirriped larvae, decapod larvae, cladocerans, copepod nauplii, copepod copepodids and adults (subdivided into five categories: freshwater copepods, brackish water copepods, euplanktonic calanoids, euplanktonic cyclopoids, euplanktonic harpacticoids), tunicates and 'other groups'.

Canonical Correspondence Analysis (CCA) (Ter Braak, 1986) was applied in data analysis. CCA allows ordination of a set of samples, each defined by a list of species abundances, under the constraint that ordination axes are a linear combination of environmental variables measured for each sample. This multivariate technique assumes that species abundance response to variables follows a unimodal, bell-shaped curve. CCA provides ordination of both samples and species points, highlighting the importance of species in determining sample composition.

A simple graphical representation (biplot) can be drawn, where environmental variables are represented by arrows in the ordination space. Then sample (and species) point projections on the arrows are estimates of the 'score' of each point with respect to environmental variables. In the case of species, this score can be thought of as species performance estimates along the corresponding environmental gradient.

All ordination techniques are by their own nature explorative, thus a significance test of the results has a purely indicative value. In the case of CCA, a crucial issue is that the species data set is fully or partly explained by the environmental variables considered. To assess the significance of ordination axes for explorative purposes, a Monte Carlo permutation test is usually performed: in each one of a number of CCAs, the order of environmental variable values is randomly permuted. From the distribution of the ordination statistic chosen (usually the eigenvalue), one can estimate the probability that environmental variables would have no significant effect on the species data set, by computing the percentage $p$ of runs yielding statistics equal to or greater than that observed.

A CCA of the Sacca di Goro data was performed using abundances of the 18 taxonomic groups, after a $\log(1 + x)$ transformation. Environmental variables were: temperature, salinity, dissolved oxygen and phytoplankton chlorophyll-$a$ content: data recorded at the surface layer of the water column were used. The excessive sensitivity of CCA to rare species was compensated by using downweighting as suggested by Hill (1979). CCA was performed by means of the CANOCO program (Ter Braak, 1988), running on a IBM-compatible PC. Separate CCAs were conducted for the two sampling stations, with Monte Carlo significance tests using first eigenvalues extracted and 1000 permutations each time. The biplot diagrams of environmental variables and taxa scores are represented separately for the periods before and after the cut of the sand-bank; in other diagrams sample scores for the entire period of investigation are given separately for each year.

## Results

The change in the lagoon hydrological conditions after the hydraulic intervention in August 1992 was reflected, first of all, by higher values of phytoplankton chlorophyll-$a$ content at St. 8 (Figure 2). In previous years, particularly in 1992, chlorophyll-$a$ content of the water ranged around values less than 5 $\mu$g $l^{-1}$ through the entire growing season: isolated phytoplankton blooms occurred only after the summer anoxic episodes, following improvised emergency interventions on the sand bank aimed to favour a temporary inflow of seawater into the lagoon (Ferrari et al., 1993). The increase of phytoplankton chlorophyll-$a$ coincided with the reduction of *Ulva* biomass. Other phenomena closely related to the decreased spreading of this nitrophilous macroalga have been detailed by Viaroli et al. (1993b) and Bartoli & Viaroli (pers. comm.): during 1993 and 1994 a shorter phase of spring depletion of dissolved inorganic nitrogen and a shorter duration of summer anoxia in a narrow easternmost area of the Valle di Gorino were observed. In addition, experiments carried out in 1993 and 1994 with the dark and light bottle method and dark core incubation showed, in comparison with the previous years, a significant reduction of the metabolic activity (production and respiration) of the macroalgal compartment compared to the planktonic compartment as well as a decrease in the sediment oxygen demand.

Zooplankton abundance and composition also underwent significant changes. Until 1992 total zooplankton density at St. 8 was characterized by sudden, very high peaks, generally linked to phytoplankton chlorophyll-$a$ maxima. In the subsequent years total density had more reduced fluctuations; furthermore, its seasonal trend showed less pronounced differences from St. 4 (Figure 2).

The copepod nauplii were, on the average, the dominant group, with the highest abundance peak in the Valle di Gorino (1299 ind $l^{-1}$ in July 1992); they followed the same general trend observed for total zooplankton (Figure 2). The species belonging to the five categories of copepodids and adults of copepods are listed in Table 1. Freshwater copepods occurred with a very low frequency. The brackish water fraction, mainly represented by epibenthic and phytal harpacticoids, showed larger abundances at St. 8 (16 and 17 ind $l^{-1}$ in July 1988 and August 1992, respectively) during the decomposition phase of macroalgal beds; at the same station, a decreasing trend was observed after August 1992: maximum density was 9 ind $l^{-1}$ in 1993 (May) and 2 ind $l^{-1}$ in 1994 (December). Abundances of neritic euplanktonic cyclopoids (chiefly belonging to the genus *Oithona*) and harpacticoids did not vary appreciably: density maxima were 7–8 and 5 ind $l^{-1}$, respectively, both before and after August 1992.

From 1987 to 1989 calanoids were characterized by relatively high abundance values during winter months, mainly sustained by neritic forms, in particular *Acartia clausi* Giesbrecht. In 1993 and 1994 they showed sharply higher densities than in previous years

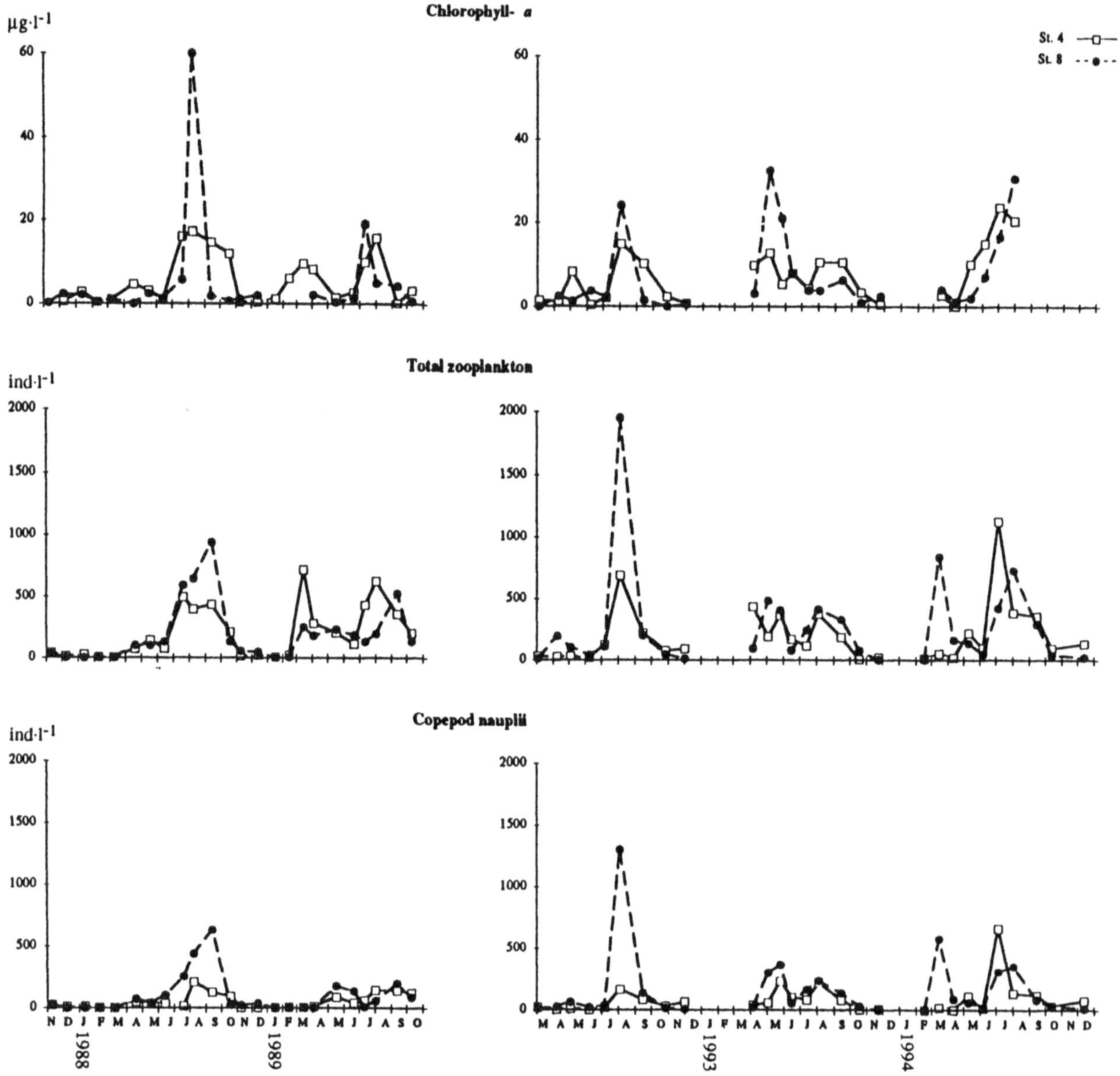


*Figure 2.* Sacca di Goro: phytoplankton chlorophyll-*a* concentration at the surface layer, density of total zooplankton and copepod nauplii at St. 4 and St. 8 from 1987 to 1989 and from 1992 to 1994.

at both stations (Figure 3): this increased abundance was due almost exclusively to *Acartia tonsa* Dana, a species typical of highly eutrophic coastal areas and harbour environments (Brylinski, 1981). *A. tonsa* was found for the first time in the Mediterranean Sea by Gaudy & Viñas (1985) in a great brackish water pond near Marseille and for the first time in the Adriatic Sea by Farabegoli et al. (1989) in the lagoons of the Po River Delta. Over the period under investigation, *A. tonsa* gradually came to be the dominant species of the plankton copepod component of the Sacca di Goro: the peaks of calanoid density in September 1989 (75 ind $l^{-1}$ at St.4) and then in July (189 ind $l^{-1}$ at St. 4) and August 1994 (198 ind $l^{-1}$ at St. 8) were mostly supported by *A. tonsa*. This led to the progressive disappearance of the congeneric species *Acartia margalefi* Alcaraz and *Acartia latisetosa* Kriczaguin, which are typical of the most confined areas of the lagoons of the Adriatic Sea: after 1989 no individuals belonging to these two species were found. The congeneric marine species *A. clausi* was found in most samples, especially at St. 4, but its abundance gradually decreased both

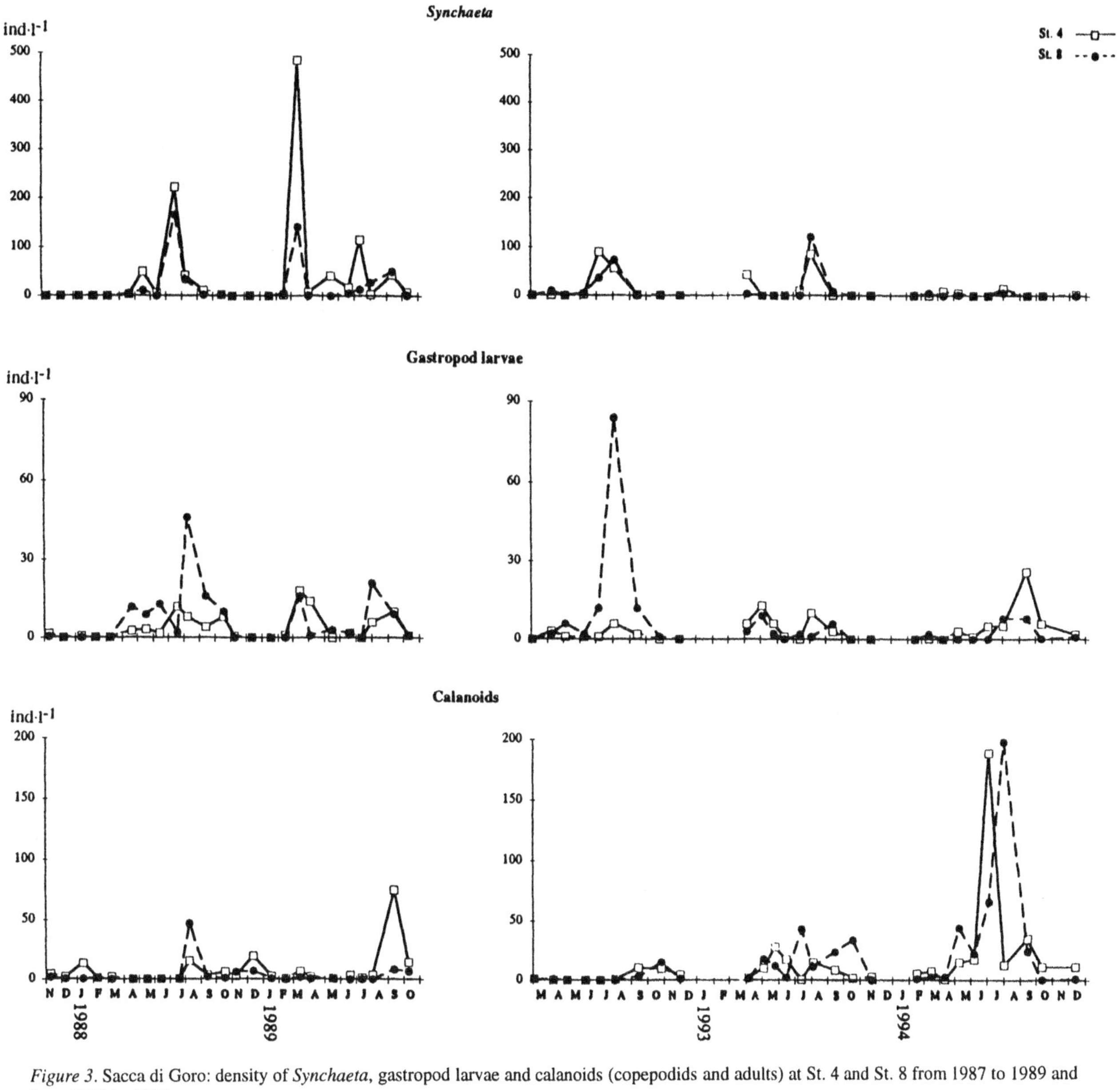


*Figure 3.* Sacca di Goro: density of *Synchaeta*, gastropod larvae and calanoids (copepodids and adults) at St. 4 and St. 8 from 1987 to 1989 and from 1992 to 1994.

in the Valle di Gorino and in the centre of the lagoon, until it represented only a slight fraction of the *Acartia* complex density in 1994.

A clear increase in the Valle di Gorino was observed for marine cladocerans as well: from 1987 to 1992 at St. 8 they never exceeded 1 ind $l^{-1}$, whereas they showed a peak of 11 ind $l^{-1}$ in both September 1993 and September 1994. The highest density value (37 ind $l^{-1}$) was recorded in June 1994 at St. 4. The most important species was *Podon polyphemoides* Leuck.

Among meroplankton, decapod larvae exceeded 2 ind $l^{-1}$ only at St. 8 from May to July 1993 and in June 1994; from 1987 to 1994 they were significantly increasing at St. 4 too. Polychaete, bivalve and cirriped larvae showed conspicuous densities at both stations, generally from March to September, throughout the research period. Density maxima for these three taxa were found at St. 4: 417 ind $l^{-1}$ in August 1989 for polychaete larvae, 196 ind $l^{-1}$ in July 1988 for bivalve larvae and 35 ind $l^{-1}$ in July 1989 for cirriped larvae. The high densities of bivalve larvae

*Table 1.* Species of copepods occurring in the zooplankton samples collected in the Sacca di Goro.

| | |
|---|---|
| Freshwaters copepods | |
| | *Acanthocyclops* sp. |
| Brackish copepods | |
| | *Calanipeda aquaedulcis* Kritz |
| | *Halicyclops rotundipes* Kiefer |
| | *Halicyclops* sp. |
| | *Cyclopina gracilis* Claus |
| | *Cyclopina* sp. |
| | Ergasilidae |
| | *Parategastes sphaericus* Claus |
| | *Canuella perplexa* Scott |
| | *Ectinosoma dentatum* Steuer |
| | *Ectinosoma* sp. |
| | *Horsiella* sp. |
| | *Microarthridion fallax* Perkins |
| | *Harpacticus flexulosus* Ceccherelli |
| | *Harpacticus flexus* Brady & Robertson |
| | *Tisbe holoturiae* Humes |
| | *Tisbe* sp. |
| | *Amphiascus parvus* Sars |
| | *Amonardia normani* Brady |
| | *Schizopera compacta* De Lint |
| | *Diosaccopsis rubeus* Brian |
| | *Ameira parvula* Claus |
| | *Nitocra typica* Boeck |
| | *Mesochra pygmaea* Claus |
| | *Asellopsis sarmantica* Jakubisiak |
| | *Robertgurneya similis* A. Scott |
| Euplanktonic calanoids | |
| | *Calanus helgolandicus* Claus |
| | *Paracalus parvus* Claus |
| | *Pseudocalanus elongatus* Boeck |
| | *Clausocalanus* sp. |
| | *Temora longicornis* Müller |
| | *Temora stylifera* Dana |
| | *Centropages ponticus* Karavaev |
| | *Centropages typicus* Kröyer |
| | *Acartia clausi* Giesbrecht |
| | *Acartia latisetosa* Kriczaguin |
| | *Acartia margalefi* Alcaraz |
| | *Acartia tonsa* Dana |
| Euplanktonic cyclopoids | |
| | *Oithona nana* Giesbrecht |
| | *Oithona similis* Claus |
| | *Oithona* sp. |
| | *Oncaea subtilis* Giesbrecht |
| | *Oncaea media* Giesbrecht |
| | *Oncaea* sp. |
| | *Corycaeus* sp. |
| Euplanktonic harpacticoids | |
| | *Microsetella norvegica* Boeck |
| | *Euterpina acutifrons* Dana |

can be related to the presence in the southern area of the lagoon of extensive banks of the cultivated clam *Tapes philippinarum* Adams & Reeve. The considerable cirriped larvae abundance seems to be dependent on the large number of wood and cement artificial substrata widespread in all the lagoon. Gastropod larvae, probably belonging to the genus *Hydrobia*, a small snail closely associated with the *Ulva* thalli, at St. 8 decreased sharply from the peak of 84 ind $l^{-1}$ in August 1992 to maximum densities of 8 to 9 ind $l^{-1}$ in 1993 and 1994 (Figure 3).

A decreasing trend for *Synchaeta* was evidenced at both the stations: the peaks observed in 1988 (222 ind $l^{-1}$ at St. 4 and 165 ind $l^{-1}$ at St. 8 in July) and 1989 (485 ind $l^{-1}$ at St. 4 and 141 ind $l^{-1}$ at St. 8 in March) progressively decreased; in 1994 a remarkably lower density was noted (Figure 3). Other rotifer species rarely occurred, although in August 1992 isolated peaks (255 ind $l^{-1}$ at St. 4 and 43 ind $l^{-1}$ at St. 8), mainly sustained by *Brachionus plicatilis* O.F. Müller, were recorded.

*Noctiluca* and tintinnids were generally found in low density, except for a tintinnid abundance peak of 260 ind $l^{-1}$ at St. 8 in August 1992.

Both the occurrence frequency and abundance of tunicates (appendicularians), a typical marine group generally associated with neritic species of copepods, were quite low throughout the research period; at St. 4, however, their frequency and abundance (with a maximum of 6 ind $l^{-1}$ in September 1992) appeared more relevant than at St. 8, where tunicates were found only in three samples with a maximum of 1 ind $l^{-1}$ in September 1993.

The main results of CCA are presented in the biplot diagrams of the environmental variables and taxa scores for St. 4 and St. 8, separately for the pre- and post-cut period (Figure 4). The statistical significance of the effects of the environmental variables, tested by the Monte Carlo permutation procedure, should be noted. Before the channel was cut, at both stations the community composition could be accounted for by environmental variables: the $p$ value is $< 0.01$ for St. 4 and 0.01 for St. 8. In contrast, after August 1992 the environmental variables do not significantly explain the zooplanktonic community structure: $p$ values are 0.20 and 0.28 for St. 4 and St. 8, respectively.

CCA ordination before August 1992 at St. 4 is characterized by the importance of salinity which contributes particularly to the first axis in contrast both to chlorophyll concentration and temperature. The typically marine taxa, tunicates, cyclopoids, harpacti-

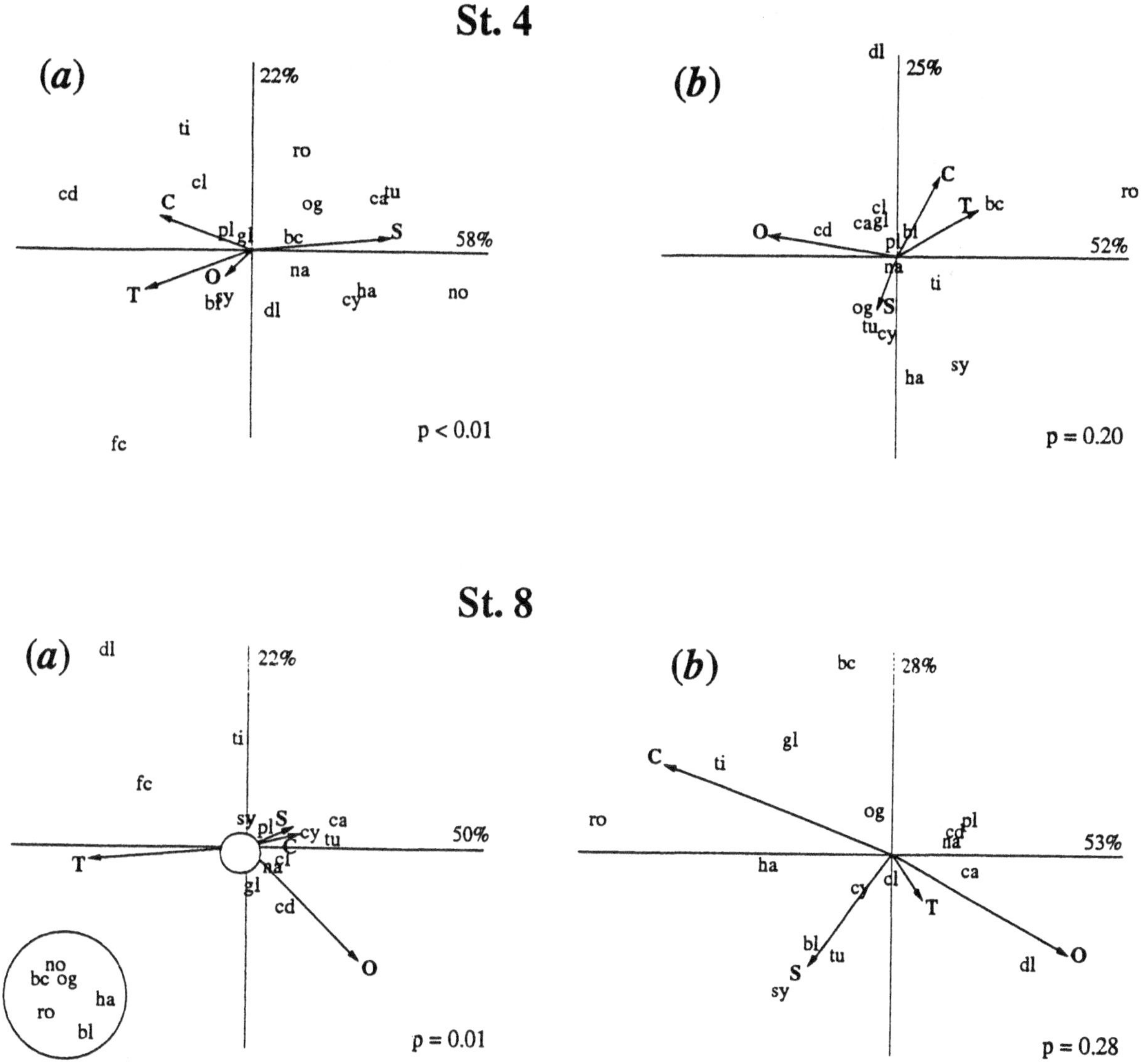


| | | | | | | | | |
|---|---|---|---|---|---|---|---|---|
| no | = | *Noctiluca* | bl | = | Bivalve larvae | bs | = | Brackish copepods |
| ti | = | Tintinnids | cl | = | Cirriped larvae | ca | = | Euplanktonic calanoids |
| ro | = | Rotifers | dl | = | Decapod larvae | cy | = | Euplanktonic cyclopoids |
| sy | = | *Synchaeta* | cd | = | Cladocerans | ha | = | Euplanktonic harpacticoids |
| pl | = | Polychaete larvae | na | = | Copepod nauplii | tu | = | Tunicates |
| gl | = | Gastropod larvae | fc | = | Freshwater copepods | og | = | Other groups |

*Figure 4.* CCA of the zooplankton in the Sacca di Goro: scores of environmental variables and taxonomic groups in the plane of the first two axes of the ordination for the pre- and post-cut periods (*a* and *b* respectively) at St. 4 and St. 8.

coids, and especially calanoids whose relative abundance peaked during the winter months, are associated with the highest salinity values. At St. 8 ordination space shows the larger part of taxa aggregated close to the origin; tunicates, cyclopoids and calanoids, however, are separated from the other groups and located on the upper right quadrant where salinity and chlorophyll content are at their maxima: indeed, in 1988 and 1989 cyclopoids and calanoids occurred with isolated summer peaks after the dystrophic crisis.

After August 1992 ordination at both stations still points to a close relationship between salinity and the two typically marine taxa represented by tunicates and cyclopoids, whereas calanoids, which were more and more dominated by *Acartia tonsa*, are associated with autochthonous meroplankton taxa (gastropod, cirriped and bivalve larvae at St. 4, polychaete larvae at both St. 4 and St. 8). Calanoids, in contrast to what was observed before the channel was cut, appear to be inversely related to chlorophyll content.

Environmental variable plots calculated over the entire investigation period, for an easier graph interpretation, are presented only for the first year (Figure 5). At St. 8 the first axis is clearly defined by the opposition between dissolved oxygen and temperature; such an ordination underscores the central importance of summertime anoxia for this station. After 1992 a trend of samples (summer ones included) to move towards the left quadrants, where dissolved oxygen is at its maximum, is quite evident. At St. 4, on the contrary, the key variable appears to be salinity, which is in opposition to chlorophyll content, thus confirming what had been observed in biplots of taxa scores for the same station.

## Discussion

The case of the Sacca di Goro is representative of a condition of eutrophy common to many coastal lagoons and not only those in the Northern Adriatic (Viaroli et al., 1992; Ceccherelli et al., 1994). The eutrophication process accelerated rapidly during the 1980's, leading to the enormous growth of *Ulva* biomass and the consequent severe summer anoxia due to its decomposition. The first dystrophic episode in the lagoon took place in 1987, and every summer up to 1992 was characterized by increasingly more severe anoxic crises. The hydraulic intervention in August 1992 did not result in a substantial reduction of water column trophism, inasmuch as it remained elevated due to the heavy nutrient loading from both the catchment area and the sediment. It did, however, attenuate environmental stress which was particularly severe in the Valle di Gorino and by then also evident in the centre of the lagoon (Ferrari et al., 1993; Viaroli et al., 1993a, 1993b).

The series of data examined in this study shows the transition from an extreme condition towards an environment which allows the recovery of a highly resilient ecosystem. The hydrodynamic and hydrological conditions created by cutting of the channel resulted first of all in the reduction of the area covered by the *Ulva* beds and a shift from the dominance of macroalgae and epiphytes to the dominance of phytoplankton within the primary producer community.

The response of zooplankton to the environmental changes was analysed on an incomplete set of samples. Moreover, data analysis was done on mixed groups identified according to taxonomic and ecological analogies among species, and statistical processing was carried out for principally explorative purposes. Nevertheless, these results seem promising.

CCA ordination of both taxa and samples suggests two lines of systemic interpretation of the evolution of the lagoon system we have investigated. First, the significance of the differences in zooplankton structure before and after the channel was cut is demonstrated: a biocoenosis strongly influenced by the forcing of environmental variables is replaced, after 1992, by a biocoenosis more resilient to external factors. Secondly, the main ecological differences between the two sampling areas are represented synthetically: in the centre of the lagoon zooplankton variability appears to be dependent on the action of two factors, autochthonous production (phytoplankton chlorophyll-*a*) and the influence of the sea (salinity); the Valle di Gorino is characterized chiefly by summer anoxia, but a tendency to attenuation of dystrophic events after 1992 is pointed out.

Analysis of the spatial-temporal patterns of the individual zooplankton taxa shows regularities and consistent trends, above all in the Valle di Gorino: here, after August 1992, a decrease in taxa strictly associated with *Ulva* thalli (e.g., brackish water copepods and gastropod larvae) and, more generally, in the minute zooplankton forms, in particular rotifers, was observed. Concurrently, the grazing chain appears to have been strengthened by the increased density of larger-sized forms of filter feeders: cladocerans, decapod larvae and above all calanoids represented by *Acartia tonsa*. The latter species has most probably only recently colonized the lagoon (Farabegoli et al., 1989), but within a period of a few years it has become the dominant planktonic copepod (resulting in a reduction in the diversity of the *Acartia* complex typically confined to the lagoon) and probably a key species of the entire plankton community. A short-term zooplankton sampling campaign, we carried out in the Sacca di Goro in July 1994 (unpublished data), confirmed the primary importance of *A. tonsa*: in the Valle di Gorino, at ebb tide, copepodids and adults of this species reached density peaks of more than 500 ind $l^{-1}$.

*(a)*

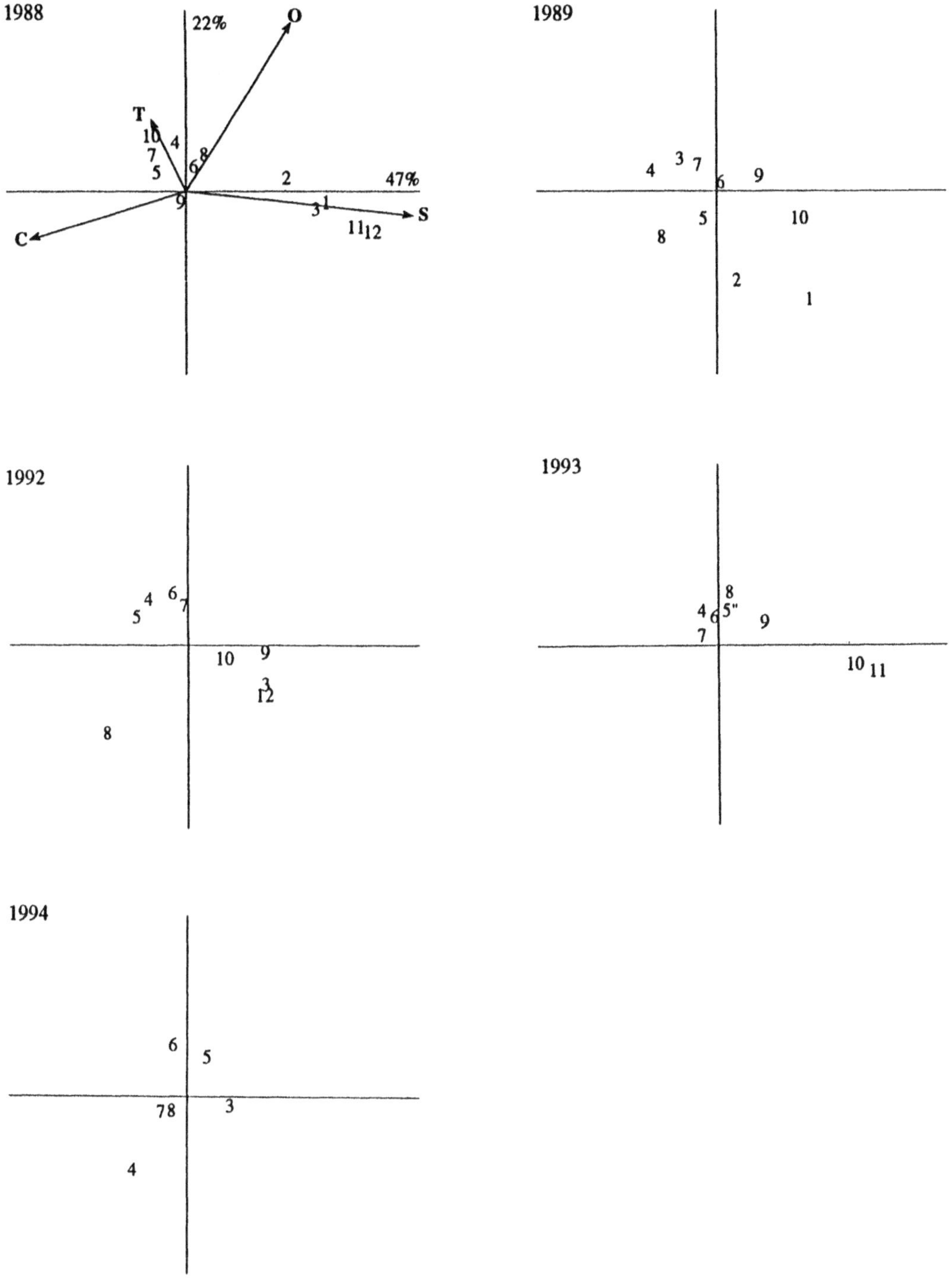


*Figure 5.* CCA of the zooplankton in the Sacca di Goro from 1988 to 1994: scores of environmental variables and samples in the plane of the first two axes of the ordination for St. 4 (*a*) and St. 8 (*b*). The axes in the yearly subplots have the same scaling; arrows of environmental variables are shown only in the first subplot. The numbers correspond to sampling months; 5′ and 5″ indicate two different samples collected in May 1993.

(*b*)

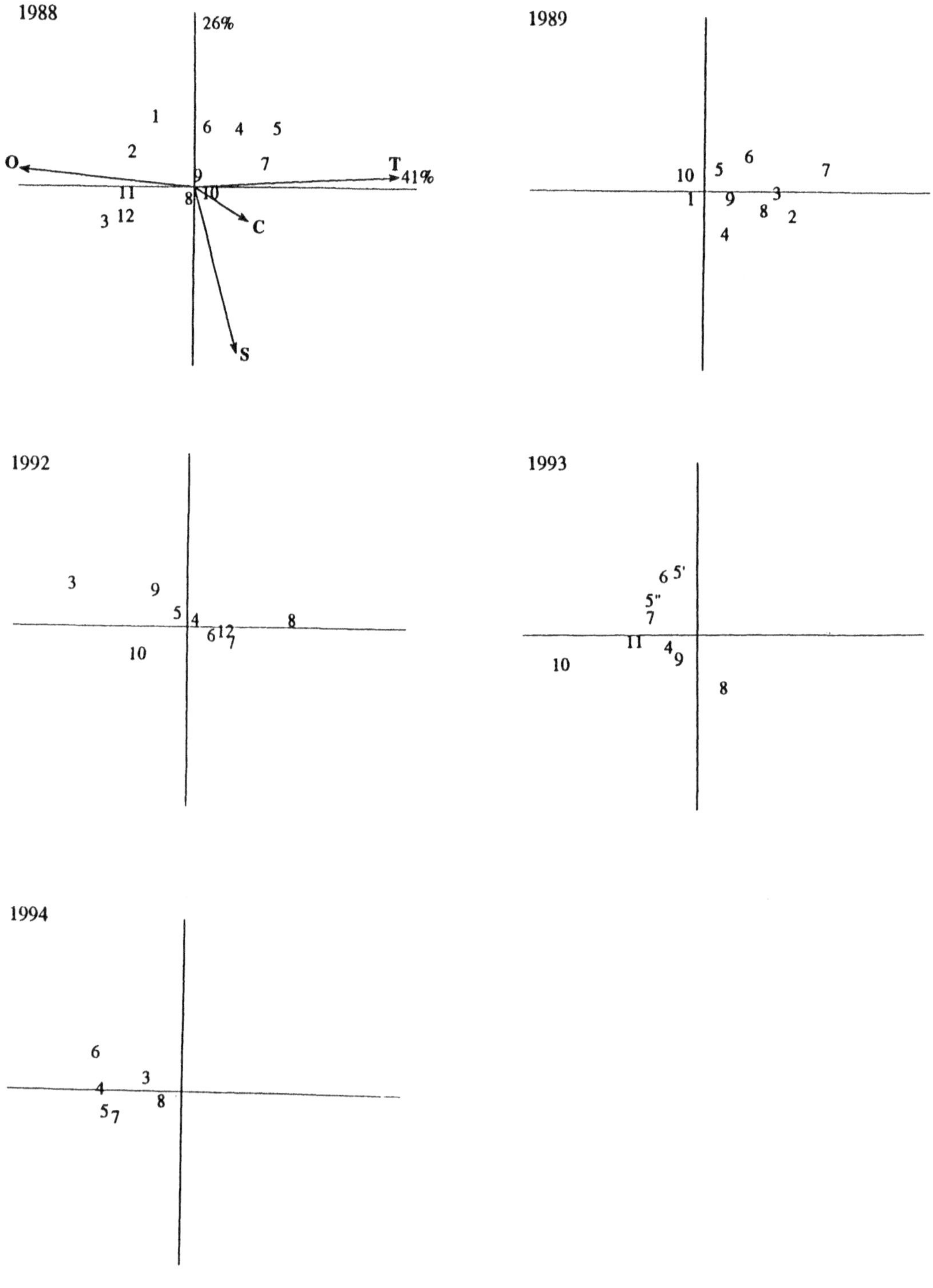
1988
26%
1
6 4 5
2
O
T
7
9
41%
11
8
10
C
3 12
S
1989
6
10 5 7
1 9 3
8 2
4
1992
3
9
5 4
8
6 12
7
10
1993
6 5'
5"
7
11
4
9
10
8
1994
6
3
4
8
5 7

*Figure 5b.*

## Acknowledgements

This research was supported by the joint EU project 'CLEAN: Coastal Lagoon Eutrophication and ANaerobic processes' (Contract NO EV5V-CT92-0080). We are grateful to the two anonymous referees for their helpful comments and criticism on the initial version of this manuscript. Part of our work was made possible thanks to Mr Vittorio Gaiani (Dipartimento di Biologia, Università di Ferrara), Mr Tony Zappata (Battello Hydra, Amministrazione Provinciale di Ferrara), Mr Mimmo Cavalca (Istituto di Ecologia, Università di Parma) and Dr Matteo Cattadori (ITIS "G. Galilei", San Secondo, Parma). Thanks are also due to Dr Giovanni Zurlini (CNR, Roma) for his assistance in the data statistical processing.

## References

Ambrogi, R., I. Ferrari & S. Geraci, 1989. Biotic exchange between river, lagoon and sea: The case of zooplankton in the Po Delta. Scient. Mar. 53: 601–608.

A.P.H.A., A.W.W.A., W.P.C.F., 1975. Standard methods for the examination of water and wastewater. 14th edition, A.P.H.A., Washington, 1114 pp.

Bakker, C., 1994. Zooplankton species composition in the Oosterschelde (SW Netherland) before, during and after the construction of a strorm-surge barrier. Hydrobiologia 282/283: 117–126.

Bakker, C. & P. van Rijswijk, 1994. Zooplankton biomass in the Oosterschelde (SW Netherland) before, during and after the construction of a strorm-surge barrier. Hydrobiologia 282/283: 127–143.

Brylinski, J. M., 1981. Report on the presence of *Acartia tonsa* Dana (Copepoda) in the harbour of Dunkirk (France) and its geographical distribution in Europe. J. Plankton Res. 3: 255–260.

Carrada, G. C., V. U. Ceccherelli & I. Ferrari, 1987. Les lagunes italiennes. Bull. Ecol. 18: 149–158.

Castel, J., 1993. Long-term distribution of zooplankton in the Gironde estuary and its relation with river flow and suspended matter. Cah. Biol. mar. 34: 145–163.

Ceccherelli, V. U., I. Ferrari & P. Viaroli, 1994. Ecological research on the animal communities of the Po River Delta lagoons. Boll. Zool. 61: 425–436.

Farabegoli, A., I. Ferrari, C. Manzoni & A. Pugnetti, 1989. Prima segnalazione nel Mare Adriatico del copepode calanoide *Acartia tonsa* DANA. Nova Thalassia 10: 207–208.

Ferrari, I. & A. R. Chieregato, 1981. Feeding habits of juvenile stages of *Sparus auratus* L., *Dicentrarchus labrax* L. and Mugilidae in a brackish embayment of the Po River Delta. Aquaculture 25: 243–257.

Ferrari, I., M. T. Cantarelli, M. G. Mazzocchi & L. Tosi, 1985. Analysis of a 24-hour cycle of zooplankton sampling in a lagoon of the Po River Delta. J. Plankton Res. 7: 849–865.

Ferrari, I., V. U. Ceccherelli, M. Naldi & P. Viaroli, 1993. Planktonic and benthic communities in a shallow-water dystrophic lagoon. Verh. int. Ver. Limnol. 25: 1043–1047.

Gaudy, R. & M. D. Viñas, 1985. Première signalisation en Méditerranée du Copépode pélagique *Acartia tonsa*. Rapp. Comm. int. Mer Médit. 29: 227–229.

Hill, M. O., 1979. DECORANA: a FORTRAN program for detrended correspondence analysis and reciprocal averaging. Section of Ecology and Systematics, Cornell University, Ithaca, New York.

Lam Hoai, T. & M. Amanieu, 1989. Spatial structure and seasonal evolution of surface zooplankton in two mediterranean lagoon ecosystems. Oceanol. Acta 12: 65–77.

Laprise, R. & J. J. Dodson, 1994. Environmental variability as a factor controlling spatial patterns in distribution and species diversity of zooplankton in the St. Lawrence Estuary. Mar. Ecol. Prog. Ser. 107: 67–81.

Menendez, M. & F. A. Comin, 1986. Variaciòn estacional del zooplancton en las lagunas del Delta del Ebro (NE España). Oecol. aquat. 8: 47–60.

Soetaert, K. & P. van Rijswijk, 1993. Spatial and temporal patterns of the zooplankton in the Westerschelde estuary. Mar. Ecol. Prog. Ser. 97: 47–59.

Ter Braak, C. J. F., 1986. Canonical Correspondence Analysis: a new eigenvector technique for multivariate direct gradient analysis. Ecology 67: 1167–1179.

Ter Braak, C. J. F., 1988. CANOCO – a FORTRAN program for canonical community ordination by (partial) (detrended) (canonical) correspondence analysis, principal component analysis and redundance analysis (version 2.1). Agricultural Mat. Group, Ministry of Agriculture and Fisheries, Wageningen, 95 pp.

Viaroli, P., A. Pugnetti & I. Ferrari, 1992. *Ulva rigida* growth and decomposition processes and related effects on nitrogen and phosphorus cycles in a coastal lagoon (Sacca di Goro, Po River Delta). In G. Colombo, I. Ferrari, V. U. Ceccherelli & R. Rossi (eds), Marine Eutrophication and Population Dynamics, Olsen & Olsen, Fredensborg: 77–84.

Viaroli, P., M. Naldi, R. R. Christian & I. Fumagalli, 1993a. The role of macroalgae and detritus in the nutrient cycles in a shallow-water dystrophic lagoon. Verh. int. Ver. Limnol. 25: 1048–1051.

Viaroli, P., M. Bartoli, C. Bondavalli, M. Naldi & I. Ferrari, 1993b. Macroalgae growth and decomposition and nutrient cycling in the Sacca di Goro (Po river delta). C.E.C.-Environmental Research Programme, C.L.E.AN Progress report: 281–307.

*Hydrobiologia* **329**: 57–67, 1996.
*P. Caumette, J. Castel & R. Herbert (eds), Coastal Lagoon Eutrophication and ANaerobic Processes (C.L.E.AN.).*



# A field experiment on the effect of two types of sediment disturbance on the rate of recovery of a meiobenthic community in a eutrophicated lagoon

M. A. Colangelo[1], T. Macrí[1] & V. U. Ceccherelli[2]
[1] *Dipartimento di Biologia, Università di Ferrara, via L. Borsari 46, 44100 Ferrara, Italy*
[2] *Dipartimento di Biologia evoluzionistica e sperimentale, Università di Bologna, via F. Selmi 3, 40126 Bologna, Italy*

*Key words:* Organic enrichment, meiobenthos, copepods, recolonization processes

## Abstract

A recolonization field experiment of two different artificially disturbed sediments (both defaunated sand and defaunated/reduced sand through organic enrichment) was carried out in the Sacca di Goro (Adriatic sea, Po river Delta, Italy). Copepods showed themselves better colonizers than nematodes. In particular, copepods, in the defaunated sand, were able to reach the same densities as the control site after only seven days from the beginning of the experiment. In the reduced-sand, copepod recolonization occurred more slowly but reached the densities found in both azoic and control sediments at the end of the experiment (15 days), when the values of total carbon content decreased. The recovery evolution of the community structures was mostly dependent on the different behaviour of the active epibenthic species of the harpacticoids (e.g. *Canuella perplexa* T. & A. Scott, 1893, *Ameira parvula* (Claus, 1866), *Robertgurneya similis* (A. Scott, 1896)) and of the passively transported endobenthic ones (e.g. *Asellopsis sarmatica* Jakubisiak, 1938, *Ectinosoma dentatum* Steuer, 1940).

## Introduction

Soft sediments of shallow water environments are affected by different kinds of disturbance, which are represented either by mechanical perturbations, due to sea storms or dredging, or by sediment organic enrichment and pollutants (Pearson and Rosenberg, 1978). Although in off-shore areas most organic matter particles are derived from phytoplankton and zooplankton faecal pellets and exuvia, in inshore waters and lagoons they are derived mainly from the breakdown of macroalgae which causes an often excessive organic enrichment. Moreover, this annual supply of organic matter is deposited over a very short time on the lagoon bottom. In fact, the vegetation mats undergo a rapid decomposition during summer months, when temperature is high and water is stagnant. The consequence is an immediate and substantial organic enrichment of the sediments.

When the primary cause of perturbation ceases to act, the resultant alterations of sediments could have a different effect on the benthic community recovery to either the original structure or a different organism assemblages (Le Guellec, 1988).

A suitable approach to the study of disturbance impact and recolonization processes is controlled field experiments with artificially disturbed or defaunated sediments (Coull & Palmer, 1984). Many works have focused on the responses of the meiofaunal components of the benthic community, because of their potential for more rapid responses to environmental change than macrofauna (Coull & Chandler, 1992). Meiofaunal organisms might be useful biological indicators of such changes because their short generation times and the fact that many of them, especially epibenthic copepods, frequently adopt movement through water column for their dispersal (Palmer, 1988) result in a faster recovery of community structure over a reasonable time-span (weeks or months) (Heip et al., 1988).

Benthic environments of the northern Adriatic coastal lagoons, particularly those of the Po river Delta (Italy), have always been subjected both to natural and man induced disturbances. Recently they are more and more affected by increased eutrophication which had caused, above all, a huge growth of macroalgal beds (Ulvaceae) whose decomposition resulted in organic enrichment of the sediments. The latter may cause complete anoxia and reduced conditions in the sedimentary habitat, which represents the main factor of acute disturbance for the biota, especially for benthic organisms (Viaroli et al., 1992).

There are very few studies in the literature on the recovery rates of sublittoral meiofauna communities with respect to experimental organic enrichment of sediment (see Coull & Chandler, 1992).

The aim of the present study was to investigate, in one of these lagoons (Sacca di Goro), whether the recolonization process and the recovery rate of a meiobenthic community in sandy patches were differently affected by two kinds of artificially induced disturbance. We compared the effects of either a merely defaunated sand or a similarly defaunated but completely reduced sand by organic enrichment with powderded *Ulva* thalli. The latter tried to simulate one of the most severe disturbance event which periodically occurs in these environments. In particular, copepod species compositions and community structure differences among the different sediment treatments and the natural environment were analyzed to determine if different species behaved differently. The meiobenthic assemblage studied was that of a fine sand shoal in the investigated lagoon.

## Materials and methods

The experiment was carried out in the Sacca di Goro (Figure 1), a subtidal shallow water embayment, one of the largest (26 km$^2$) in the Po river Delta (Northwest Adriatic coast) in June 1993. This lagoon shows wide yearly variations in salinity (10.8–32.9‰) (Colombo et al., 1991). In spite of the limited tidal range (typical of the Mediterranean sea) tidal effects are not negligible, mainly in the West-central part, owing to the wide sea mouth (2.5 km). Sediments are fine-grained (from silty clay to clayey silt) in the inner part and become coarser toward the sea, where relict and recent sand bars are located (Dal Cin & Pambianchi, 1991). Owing to the eutrophication conditions of this lagoon (Pugnetti & Zaccaria, 1991), severe anoxic crises, mainly due to decomposition of *Ulva rigida* Agardh beds, frequently occur during the summer.

The experimental site was located on the subtidal side of a sand bank facing the lagoon. From this station (CR, Figure 1) the sand (Md = 0.207 mm) was gathered which was made azoic for the experiment by repeated freezing (−20 °C) and thawing (40 °C) and dried in the air for about one month. In order to obtain reduced conditions a part of this azoic sand was enriched with powdered dry *Ulva* thalli. The amount of this addition corresponded to a total carbon content of about 100 g C m$^{-2}$ in the uppermost 3 cm of sediment. Thirty-six plastic containers (each with a surface area of 49 cm$^2$) were filled up to the rim: 18 with the simply defaunated sand (DF treatment) and 18 with the organically enriched sand (RD treatment). Before being submerged *in situ*, all of those containers were incubated for 5 days in filtered sea water until complete reduced conditions were reached in those with organically enriched sand. Containers were randomly arranged in a steel tray which was placed directly on the sandy bottom of the experimental station (TR, Figure 1). Only the sixteen internal containers were sampled in order to avoid the edge effect. Sampling was scheduled on 1–3–7–15 days from the beginning of the experiment because a previous study, carried out only with defaunated sand in the same site (Colangelo et al., 1994), had shown that the main changes in the recovery of the copepod assemblage occurred during the first two weeks. On each sampling occasion, 4 replicate containers were randomly sampled: two with defaunated sand and two with reduced sand. From each container, 2 pseudoreplicate cores (3.14 cm$^2$ each) were directly collected by hand (a total of 4 cores from DF treatment and 4 from RD treatment). At the control site (CR, Figure 1) 2 pairs of pseudoreplicate cores were sampled from the sandy bottom by means of a remotely operated multiple corer. The uppermost 3 cm of sediment were retained, stained with rose bengal and preserved in 8% buffered formalin. In the laboratory, sampled were sieved through both 1.0 and 0.03 mm mesh size. The fraction on the 0.03 mm sieve was centrifuged three times with a LUDOX-AM water solution (60%) in order to separate organisms from sediments. Meiobenthic organisms were counted to major taxa. Copepods were identified to the species, staged and sexed.

At the start of the experiment and on each sampling date, two samples of surface sediment from CR and four random marginal containers (not sampled for meiofauna), two with DF sand and two with RD sand, were analyzed for the following abiotic parameters: Eh, with a 'Radiometer' platinum combined elec-

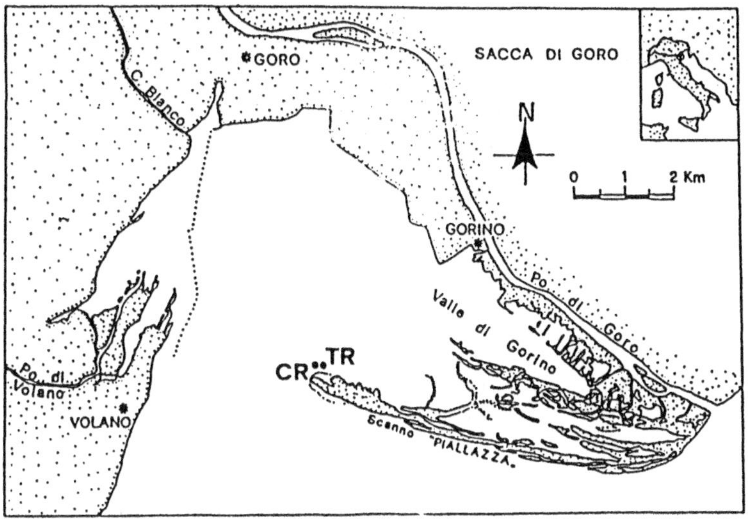


*Figure 1.* Map of the Sacca di Goro lagoon. Locations of control (CR) and experimental (TR) sites are shown.

trode; total organic carbon, estimated by chromic acid wet oxidation technique (Rofes, 1980); organic matter ash free dry weight (AFDW), by dessication at 70°C overnight and ignition at 550°C for 24 hours. The mean value for each parameter is reported because the chemical and meiofauna replicate samples were not matched pairs.

Sampling design followed a two-way ANOVA layout with treatments (CR, DF, RD) as fixed and times (1–3–7–15 days) as random factors. Calculations were made both on $\log(x + 1)$ transformed densities of taxa and on diversity index values of copepod biocoenoses (No. of species, Hill's N1 and arcsin transformed evenness values). Every time a significant difference resulted only for main effects, the Student-Newman-Keuls (SNK) post-hoc test was performed to obtain all pair-wise comparisons between factor averages (significance alpha level 0.05). When significant interaction was found, the SNK test was restricted to comparisons between levels of each significant factor within each level of the other factor (Underwood, 1981). For copepod communities, non-metric multidimensional scaling (MDS) ordination was carried out on the dissimilarity matrix of Manhattan (city block) metrics between samples calculated on double root transformed abundance data. The choice of that particular distance index was justified by the very impoverished biota (Field et al., 1982).

## Results

Table 1 shows the variations of the most important abiotic parameters at the beginning of the experiment and at each sampling dates. Values of the redox potential were recorded in a rather rough way and can be considered representative, approximately, of the first centimetre sand layer. Notwithstanding this methodological drawback, the redox values were always negative in the containers with reduced sand even if they showed an appreciable decrease of their absolute values after a few days. The per cent carbon content (as well as the total organic matter ash free dry weight) showed intermediate values in the DF treatment but the highest values in the RD treatment during the first phase of the experiment. However, the latter became equal to those of the defaunated sand after three days from the start, and all the values were the same in both the treatment containers and in the control site at the end of the experiment.

The average densities of the two major meiobenthic taxa and results of the post-hoc SNK tests are shown in Figure 2. Averages at each sampling date which share the same symbol across control and treatments are not significantly different. Symbols referring only to the legend labels mean that the SNK test comparisons were performed between treatment factor effects averaged across times, because interactions were not significant. On the other hand, the same letters mean no statistically significant differences among sampling date average values within each single treatment (CR, DF and RD respectively).

*Table 1.* Abiotic parameter mean values of temperature, redox potential, carbon content and organic matter ash free dry weight (AFDW) content of the sediments for each treatment on each day from the start of the experiment. CR: control; DF: defaunated sand; RD: reduced sand.

| | | Start | 1 | 3 | 7 | 15 |
|---|---|---|---|---|---|---|
| T (°C) | CR | 26.4 | 27.1 | 26.7 | 21.8 | 26.3 |
| | DF | – | 26.3 | 26.7 | 22.0 | 26.1 |
| | RD | – | 26.7 | 27.3 | 21.9 | 25.5 |
| Eh (mV) | CR | 102 | 90 | 102 | 110 | 10 |
| | DF | 50 | 182 | 90 | 13 | 65 |
| | RD | –125 | –73 | –26 | –34 | –21 |
| Tot. C (%) | CR | 0.64 | 0.66 | 0.60 | 0.62 | 0.66 |
| | DF | 0.75 | 0.81 | 0.74 | 0.66 | 0.57 |
| | RD | 1.83 | 1.78 | 0.72 | 0.73 | 0.40 |
| AFDW (%) | CR | 3.52 | 3.52 | 3.30 | 3.41 | 3.63 |
| | DF | 2.80 | 4.40 | 3.60 | 3.50 | 3.80 |
| | RD | 5.40 | 5.20 | 4.40 | 4.10 | 3.30 |

*Table 2.* Two-way ANOVA results for treatments (control; defaunated; reduced) and times (days from the beginning of the experiment) relative to total meiofauna, nematodes and copepods (for details of experimental design see 'Materials and Methods').

| Source of variation | df | F-value | | |
|---|---|---|---|---|
| | | Total meiofauna | Nematodes | Copepods |
| Treatment (TR) | 2 | 20.46** | 63.71** | 9.14* |
| Time (D) | 3 | 8.36** | 1.05 | 18.60** |
| TRxD | 6 | 5.63** | 1.43 | 10.88** |
| Error | 36 | | | |

* $p < 0.05$; ** $p < 0.01$

Densities of nematodes were always statistically lower in both treatment containers than in the control (Table 2). Their constantly low densities in the treatment containers resulted in no significant interaction with respect to time. The recolonization behaviour of copepods appeared to be different. In the defaunated sand, they reached the same densities as the control after seven days from the beginning of the experiment. In the reduced sediments the copepod recolonization occurred more slowly. Density values of this taxon were not significantly different among ambient sediment and treatments only at the end of the experimental period.

Within the copepod assemblage, four species of harpacticoid played a major role in the recolonization process: two epibenthic harpacticoids, the muddy-sand dweller *Canuella perplexa* T. & A. Scott, 1893 and the eurytopous *Ameira parvula* (Claus, 1866), and two endobenthic species, the sand dwellers: *Ectinosoma dentatum* Steuer, 1940 and *Asellopsis sarmatica* Jakubisiak, 1938. The behaviour of each of them appeared to be different in relation to their functional characteristics (Appendix Table 1). The ANOVA F-values for the two epibenthic species are reported in Table 3. Their average densities and the SNK test results are shown in Figure 3. *Canuella perplexa* seems to be a slow colonizer, its abundance values were significantly lower in the treatment containers, particularly in those with reduced sediments, until day 15. By contrast, *Ameira parvula* confirmed itself as an opportunistic,

*Table 3.* Two-way ANOVA results for treatments (control; defaunated; reduced) and times (days from the beginning of the experiment) relative to the major copepod species (for details of experimental design see 'Material and Methods').

| Source of variation | df | F-value | | | | |
|---|---|---|---|---|---|---|
| | | *Canuella perplexa* | *Ameira parvula* | *Robertgurneya similis* | *Ectinosoma dentatum* | *Asellopsis sarmatica* |
| Treatment (TR) | 2 | 12.42** | 4.04 | 2.09 | 31.87** | 78.55** |
| Time (D) | 3 | 20.33** | 10.56** | 4.82** | 2.25 | 0.94 |
| TRxD | 6 | 8.74** | 7.59** | 2.64* | 1.29 | 1.35 |
| Error | 36 | | | | | |

* $p < 0.05$; ** $p < 0.01$.

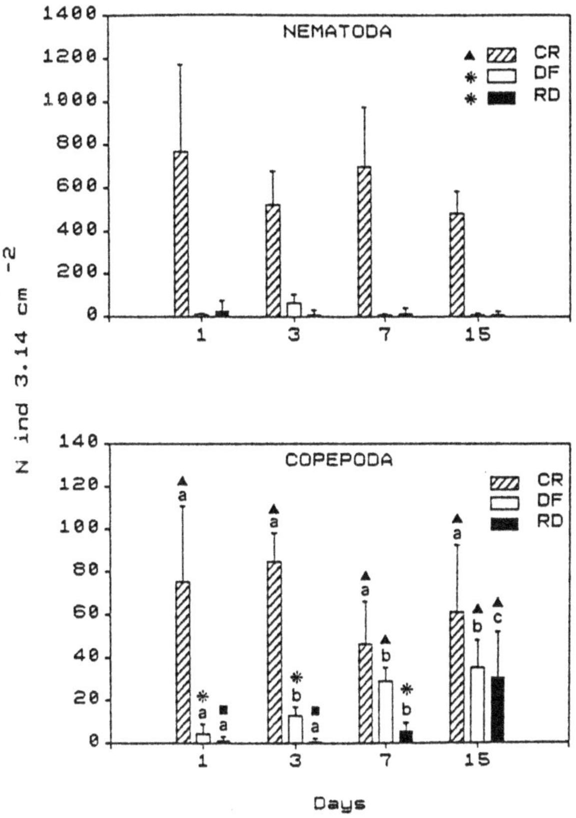


*Figure 2.* Average densities (± SD) per core (3.14 cm$^2$) of nematodes and copepods in each treatment on each sampling date. Averages with the same symbols among treatments on the same sampling date are not significantly different. Symbols referring only to the legend labels in the Nematoda graph are relative to significant differences between treatment factor averages. The same letters mean no significant differences among sampling date average values within each treatment. CR: control; DF: defaunated sand; RD: reduced sand (see text as for detail).

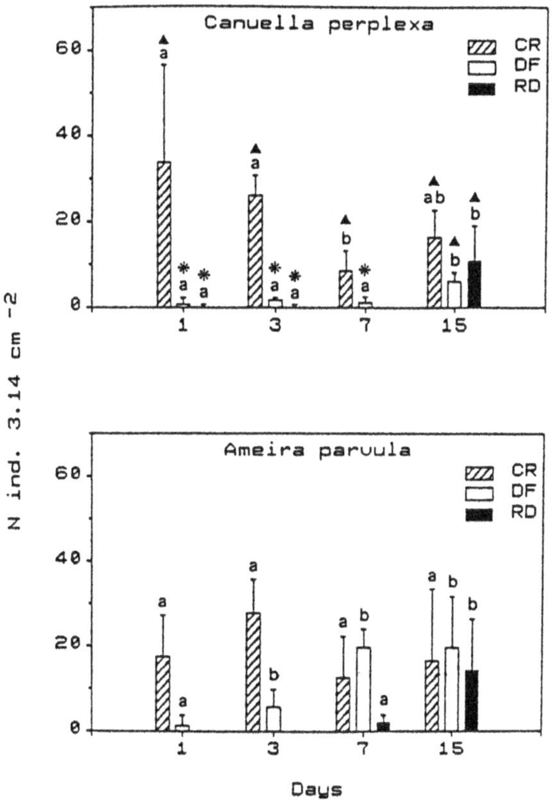


*Figure 3.* Average densities (± SD) per core (3.14 cm$^2$) of *Canuella perplexa* and *Ameira parvula* in each treatment on each sampling date. Averages with the same symbols among treatments on the same sampling date are not significantly different. The same letters mean no significant differences among sampling date average values within each treatment. CR: control; DF: defaunated sand; RD: reduced sand (see text as for detail).

fast colonizer (Ceccherelli et al., 1992). In fact, the treatment main factor, averaged across sampling dates, was not statistically significant (Table 3), because of the high abundance values which *A. parvula* reached in the DF containers by day 7. Only density differences over time were statistically significant, with the abundances in the reduced sand increasing more slowly than in the defaunated sand (Figure 3). The evidence that epibenthic species play a major role in the colonization

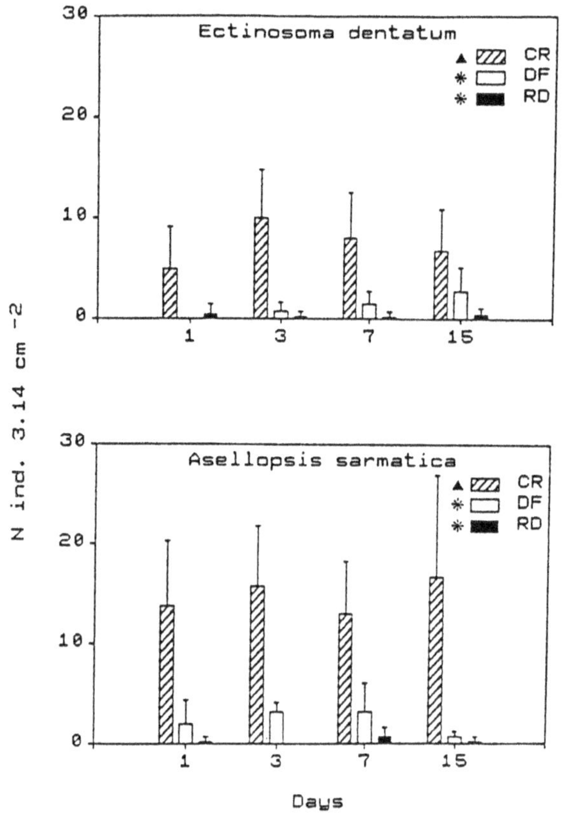


*Figure 4.* Average densities (± SD) per core (3.14 $cm^2$) of *Ectinosoma dentatum* and *Asellopsis sarmatica* in each treatment on each sampling date. Treatment legend labels with the same symbols indicate no significant difference between treatment factor averages. CR: control; DF: defaunated sand; RD: reduced sand (see text as for detail).

process was confirmed by the results for *Robertgurneya similis* (A. Scott, 1896) (Table 3). This species was rare in the CR but it showed no abundance differences between CR and RD treatment and it exceeded significantly the environment densities in the DF treatment by day 7 (Appendix Table 1). The endobenthic species, *Asellopsis sarmatica* and *Ectinosoma dentatum*, were present in the treatments from day 1 (Figure 4), but their average densities in the experimental sand containers were always significantly lower than in the control site (Table 3).

The overall pattern of community changes of the copepod assemblage is shown by the results of the MDS ordination (Figure 5). The date points of the control (CR) form a separate and homogeneous cluster which indicates the relatively constant copepod assemblage in the environment. On the contrary, the date points of both DF and RD treatment communities are more scattered, indicating different recolonization processes. They can be better analyzed in term of trajectories than clusters. The trend of the development in community structure in the azoic sand appeared to be a constant progression. On the other hand, the date point trend of the reduced sand community has a different trajectory because the recolonization started later and its pioneer phase showed marked differences from the defaunated sand assemblage. However, the structure of the two communities converged toward a very similar community in both the azoic and reduced sand by day 15.

The ANOVA results of the diversity index values are shown in Table 4, trends of their averages in Figure 6 and SNK test comparisons in Table 5. It is easy to see a faster increase of the diversity values in the defaunated sand than in the reduced sand. However, the values of all diversity indexes became not significantly different between control and treatments toward the end of the experimental period. The univariate measures of diversity take into account the distribution of individuals amongst species but do not utilize species identity. The fact that diversity was the same for the communities in the three types of sediments by day 15 means that, in general, the copepod assemblages in the two treatments reached an organized structure similar to that of the control. Instead the multivariate species-dependent MDS method was much more sensitive in discriminating the three communities. This means that the dominance relationships among species changed in the defaunated and reduced sand assemblages compared to the natural environment. In particular the dominance of the epibenthic species increased in the experimental sand containers (Appendix Table 1).

## Discussion

The present investigation shows meiobenthic organisms rapidly colonized the experimental containers. In particular, the recovery of meiobenthic community structures seems to occur after just two weeks, irrespective of the experimental conditions which simulated two different kinds (and levels) of disturbance impact on the sedimentary environment. Fast colonization rates have been reported also in other studies on recolonization of disturbed habitats. After a mechanically induced disturbance, Sherman & Coull (1980) observed that densities of most meiobenthic taxa reached the same levels as those at the control site after just 12 hours. Le Guellec (1988), working

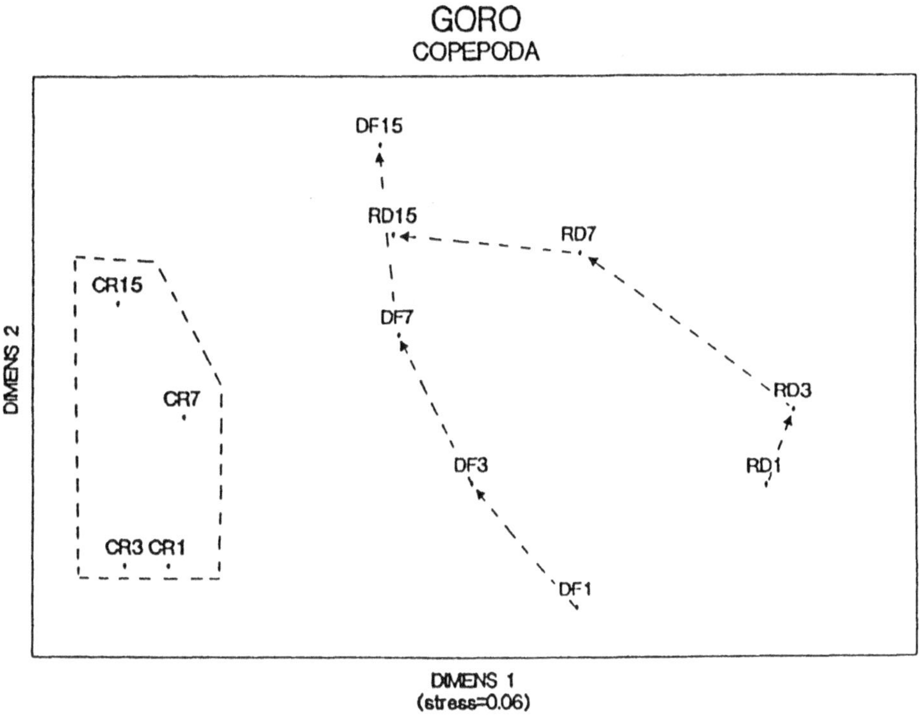


*Figure 5.* MDS ordination plot of date points for the three treatments. CR: control; DF: defaunated sand; RD: reduced sand. 1–3–7–15: days from the beginning of the experiment.

*Table 4.* Two-way ANOVA results for treatments (control; defaunated; reduced) and times (days from the beginning of the experiment) relative to the diversity index values for the copepod assemblage (for details of experimental design see 'Material and Methods').

| Source of variation | df | F-value | | |
|---|---|---|---|---|
| | | No species | Hill's N1 | Evenness |
| Treatmnet (TR) | 2 | 5.57* | 7.90* | 1.40 |
| Time (D) | 3 | 9.06** | 6.59** | 6.04** |
| TRxD | 6 | 6.78** | 3.40** | 3.00** |
| Error | 36 | | | |

* $p < 0.05$; ** $p < 0.01$.

with exogenous azoic sand, reported a similar result after only two tidal cycles. Thistle (1980) also found that the recolonization of defaunated enteropneust faecal mounds by subtidal harpacticoids occurred in less than 24 hours. Sun & Fleeger (1994), even though they were mainly interested in investigating meiofaunal recolonization of azoic sediment mimicking bottom depressions, remarked how the abundances of the dominant copepods showed no significant differences between experimental and ambient sediments after 24–48 hours.

As for the effect of physicochemical alterations of sediments induced by organic enrichment, the present investigation showed that, in general, colonization by most meiofaunal populations and recovery of meiobenthic assemblage structure seemed to be delayed in reduced sand compared to the defaunated sand. The recovery rates of different meiobenthic taxa were related both to their different movement mechanism, and consequent ability to recruit to sediment, and to the degree of disturbance of the experimental sediments. Nematodes were the least active colonizers, but this

*Appendix 1.* Average densities (± SD) per core (3.14 $cm^2$) of meiobenthic copepod species recorded in each treatment on each sampling date from the beginning of the experiment. Functional habits: M = mud; S = sand; Ph = phytal; E = eurytopous; epi = epibenthic; end = endobenthic.

| | | CONTROL | | | | DEFAUNATED | | | | REDUCED | | | |
|---|---|---|---|---|---|---|---|---|---|---|---|---|---|
| | | 1 | 3 | 7 | 15 | 1 | 3 | 7 | 15 | 1 | 3 | 7 | 15 |
| *Canuella perplexa* | (MS-epi) | 33.8 (±22.7) | 26.0 (± 4.8) | 8.5 (± 4.7) | 16.5 (± 6.2) | 0.8 (±1.5) | 1.8 (±0.5) | 1.3 (±1.3) | 6.3 (± 2.1) | 0.3 (±0.5) | 0.3 (±0.5) | | 11.0 (± 8.1) |
| *Ectinosoma dentatum* | (S-end) | 5.0 (± 4.2) | 10.0 (± 4.8) | 8.0 (± 4.6) | 6.8 (± 4.1) | | 0.8 (±1.0) | 1.5 (±1.3) | 2.8 (± 2.4) | 0.5 (±1.0) | 0.3 (±0.5) | 0.3 (±0.5) | 0.5 (± 0.6) |
| *Microarthridion fallax* | (MS-end) | 2.5 (± 1.0) | 2.3 (± 1.3) | 0.5 (± 0.6) | 0.8 (± 1.0) | | | | | | | | |
| *Harpacticus flexus* | (S-epi) | 1.8 (± 1.0) | 1.0 (± 0.8) | 0.3 (± 0.5) | | | | | | | | 0.5 (±0.6) | |
| *Tisbe holothuriae* | (Ph) | | | 0.8 (± 1.5) | 3.0 (± 4.8) | | | 0.3 (±0.5) | 0.5 (± 1.0) | | | 0.8 (±1.0) | 2.0 (± 2.7) |
| *Dactylopusia tisboides* | (E-epi) | 0.3 (± 0.5) | 0.5 (± 1.0) | | | | | | | 0.3 (±0.5) | 0.3 (±0.5) | | |
| *Amphiascus parvus* | (Ph) | | | | | | | | 0.3 (± 0.5) | | | | 0.3 (± 0.5) |
| *Robertgurneya similis* | (E-epi) | 0.8 (± 1.0) | 0.8 (± 1.0) | 0.5 (± 0.6) | 0.8 (± 1.0) | | 0.8 (±1.5) | 2.5 (±0.6) | 4.8 (± 3.9) | | 0.3 (±0.5) | 0.3 (±0.5) | 1.0 (± 1.2) |
| *Schizopera compacta* | (E-epi) | | | | | | | | | | | 0.8 (±1.5) | |
| *Ameira parvula* | (E-epi) | 17.5 (± 9.6) | 27.8 (± 7.9) | 12.8 (± 9.6) | 16.8 (±16.8) | 1.3 (±2.5) | 5.8 (±4.1) | 19.8 (±4.3) | 19.8 (±12.1) | | | 2.3 (±1.7) | 14.5 (±11.9) |
| *Mesochra pontica* | (E-epi) | | | 0.3 (± 0.5) | | | | | | 0.3 (±0.5) | | | 0.5 (± 1.0) |
| *Enhydrosoma gariene* | (M-end) | | | | | | | | 0.3 (± 0.5) | | | | |
| *Paralaophonte congenera* | (E-epi) | 0.3 (± 0.5) | 0.8 (± 1.0) | 1.3 (± 1.5) | | 0.3 (±0.5) | 0.3 (±0.5) | 0.5 (±0.6) | | | | 0.5 (±0.6) | 0.5 (± 0.6) |
| *Asellopsis sarmatica* | (S-end) | 13.8 (± 6.6) | 15.8 (± 6.1) | 13.0 (± 5.3) | 16.8 (±10.2) | 2.0 (±2.5) | 3.3 (±1.0) | 3.3 (±2.9) | 0.8 (± 0.5) | 0.3 (±0.5) | | 0.8 (±1.0) | 0.3 (± 0.5) |
| *Cyclopina gracilis* | (E-epi) | | | 0.3 (± 1.0) | | 0.3 (±0.5) | 0.5 (±0.6) | | | | | | 0.5 (± 0.6) |
| TOTAL | | 75.5 (±35.5) | 84.8 (±13.4) | 46.5 (±19.8) | 61.3 (±31.4) | 4.5 (±4.4) | 13.0 (±3.9) | 29.0 (±6.3) | 35.3 (±13.3) | 1.5 (±1.7) | 1.0 (±1.4) | 6.0 (±3.5) | 31.0 (±21.4) |

*Table 5.* Studentized-Newman-Keuls post hoc test for multiple comparisons of diversity index values for the copepod assemblage, relative to the significant interaction. Average values with the same symbols among treatments on the same sampling date are not significantly different. The same letters indicate no statistically significant differences among sampling date average values within each single treatment. CR: control; DF: defaunated sand; RD: reduced sand (see text as for detail)

| Treatments | | No species | Hill's N1 | Evenness |
|---|---|---|---|---|
| CRxTimes | 1 | ∇ 6.75 a | ∇ 4.08 a | 59.41 a |
| | 3 | ∇ 7.00 a | ∇ 4.47 a | 61.59 a |
| | 7 | ∇ 6.75 a | ∇ 4.93 a | 65.90 a |
| | 15 | ∇ 5.50 a | ∇ 4.03 a | 65.50 a |
| DFxTimes | 1 | * 1.50 a | * 1.39 a | 16.82 a |
| | 3 | ∇ 4.50 b | ∇ 3.80 b | 69.98 b |
| | 7 | *∇ 5.25 b | * 2.81 ab | 52.30 ab |
| | 15 | ∇ 5.50 b | ∇ 3.51 b | 58.89 ab |
| RDxTimes | 1 | * 1.25 a | * 1.46 a | 19.15 a |
| | 3 | * 1.00 a | * 1.50 a | 22.50 a |
| | 7 | * 3.50 b | *∇ 3.07 b | 72.73 b |
| | 15 | ∇ 5.50 b | ∇ 3.23 b | 56.77 ab |

result was probably due to our experimental set-up. This did not allow lateral invasion by crawling through sediment interstices which is the main route by which nematodes recolonize azoic sediment patches (Chandler & Fleeger, 1983). Nematodes are known to occur deeper in sediments (Jacobs, 1984) and to be passively suspended in the water column only when shear-currents exceed a certain threshold (Sherman & Coull, 1980; Palmer, 1988). So, in our experiment, they could arrive in the containers only by passive transport via water column. That is probably the reason why we observed a poor recolonization of experimental sand by those organisms. The same thing probably applied, to a certain extent, to endobenthic copepod species, *Ectinosoma dentatum* and *Asellopsis sarmatica*. Their abundances in the treatments never reached those of the control site, not even at the end of the experimental period. These endobenthic species seem to be less good colonizer but only because, probably, they are able to burrow into the sediment as flow increases (e.g. at high water) and thus avoid transport (Palmer, 1988). Consequently, few of them randomly encounter the containers. On the other hand, even if no statistical differences were found between densities in defaunated and reduced sand, in the latter the abundances of these two species always seemed to be lower (Figure 4). At high levels of organic enrichment (RD treatment) the permanence of reduced conditions in deeper layers of sediments (Table 1) prevents those species living beneath the surface of the sediment from colonizing successfully (Gee et al., 1985). Thus endobenthic species may take much longer time to recover. Probably, the short duration of this experiment might have been critical in masking the time settlement potential of these slow colonizing species (Sun & Fleeger, 1994).

Epibenthic copepods (e.g. *Canuella perplexa, Ameira parvula* and *Robertgurneya similis*) seem to be very good colonizers. This depends on their ability to enter actively the water column to be transported (Kern, 1990). In fact, their densities in both experimental treatments fully reached ambient sediment levels toward the end of the experimental period (Figure 3; Table 3). However, notwithstanding their dispersal capability, the recolonization of the reduced sand appeared to be delayed with respect to the azoic sand. Probably, the epibenthic species show a selective settlement behaviour in relation to the chemical status of the sediments. Although they are able to settle rapidly in the merely defaunated sand, they might postpone their settlement in the reduced sand perhaps waiting for more oxygenated conditions of surface sediment.

Gee et al. (1985) carried out a mesocosm experiment on the effects of organic enrichment on meiofauna by adding powdered macroalgae (*Aseophyllum*) to experimental containers of mud in two different quantities (50 and 200 g C m$^{-2}$ respectively). In their high dose treatments, after 56 days, all the burrowing species had disappeared and the fauna was dominated by only five species of the *Tisbe* guild, even though no specimens of this genus were found in the field samples. Since members of the genus *Tisbe* are opportunist organisms this may explain their presence in these organically stressed environments. In the present study, at intermediate quantity of powdered macroalgae addition (100 g C m$^{-2}$), the opportunistic role seemed to be taken by *Ameira parvula* (Figure 3). The adults of this species have an active migratory behaviour (Kern, 1990) and are able to settle massively (mainly with ovigerous females) during the first phase of recolonization of defaunated sediments (Colangelo et al., 1994). Also Warwick et al. (1990) found an *Ameira* sp. which seemed to swim actively at high water slack and to be capable or redistributing itself in the sandy habitat, in order either to avoid predation or to search out sediment patches more attractive in terms of available food resources.

The ordination plot of the copepod assemblage (Figure 5) and the trends of the diversity index values

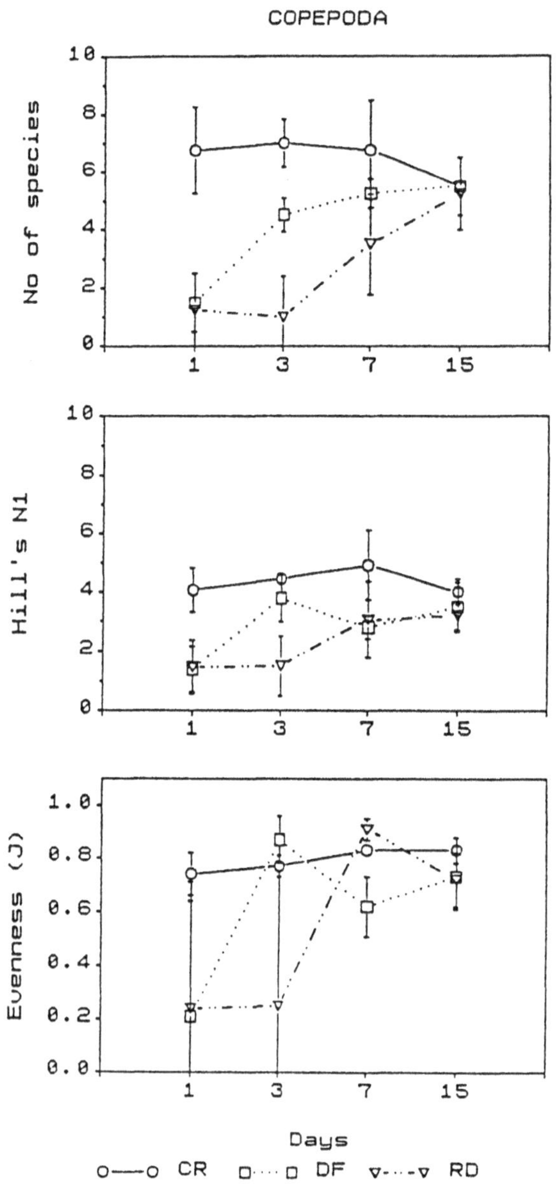


*Figure 6.* Trends of diversity indices average values (± SD) for the copepod assemblage in each treatment. CR: control; DF: defaunated sand; RD: reduced sand.

(Figure 6) demonstrated the timing and pattern of the community recovery. During the first recolonization phase (up to day 3), when higher values of both carbon content and negative redox potential were recorded in the reduced sediments (Table 1), the greatest differences in community composition were observed. Afterwards, when the levels of the organic enrichment decreased, the biotic structuring forces of the communities tend to play an increasing role and all copepod assemblages approached a similar structure. Altered sediments have to recover suitable chemical composition in order to allow the development of colonizing populations.

It is clear that, besides the granulometric characteristic of sediments, settlement of meiobenthic organisms (that of harpacticoid at least) seems to be controlled by the chemical status of the sedimentary habitat. We have still to understand the complicated dynamics in the sediment strata. Notwithstanding the permanent negative redox potentials, the oxidative processes of the sediment carbon content seem to have been set up quickly in the investigated reduced sand (Table 1). This probably means that oxic/anoxic and sulphidic chemoclines become established in the sediment (Revsbech et al., 1980). Oxic microenvironments could exist amid reduced sediment and enable co-occurring aerobic and anaerobic processes on a millimetre scale (Fenchel, 1992). For meiobenthos, the eco-physiological adaptive mechanism involved in resistance to reduced conditions are still largely unknown. However, the present results seem to emphasise that meiobenthic populations, living in the complicated systems around the above-said chemoclines, are probably adapated to exploit a set of chemical and ecological microniches.

## Acknowledgements

The authors wish to thank Mr Vittorio Gaiani for his precious collaboration in field work. The research was supported by a grant from the ECC-CLEAN Project (contract No. EV5V CT92 0080).

## References

Ceccherelli, V. U., G. Caramori, M. A. Colangelo, D. Fornasari, V. Gaiani, C. Mantovani & G. Reggiani, 1992. La successione ecologica nelle comunità bentoniche di ambienti lagunari in relazione ai fenomeni di disturbo. S.It.E./ATTI 15: 135–153.

Chandler, G. T. & J. W. Fleeger, 1983. Meiofaunal colonisation of azoic estuarine sediment in Louisiana: mechanisms of dispersal. J. exp. mar. Biol. Ecol. 69: 175–188.

Colangelo, M. A., V. Gaiani & V. U. Ceccherelli, 1994. Evoluzione del popolamento meiobentonico su sabbia artificialmente defaunata. Biol. Mar. Medit. 1: 243–247.

Colombo, G., C. Camerota, R. Bisceglia, V. Zaccaria, V. Gaiani & A. Carrieri, 1991. Variazioni spaziali e temporali delle caratteristiche fisico-chimiche delle acque e della biomassa fitoplanctonica della Sacca di Goro. In S. Bencivelli & N. Castaldi (eds), Studio integrato sull'ecologia della Sacca di Goro. Franco Angeli, Milano: 9–38.

Coull, B. C. & M. A. Palmer, 1984. Field experimentation in meiofaunal ecology. Hydrobiologia 118: 1–19.

Coull, B. C. & G. T. Chandler, 1992. Pollution and meiofauna: field, laboratory, and mesocosm studies. Oceanogr. Mar. Biol. ann. Rev. 30: 191–271.

Dal Cin, R. & P. Pambianchi, 1991. I sedimenti della Sacca di Goro (Delta del Po). In S. Bencivelli & N. Castaldi (eds), Studio integrato sull'ecologia della Sacca di Goro. Franco Angeli, Milano: 253–263.

Fenchel, T., 1992. What can ecologists learn from microbes: life beneath a square centimetre of sediment surface. Funct. Ecol. 6: 499–507.

Field, J. G., K. R. Clarke & R. M. Warwick, 1982. A practical strategy for analysing multispecies distribution patterns. Mar. Ecol. Prog. Ser. 8: 37–52.

Gee, J. M., R. M. Warwick, M. Schaanning, J. A. Berge & W. G. Ambrose Jr, 1985. Effects of organic enrichment of meiofaunal abundance and community structure in sublittoral soft sediments. J. exp. mar. Biol. Ecol. 91: 247–262.

Heip, C., R. M. Warwick, M. R. Carr, P. M. J. Herman, R. Huys, N. Smol & K. van Holsbeke, 1988. Analysis of community attributes of the benthic meiofauna of Frierfjord/Langesundfjord. Mar. Ecol. Prog. Ser. 46: 171–180.

Jacobs, L. J., 1984. The free-living inland aquatic nematodes of Africa – a review. Hydrobiologia 113: 259–291.

Kern, J. C., 1990. Active and passive aspects of meiobenthic copepods dispersal at two sites near Mustang Island, Texas. Mar. Ecol. Prog. Ser. 60: 211–223.

Le Guellec, C., 1988. Colonisation d'un sable azoïque exogène par la méiofaune. Cah. Biol. Mar. 29: 469–481.

Palmer, M. A., 1988. Dispersal of marine meiofauna: a review and conceptual model explaining passive transport and active emergence with implications for recruitment. Mar. Ecol. Prog. Ser. 48: 81–91.

Pearson, T. H. & R. Rosenberg, 1978. Macrobenthic succession in relation to organic enrichment and pollution of the marine environment. Oceanogr. Mar. Biol. ann. Rev. 16: 229–311.

Pugnetti, A. & V. Zaccaria, 1991. Ricerche sul fitoplancton della Sacca di Goro. In S. Bencivelli & N. Castaldi (eds), Studio integrato sull'ecologia della Sacca di Goro. Franco Angeli, Milano: 59–71.

Revsbech, N. P., J. Sørensen, T. H. Blackburn & J. P. Lomholt, 1980. Distribution of oxygen in marine sediments measured with microelectrodes. Limnol. Oceanogr. 25: 403–411.

Rofes, G., 1980. Etudes des sédiments. Méthodes de prélèvement et d'analyse pratiquées au laboratoire de sédimentologie. CTGREF, DIVISION QEPP, Etude 47, 50 pp.

Sherman, K. M. & B. C. Coull, 1980. The response of meiofauna to sediment disturbance. J. exp. mar. Biol. Ecol. 46: 59–71.

Sun, B. & J. W. Fleeger, 1994. Field experiments of the colonisation of meiofauna into sediment depressions. Mar. Ecol. Prog. Ser. 110: 167–175.

Thistle, D., 1980. The response of a harpacticoid copepod community to a small scale natural disturbance. J. mar. Res. 38: 381–395.

Underwood, A. J., 1981. Techniques of analysis of variance in experimental marine biology and ecology. Oceanogr. Mar. Biol. ann. Rev. 19: 513–605.

Viaroli, P., A. Pugnetti & I. Ferrari, 1992. *Ulva rigida* C. AGARDH growth and decomposition processes and related effects on nitrogen and phosphorus cycles in a coastal lagoon (Sacca di Goro, Po River Delta). In: G. Colombo, I. Ferrari, V. U. Ceccherelli, R. Rossi (eds), Marine eutrophication and population dynamics. Olsen & Olsen, Freedenborg: 77–84.

Warwick, R. M., K. R. Clarke & J. M. Gee, 1990. The effect of disturbance by soldier crabs *Mictyris platycheles* H. Milne Edwards on meiobenthic community structure. J. exp. mar. Biol. Ecol. 135: 19–33.

*Hydrobiologia* **329**: 69–78, 1996.
*P. Caumette, J. Castel & R. Herbert (eds), Coastal Lagoon Eutrophication and ANaerobic Processes (C.L.E.AN.).*



# Diel and seasonal vertical distribution of meiobenthic copepods in muddy sediments of a eutrophic lagoon (fish ponds of Arcachon Bay)

Evelyne Buffan-Dubau & Jacques Castel
*Laboratoire d'Océanographie Biologique, Université Bordeaux I, 2 rue du Professeur Jolyet, F-33120 Arcachon, France*

*Key words:* Meiobenthos, harpacticoid copepods, vertical distribution, emergence, *Canuella perplexa*, *Amphiascus parvus*

## Abstract

The vertical distribution of meiobenthic copepods was investigated within muddy sediments of a eutrophic lagoon (fish ponds of Arcachon Bay, France). The aim of the study was to determine if in muddy sediments, as previously established in sandy sediments, meiobenthic copepods migrate vertically according to the seasons or diel periods. Two experimental approaches were used, viz: a three-season comparison was made of the diel vertical distribution of the harpacticoid *Canuella perplexa* T. & A. Scott (1893) and secondly the depth distribution of a meiobenthic copepod assemblage was followed for a 24 h period, in shallow water subtidal locations. The harpacticoid *C. perplexa* vertically migrated through the top three centimeters of the sediment, showing diel and seasonal variations in depth distribution. The differential vertical distributions shown by the dominant meiobenthic populations suggest that emergence into the water column may mainly concern surface dwelling copepods. The physical and biological factors affecting seasonal and diel changes in the copepod assemblage of the fish ponds are discussed.

## Introduction

Vertical distribution of meiobenthic copepods has been intensively investigated in a wide variety of habitats. Earlier studies indicated that the majority of the meiofauna were restricted to the upper few centimeters in sandy habitats, and to the upper few millimeters in muddy sediments (see review by Huys et al., 1986). In general, vertical distribution patterns may be governed by a large number of factors. The penetration depth of oxygen (McLachlan, 1978), the presence of biogenic structures such as seagrass roots and macro-infauna tubes (Meyers et al., 1987; Escaravage et al., 1989) and the vertical distribution pattern of food sources (Montagna et al., 1989) may all influence the vertical distribution of meiobenthic copepods in both sandy and muddy sediments.

In sandy sediments, several studies have established that meiobenthic organisms, and especially harpacticoid copepods, generally colonize the top 10 centimeters and have been found at a maximum depth of 30–50 cm, (Harris, 1972a; McIntyre & Murison, 1973; McLachlan et al., 1977; Joint et al., 1982). Seasonal and diurnal changes in depth distributions of meiobenthic copepods have mostly been investigated in sandy habitats. Pronounced seasonal changes were shown indicating that, in general, harpacticoids migrate down and live deeper in the sediment during winter, whereas they migrate and live closer to the surface during the summer. These seasonal changes have been related to a number of factors such as the seasonal pattern of temperature in intertidal sandy areas (Harris, 1972b), seasonal fluctuations of oxygen penetration and breeding activity in both tidal and subtidal sandy areas (Harris, 1972c; Huys et al., 1986). Diurnal changes in depth distribution have been related to tide periods indicating that harpacticoids from intertidal sand flats may display downwards migrations at high tide (Joint et al., 1982).

In muddy sediments, most meiobenthic taxa and particularly copepods are usually concentrated in the uppermost centimeter. Arlt (1973) determined that

67% of all meiofauna from shallow (50 cm depth) muddy areas were restricted to the top first cm. Yingst (1978) counted 88% of harpacticoids within the uppermost cm of muddy sediments from a 14 m deep bottom. Similarly Jensen (1983) found 98% of the total number of harpacticoids in the top first cm of a sublittoral soft bottom. Smith & Coull (1987) determined that mud flats contain in the top first cm approximately twice as many meiofauna as the first 10 cm of sandy sediments. Furthermore, investigations using flume experiments could not find evidence that meiobenthic copepods vertically migrate in muddy sediments (Palmer & Molloy, 1986; Coull et al., 1989). However, preliminary field routine investigation carried out in 1993 in Arcachon Bay showed a surprising vertical distribution of meiobenthic copepods (Buffan & Castel, 1993). In the muddy sediment sampled in the fish ponds, 74% of the copepod assemblage were recorded in the 2–5 cm depth horizon, in spring, when the assemblage was numerically dominated by *Canuella perplexa* T. & A. Scott, 1893, a common harpacticoid of coastal environments (Heip, 1973; Castel, 1992; Ceccherelli et al., 1994). These data suggested that in these muddy sediments, migrations of meiobenthic copepods from the surface to the deeper layers may be significant.

A number of studies have shown that meiofauna and especially harpacticoids display upward migrations from sediment into the overlaying water column (Hicks, 1986; Walters, 1988; Armonies, 1989). Such movements are related to a number of factors, including light levels (Armonies, 1988), diel periods (Walters, 1988; 1991; Walters & Bell, 1994), tidal periods (Bell et al., 1988), copepod densities (Service & Bell, 1987; Walters, 1991), and species behaviour (Walters, 1991). Therefore in the present study, we have investigated the vertical distribution of meiobenthic copepods in a eutrophic lagoon using both an estimate of the density distribution according to depth in the sediment and a quantitative estimate of emerging copepods in the water column.

Field investigations were conducted to answer the following 3 questions concerning the vertical distribution patterns of meiobenthic copepods in subtidal muddy sediments.

1. Does the pattern vary over diel periods and are these variations linked to the night-day cycles?
2. Can the diel patterns be related to the emerging behaviour of harpacticoids?
3. Does the diel pattern vary according to seasons?

Diurnal changes were investigated by comparing depth distribution of a meiobenthic copepod assemblage over a diel period. Seasonal changes were investigated by comparing diel depth distribution patterns of *C. perplexa* within 3 diel samplings carried out during different seasons.

## Materials and methods

### *Study site*

Sampling sites (stations C1 and G) were located in the fish ponds of Arcachon Bay which are semi-enclosed and shallow lagoonal impoundments (see description by Castel et al., 1996). Sediments of the fish ponds which are partly covered by canopies of the seagrass *Ruppia cirrhosa*, are rich in organic detritus and silt (36–58.4%), (Escaravage & Castel, 1989). In the sampling area POC content generally varies between 4.5% to 6.5% (Etcheber, pers. comm.). The water depth at the sampling stations is 0.4 to 0.6 m. Station G was located near a deeper channel allowing renewal of water whereas station C1 was more isolated.

### *Sample collection*

Sediment was hand-collected using plastic core tubes (3 cm internal diameter and 5 cm length) and immediately preserved in liquid nitrogen. In the laboratory, frozen cores were sliced every 1 cm from 0 to 5 cm depth (in June, this was made immediately after the sampling of sediment cores). Slices were then washed through a 63 $\mu$m sieve to collect the total copepod assemblage or through a 200 $\mu$m sieve to selectively collect *Canuella perplexa.* Samples were preserved in 5% formalin and stained with Rose Bengal. The fauna were extracted from the sediment by flotation in Ludox HS-40 (De Jonge & Bouwman, 1977). Meiobenthic copepods were sorted under a dissection microscope and identified by microscopic observations. Physicochemical variables (temperature of the sediment surface, dissolved oxygen, and salinity) were measured during the sampling periods.

Copepods emerging from the sediment were collected in the water column using emergence traps made as previously described by Walters & Bell (1986). Emergence traps (Figure 1) collect any organism actively moving at least 12 cm above the sediment surface. Each trap was closed using an aluminium cap, filled with filtered seawater (63 $\mu$m), carefully positioned on a cylindrical support, and the cap was removed. The supports (plexiglass cylinders of

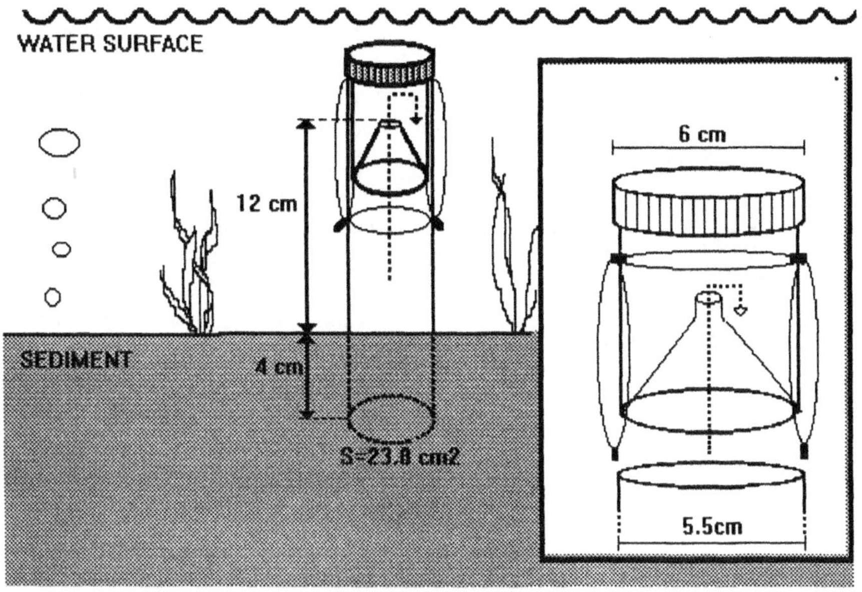


*Figure 1.* Emergence trap positioned between *Ruppia* patches in the fish ponds (station G, October). Modified from Walters & Bell (1986).

23.8 $cm^2$ surface) were carefully inserted into the sediment to a depth of 4 cm, at least 1 hour before trap placement. At the end of the field incubation, the trap contents were washed through a 63 $\mu$m sieve and preserved in 5% v/v formalin until sorting of copepods. Only adult copepods were identified to the species level.

*Sampling designs*

Station C1 was investigated for 21 h in June 1993 (03–04 June) and for 45 h in March 1994 (27-28-29 March). The aim of both samplings was to specifically study the vertical distribution patterns of *C. perplexa* over diel periods under different seasonal conditions. Every 3 hours, 3 (in June) to 8 (in March) replicate cores were collected over the diel periods.

Station G was investigated for 26 hours in October 1994 (15–16 October). Preliminary samplings have indicated both that *C. perplexa* was abundant at station G and that the copepod assemblage was more diversified than those from station C1. Thus, we decided to sample station G to study a possible difference in vertical distribution pattern between co-occurring species. Furthermore, the sampling was conducted to evaluate the emerging fraction of copepods and to examine the possible relationship between emerging behaviour and the vertical distribution patterns of copepods. Sampling points were concentrated around sunset and sunrise which may be the periods of maximum emergence over a diel period (Armonies, 1988; Walters, 1991). Four replicate cores were collected on the $15^{th}$ October at 17:00, 18:00 (sunset), 19:00, 24:00, and on the $16^{th}$ October at 01:00, 06:00, 07:00 (sunrise), 08:00, 13:00, 14:00, 17:00, 18:00 (sunset), 19:00. Two cm of overlaying water were sampled for each core. These water aliquots were then separated from sediment and considered as a slice in the results. From 19:00 on the $15^{th}$ October to 19:00 on the $16^{th}$ October, three replicate emergence traps were placed between *Ruppia* patches at each sampling time to be collected at the following sampling time.

*Statistical analysis*

Friedman two-way analysis of variance was used to test for significant differences between averages of copepod densities between sampling times over the diel periods, or according to depth. The Kolmogorov-Smirnov two-sample test was used to compare average densities of *C. perplexa*, at a depth horizon, between two sampling periods or between two depth horizons. The significance level of probability ($P$) was 0.05.

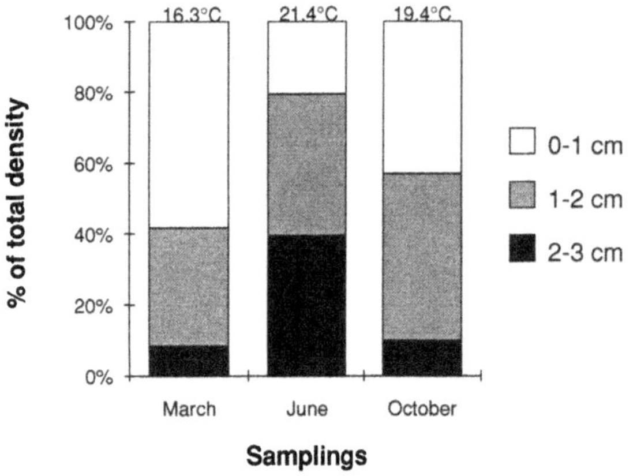


*Figure 2.* Mean vertical distribution of the harpacticoid *C. perplexa* in the top 3 cm of the sediment in March (station C1), June (station C1), and October (station G).

## Results

During the sampling programme the salinity varied between 10‰ (in March), 17‰ (in June), and 29‰ (in October). Dissolved oxygen similarly varied between the 3 diel periods, with, on average, a minimum of 2.3 mg $l^{-1}$ (early morning) and a maximum of 13.8 mg $l^{-1}$ (sunset). Generally, the copepods (79.4% to 98%) were restricted to the top 3 cm of the sediment. Consequently the vertical distribution of copepods was studied in the 3 cm depth column.

*Vertical distribution of* C. perplexa *over diel periods and between seasons (June 1993, March 1994, and October 1994, samplings)*

Between sampling periods (Figure 2), the mean temperature of the sediment surface varied from 16.3 °C (March) to 19.4 °C (October) and 21.4 °C (June). On average 60% of the copepods were concentrated in the top first cm in March ($P<0.05$, Friedman test), whereas in June 80% of the population was found in the 1–3 cm depth horizon ($P=0.05$, Friedman test). In October 89.7% of individuals were concentrated in the uppermost 2 cm ($P<0.001$, Friedman test) and equally distributed in the 0–1 cm and 1–2 cm depth horizons.

Over the diel periods (Figure 3), the temperature variation at the sediment surface was greater in March and June (around 10 °C) than in October (1 °C). In March most individuals (74–87%) were found in the top first cm at higher temperatures whereas at lower temperatures most individuals (51–59%) were found in the 1–2 cm depth horizon. The percentage of the total population in the top first cm showed increases averaging 40% and 29% for the first and the second diurnal peaks of surface sediment temperature respectively (Figure 3). In contrast, during the June sampling copepods were mostly found in the 2–3 cm depth horizon at higher temperatures (43–61% between 16:00 and 19:00), whereas at lower temperatures most individuals were found in the 1–2 cm depth horizon (41–67% at 01:00–04:00) (Figure 3). In October the temperature did not show a significant diurnal fluctuation, however, copepods tended to migrate down, from the first to the second and the third centimeters below the surface, between 06:00 and 08:00 and between 17:00 and 19:00, (Figure 3). These results indicate that factors other than surface sediment temperature influence the depth distribution of *C. perplexa.*

*Emergence into the water column of meiobenthic copepods from station G (October 1994 sampling)*

On average harpacticoids represented 87% of the meiobenthic copepod community, with copepodites and adults representing 30% and 57% of the community respectively. The most abundant species, *Canuella perplexa* and *Amphiascus parvus* Sars 1906 comprised 72% of the harpacticoid community. The remainder of the assemblage comprised 8 species: *Schizopera* sp. Sars 1905 (7.3%); *Paralaophonte brevirostris* Claus 1863 (7.06%); *Mesochra pygmea* Claus 1863 (5.8%); *Robertgurneya* sp. (2.7%); *Cletocamptus confluens* Schmeil 1894 (1.7%); unidentified species (1.3%); *Amonardia normani* Brady 1872 (0.82%); and *Paradactylopodia* sp. (0.68%). No ovigerous females were observed among the population of *C. perplexa.* In contrast, *A. parvus* ovigerous females were found.

The abundances of copepods in the emergence traps differed over the 24 h period, (Figure 4). Friedman analysis of variance was performed on data sets concerning traps incubated in the field for equivalent durations (i.e. a set pooling abundances found in traps incubated in the field for 1 h, and a set pooling abundances found in traps incubated in the field for >1 h duration). Average abundances were significantly higher ($P<0.05$) in emergence traps incubated between 19:00 and 24:00 ($291 \pm 105$ ind $trap^{-1}$ or 23.8 $cm^2$ sediment surface) and between 06:00 and 07:00 ($P=0.05$, $161 \pm 87$ ind $trap^{-1}$) than in other traps incubated for equivalent durations respectively. Considering that the total copepod abundance in the sediment reached on average $1205 \pm 712$ ind 23.8 $cm^{-2}$ over the sampling campaign, the emerging fraction can be estimated to

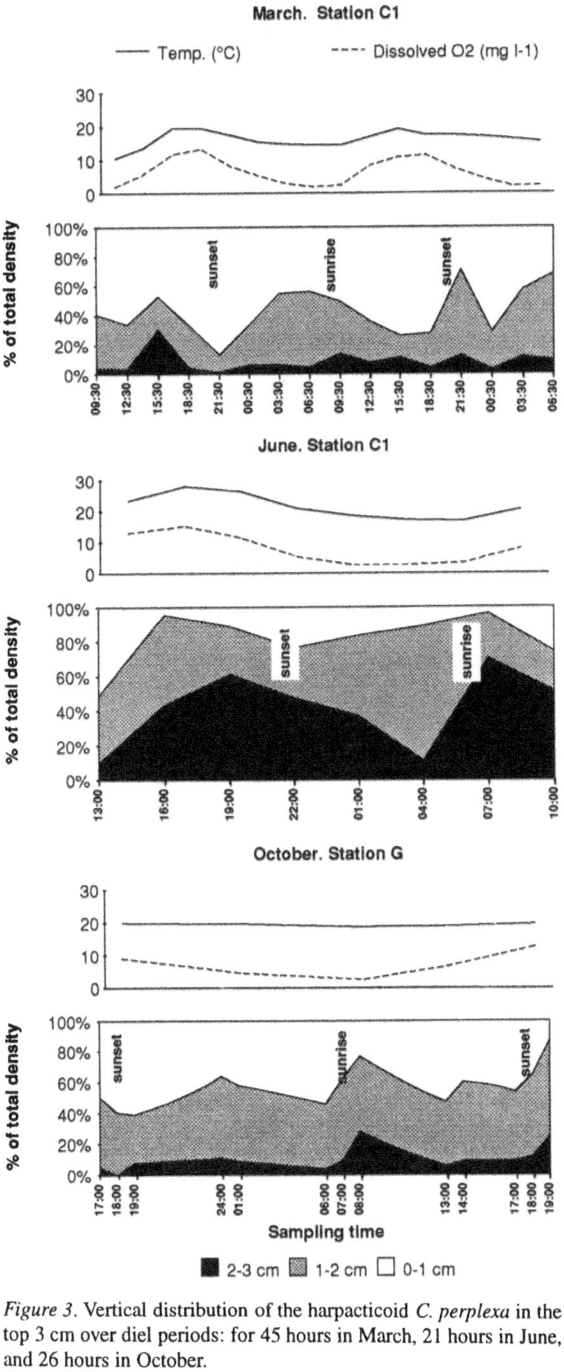


*Figure 3.* Vertical distribution of the harpacticoid *C. perplexa* in the top 3 cm over diel periods: for 45 hours in March, 21 hours in June, and 26 hours in October.

be 15–33% and 6–21% of the total copepod assemblage for the 19:00–24:00 and the 06:00–07:00 emergence periods respectively. However, since our sampling method does not allow the collecting of emerging copepods which were swimming between 2 cm and 12 cm above the sediment surface, these percentages should be considered as minimum estimates of the emerging assemblage. These results indicate that emergence of meiobenthic copepods into the water column increased around sunset (18:00) to become very active during the first half of the night (19:00–24:00). The copepods were not emerging during the second half of the night. A second emergence period was detected between 06:00–07:00, before sunrise (07:00), which was more variable and less important, in terms of abundance. Copepodites were the most abundant in trap contents and comprised 52% and 47% of the total abundances at 19:00–24:00 and 06:00–07:00 respectively. Adult harpacticoids mainly comprised *A. parvus*, 39% and 81%; *A. mormani*, 28% and 10%; and *Paradactylopodia sp.*, 23% and 6%, at 19:00–24:00 and 06:00–07:00 respectively. *C. perplexa* was not found in trap contents. The species composition was thus different between the emergence periods, with the second emergence phase mainly attributable to a single species, *A. parvus*.

### *Vertical distribution patterns of meiobenthic copepods from station G (October 1994 sampling)*

Most copepods (79.4%) were distributed in the top 3 cm of the sediment, 4% were found deeper, and 16.6% were found in the overlaying water (i.e. in the 2 cm water column above the sediment surface). On average, the density of copepods which showed an emerging behaviour (i.e. density of the non-emerging copepod *C. perplexa* was removed), significantly decreased with depth ($P<0.001$, Friedman test) with $104.5 \pm 24$ ind 7 $cm^{-3}$, $50 \pm 15$ ind 7 $cm^{-3}$, $26.5 \pm 7$ ind 7 $cm^{-3}$, at 1 cm, 2 cm, and 3 cm below the surface respectively.

Over the diel period (Figure 5) the vertical distribution pattern of copepods (*C. perplexa* excepted) tended to change around sunset. Before sunset (between 13:00 and 18:00), densities in the overlaying water and in the first cm of the sediment showed an increase averaging 14% and 13% of the total assemblage respectively, whilst in the deeper sediment strata densities tended to decrease by 13.5% and 12.5% of the total assemblage in the 1–2 cm and the 2–3 cm depth horizons respectively. These results indicate that before sunset copepods tended to migrate upwards and to regroup at the sediment water interface. At sunset (between 17:00 and 19:00), densities in the overlaying water (0–2 cm above sediment) tended to decrease. This decrease averaged between 7 to 37% of the total assemblage. This change occurred at the beginning of the emergence period sug-

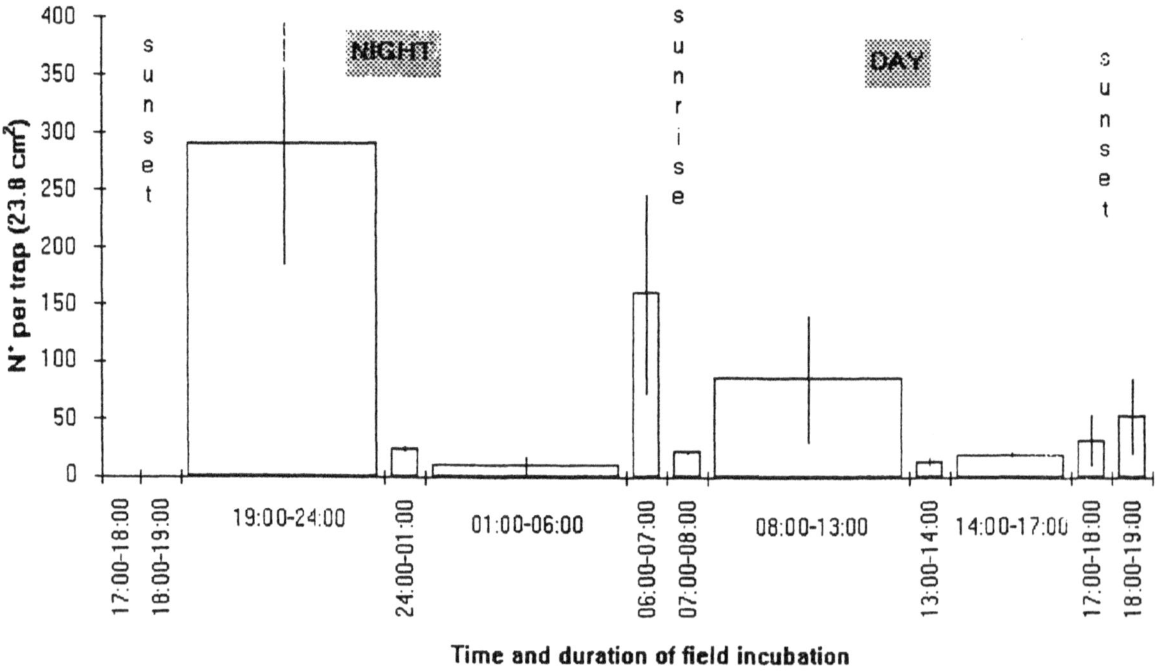


*Figure 4.* Mean abundance (± sd) of meiobenthic copepods in emergence traps collected at station G in October. Traps were positioned for different time intervals (from 1 hour to 5 hours) over a 26 hour time period.

gesting that a number of copepods began to swim up in the water column (above the overlaying water). During the night, copepods were mainly found in the first cm of the sediment.

During both emergence periods (19:00–24:00 and 06:00–07:00), the vertical distribution patterns of the emerging copepod *A. parvus* and the non-emerging copepod *C. perplexa* were compared (Figure 6). During both periods the depth distribution of *C. perplexa* was consistent with the average distribution pattern observed over the whole diel period (Figure 2). In contrast, the depth distribution of *A. parvus* tended to change during the second emergence phase (06:00–07:00). Density of the copepod in the first cm tended to decrease, inducing a more uniform depth distribution in the top 3 cm (Figure 6).

## Discussion

Generally, copepods were distributed in the uppermost 3 cm of the sediment. These data contradict the distribution generally assumed to occur in muddy sediments, where most benthic copepods (and meiofauna in general) are restricted to the first cm of the sediment (Arlt, 1973; Jensen, 1983; Coull et al., 1989). Vertical distribution of metazoans is generally related to the oxygen concentration in pore water. In the muddy sediments of the fish ponds, the maximum depth of oxygen penetration was 5 mm (Schaub & van Gemerden, 1996). However, infaunal activity (e.g. tube building by polychaetes and crustaceans) provides micro-oxygenated halos (Meyers et al., 1987) and phanerogam roots may enhance the oxygen penetration at depth. In addition, potential food sources for copepods such as phototrophic microorganisms are also present through the top 3 cm of the sediment (Buffan et al., 1993 and unpublished data), indicating that copepods may find food sources below the sediment surface.

### *Diel and seasonal variations of the vertical distribution pattern of the harpacticoid* C. perplexa

Our results agree with previous observations indicating that *C. perplexa* is a deep burrowing copepod (Thielemans & Heip, 1984). The harpacticoid *C. perplexa* showed diel changes in depth distribution which followed the diel variation of the surperficial sediment temperature. When the temperature of the sediment surface increased copepods tended to migrate up in

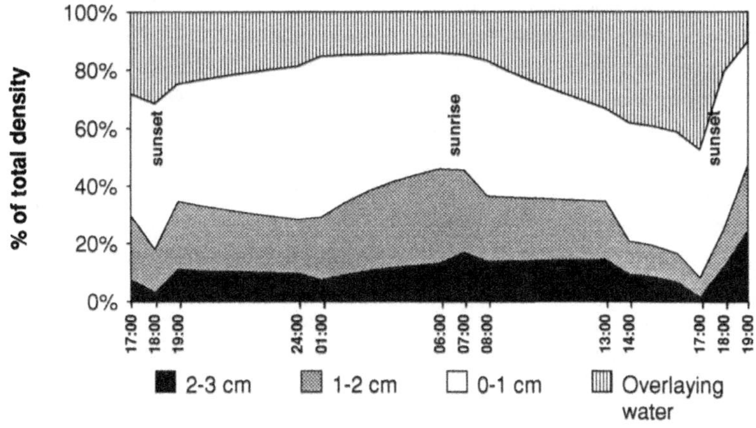


*Figure 5.* Vertical distribution of the copepod assemblage at station G in October; *C. perplexa* population was removed. Samples were taken at different time intervals over a 26 hour time period.

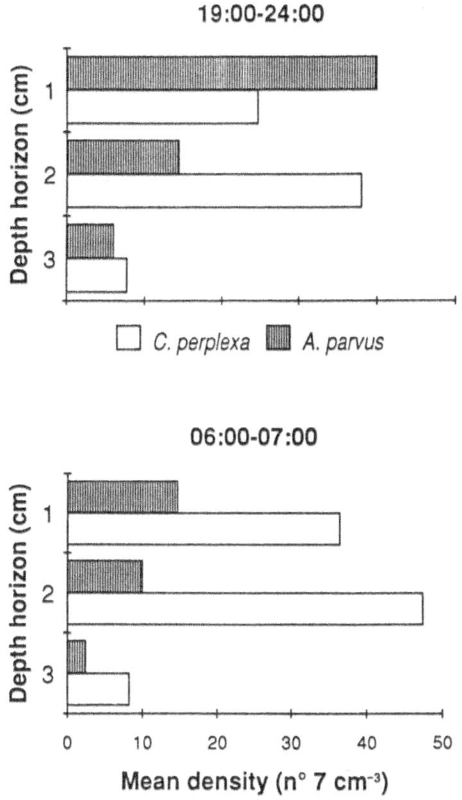


*Figure 6.* Mean vertical distribution of the most abundant harpacticoids, *C. perplexa* and *A. parvus*, in the top 3 cm of the sediment, during both emergence periods (October, station G).

March whereas in June copepods tended to migrate down. Considering that in March the maximum temperatures (17.6 –19.6 °C) were equal to the minimum temperatures (17–19 °C) measured in June, these results indicate that *C. perplexa* tended to migrate down when temperature of the sediment surface deviated (by increasing or decreasing) from between a 17–19 °C temperature interval. This suggests that in the fish ponds *C. perplexa* has a temperature optimum of 17 – 19 °C. Considering that the diel variations of temperature are slower and smaller at depth than at the sediment surface, it is likely that the copepod migrated down to avoid sudden variations of temperature and to stay close to its optimal temperature. This is consistent with the different vertical distribution patterns observed between seasons (Figure 2). The highest percentages of individuals were found in the top cm in March and October when average sediment surface temperatures approached 17 –19 °C. Furthermore, the uniform distribution of copepods in the uppermost 2 cm observed in October was tied to a vertical homogenization of temperatures in the sediment due to the stability of the diurnal temperature. However, it should be taken into account that the change of sampling station, from station C1 (March and June) to station G (October) may also have influenced the observed differences in depth distribution between seasons. These vertical distribution patterns are not totally consistent with those of Huys et al. (1986) who studied a subtidal sandy station located in the Southern Bight of the North

Sea. These authors showed that *C. perplexa* migrated towards the sediment surface in March and stayed close to the surface throughout spring and summer (90% of the population in the upper 2 cm), and migrated deeper in autumn. In the present study, the mean vertical distribution patterns observed in March and in October (Figure 2) support these results whereas the June vertical distribution does not. This was probably due to the difference of summer temperatures between the Southern Bight of the North Sea and Arcachon Bay. The authors related vertical movements of *C. perplexa* with breeding periods, indicating that during extensive reproductive activity (spring and summer) the copepod was mainly found in the upper millimeters of the sediment. Considering that *C. perplexa* shows 2 or 3 reproductive periods occurring in spring, summer and late summer to early autumn (Castel & Lasserre, 1977; Ceccherelli & Mistri, 1991), and considering that breeding activity is related to temperature (Harris, 1972c), it is likely that in the present study, reproductive activity also influenced the vertical migration of *C. perplexa* through the sediment. This is supported by the fact that the nauplius of *C. perplexa* is planktonic (Vincx & Heip, 1979), suggesting that ovigerous females migrate to the sediment surface for laying.

Oxygen concentrations also showed diel variations which were linked to temperature variations which both responded to the diel cycle of light intensity (Figure 3). Indeed, oxygen concentrations measured in the water column reflected the photosynthetic activity of microphytes which was obviously more intensive during days than during nights. There were no marked changes in mean oxygen concentrations between seasons (March: 4.4–8.3 mg $l^{-1}$, June: 3.4–11.9 mg $l^{-1}$, October: 5.4–9.7 mg $l^{-1}$), thus, seasonal changes in the vertical distribution of *C. perplexa* could not be attributed to this factor. Over diel periods we found no evidence for an effect of diel variations in oxygen concentrations on the vertical distribution of this copepod even if the copepod tended to move up during (June) or after (March and October) the decrease of oxygen concentration at night. However, considering that oxygen is a key factor for the depth distribution of meiofauna (McLachlan, 1978; Meyers et al., 1987), it is possible that the diel vertical distribution of *C. perplexa* responded to the combined effects of diel variations of temperature and oxygen concentrations.

*Emergence into the water column and vertical distribution patterns of meiobenthic copepods*

During the October sampling, the emerging fraction was estimated to be 15 to 33% (postsunset emergence) and 6 to 21% (pre-sunrise emergence) of the copepod assemblage. These results are consistent with those of Walters & Bell (1994) who observed that 7 to 31% of the copepod assemblage emerged during the post-sunset period.

As expected, active emergence of meiobenthic copepods into the water column was associated with light changes over the diel period (Figure 4). Emergence throughout the night and particularly around sunset has previously been well established (Walters, 1988, 1991; Walters & Bell, 1994). In the present study the time of emergence was species-specific. Indeed, the migrating species *A. normani* and *Paradactylopodia sp.* were found in traps sampled only after sunset, whereas *A. parvus* was found in traps sampled during both emergence phases. It may be deduced that *A. normani* and *Paradactylopodia* emerged only during the first part of the night. Another possibility was that both copepods swam for a long period (from around 19:00–24:00 to 06:00) in the water column. Traps placed above the sediment at 06:00 would not capture them if the copepods were already swimming through the water column. Such long excursions of meiobenthic copepods in the water column have been observed. For example, Bell et al. (1989) determined that an emerging copepod (*Zausodes arenicolus*) remained in the water column of field aquaria for more than 9 hours. Using experimental field aquaria, these authors also showed that the composition of the migrating species assemblage determines the time, the duration and/or the number of emergence periods occurring during the night.

We found evidence which related the emerging behaviour with the vertical distribution pattern of copepods in the sediment. At first, emerging copepods (Figure 5) and non-emerging copepods (population of *C. perplexa*) (Figure 2) showed different vertical distribution patterns over the diel period (October, station G). The emerging assemblage was in general found mainly in the first cm of the sediment, whereas *C. perplexa* showed a deeper vertical distribution. This suggests that emergence into the water column may mainly concern surface dwelling copepods. Secondly, the depth distribution of *A. parvus* only (the dominant species emerging at sunrise) tended to change during emergence phases. Density of this copepod tended to decrease in the top first cm while it was emerging into the water column (Figure 6, second emergence phase, 06:00–07:00). Consequently, it seems that both the vertical occupation of space and diel changes of the vertical distribution, particularly between day and night, may be related in part to the swimming behaviour of harpacticoids.

## Acknowledgments

We are grateful to Dr I. Auby, L. P. Souza-Santos, E. Antajan, and Y. Dubau for assistance collecting the samples throughout mosquito-infested nights. To M. Parra for making emergence traps and to A. M. Castel for assistance in sorting the trap contents. To S. Bourguès, L. P. Souza-Santos and P. J. P. Santos for their advice and discussions concerning sampling, laboratory processing, and earlier draft of the manuscript. We thank Dr W. Armonies and Dr I. Ferrari for critical comments which were helpful to improve the quality of the manuscript. We are also very grateful to G. Hello and Dr D. T. Welsh for their assistance with the English version. This work was supported by a E.U. Environment programm (CLEAN contract EV5V-CT92-0080).

## References

Arlt, G., 1973. Vertical and horizontal microdistribution of the meiofauna in the Greifswalder Bodden. Oikos Suppl. 15: 105–111.

Armonies, W., 1988. Hydrodynamic factors affecting behaviour of intertidal meiobenthos. Ophelia 28: 183–193.

Armonies, W., 1989. Meiofaunal emergence from intertidal sediment measured in the field: significant contribution to nocturnal planktonic biomass in shallow waters. Helgoländer wiss. Meeresunters. 43: 29–43.

Bell, S. S., G. R. F. Hicks & K. Walters, 1988. Active swimming in meiobenthic copepods of seagrass beds: species patterns and role in reproductive behavior. Mar. Biol. 98: 351–358.

Bell, S. S., G. R. F. Hicks & K. Walters, 1989. Experimental investigations of benthic re-entry by migrating meiobenthic copepods. J. exp. mar. Biol. Ecol. 130: 291–303.

Buffan, E. & J. Castel, 1993. Distribution of benthic microbiota in Arcachon Bay and Prévost lagoon. C.L.E.A.N. progress report part II: 73–89.

Buffan, E., L. P. Souza-Santos & J. Castel, 1993. Daily variations of vertical distribution and pigment gut content of the meiobenthic copepod *Canuella perplexa* in a brackish lagoon. C.L.E.A.N. progress report part II: 133–148.

Castel, J., 1992. The meiofauna of coastal lagoon ecosystems and their importance in the food weeb. Vie Milieu 42: 125–135.

Castel, J. & P. Lasserre, 1977. Colonisation et distribution spatiale des copépodes dans des lagunes semi-artificielles. In B. F. Keagan, P. O. Ceidigh & P. J. S. Boaden (eds), Biology of benthic organisms. Pergamon Press: 129–146.

Castel, J., P. Caumette & R. Herbert, 1996. Eutrophication gradients in coastal lagoons as exemplified by the Bassin d'Arcachon and the Etang du Prévost. Hydrobiologia 329: xi–xxx.

Ceccherelli, V. U. & M. Mistri, 1991. Production of the meiobenthic harpacticoid copepod *Canuella perplexa*. Mar. Ecol. Prog. Ser. 68: 225–234.

Ceccherelli, V. U., M. Mistri & P. Franzoi, 1994. Predation impact on the meiobenthic harpacticoid *Canuella perplexa* in a lagoon of the Po River Delta, Italy. Estuaries 17: 283–287.

Coull, B. C., M. A. Palmer & P. E. Myers, 1989. Controls on the vertical distribution of meiobenthos in mud: field and flume studies with juvenile fish. Mar. Ecol. Prog. Ser. 55: 133–139.

Escaravage, V. & J. Castel, 1989. Application de la notion de confinement aux peuplements méiobenthiques des lagunes endiguées du Bassin d'Arcachon (Côte atlantique). Acta Oecol. Gener. 10: 1–17.

Escaravage, V., M. E. Garcia & J. Castel, 1989. The distribution of meiofauna and its contribution to detritic pathways in tidal flats (Arcachon Bay, France). In J. D. Ros. (ed.), Topics in marine Biology, Scient. Mar. 53: 551–559.

Harris, R. P., 1972a. The distribution and ecology of the interstitial meiofauna of a sandy beach at Whitsand Bay, east Cornwall. J. mar. biol. Ass. U.K. 52: 1–18.

Harris, R. P., 1972b. Seasonal changes in the meiofauna population of an intertidal sand beach. J. mar. biol. Ass. U.K. 52: 389–403.

Harris, R. P., 1972c. Reproductive activity of the interstitial copepods of a sandy beach. J. mar. biol. Ass. U.K. 52: 507–524.

Heip, C., 1973. Partitioning of the brackish water habitat by copepod species. Hydrobiologia 41: 189–198.

Hicks, G. R. F., 1986. Distribution and behaviour of meiofaunal copepods inside and outside seagrass beds. Mar. Ecol. Prog. Ser. 31: 159–170.

Huys, R., R. L. Herman & C. Heip, 1986. Seasonal fluctuations in vertical distribution and breeding activity of a subtidal harpacticoid community in the southern bight, North Sea. Neth. J. Sea Res. 20: 375–383.

Jensen, P., 1983. Meiofauna abundance and vertical zonation in a sublittoral soft bottom, with a test of the Haps corer. Mar. Biol. 74: 319–326.

Joint, I. R., J. M. Gee & R. M. Warwick, 1982. Determination of fine-scale vertical distribution of microbes and meiofauna in an intertidal sediment. Mar. Biol. 72: 157–164.

Jonge, V. N. De & L. A. Bouwman, 1977. A simple density separation technique for quantitative isolation of meiobenthos using the colloidal silica LUDOX-TM. Mar. Biol. 42: 143–148.

McIntyre, A. D. & D. J. Murison, 1973. The meiofauna of a flatfish nursery ground. J. mar. biol. Ass. U.K. 53: 93–118.

McLachlan, A., 1978. A quantitative analysis of the meiofauna and the chemistry of the redox potential discontinuity zone in a sheltered sandy beach. Estuar. coast. mar. Sci. 7: 275–290.

McLachlan, A., P. E. D. Winter & L. Botha, 1977. Vertical and horizontal distribution of sublittoral meiofauna in Algoa Bay, South Africa. Mar. Biol. 40: 355–364.

Meyers, M. B., H. Fossing & E. N. Powell, 1987. Microdistribution of interstitial meiofauna, oxygen and sulfide gradients, and the tubes of macro-infauna. Mar. Ecol. Prog. Ser. 35: 223–241.

Montagna, P. A., J. E. Bauer, D. Hardin & R. B. Spies, 1989. Vertical distribution of microbial and meiofaunal populations in sediments of a natural coastal hydrocarbon seep. J. mar. Res. 47: 657–680.

Palmer, M. A. & R. M. Molloy, 1986. Water flow and the vertical distribution of meiofauna: a flume experiment. Estuaries 9: 225–228.

Schaub, B. & H. van Gemerden, 1996. Sulfur bacteria in sediments of two coastal ecosystems: the Bassin d'Arcachon and the Étang du Prévost, France. Hydrobiologia, this volume.

Service, S. K. & S. S. Bell, 1987. Density-influenced active dispersal of harpacticoid copepods. J. exp. mar. Biol. Ecol. 114: 49–62.

Smith, L. D. & B. C. Coull, 1987. Juvenile spot (Pisces) and grass shrimp predation on meiobenthos in muddy and sandy substrata. J. exp. mar. Biol. Ecol. 105: 123–136.

Thielemans, L. K. T. & C. Heip, 1984. The response of a harpacticoid copepod community to sediment disturbance in a semi-enclosed lagoon. Hydrobiologia 118: 127–133.

Vincx, M. & C. Heip, 1979. Larval development and biology of *Canuella perplexa* T. and A. Scott, 1893 (copepoda harpacticoida). Cah. Biol. Mar. 20: 281–299.

Walters, K., 1988. Diel vertical migration of sediment-associated meiofauna in subtropical sand and seagrass habitats. J. exp. mar. Biol. Ecol. 117: 169–186.

Walters, K., 1991. Influences of abundances, behavior, species composition, and ontogenic stage on active emergence of meiobenthic copepods in subtropical habitats. Mar. Biol. 108: 207–215.

Walters, K. & S. S. Bell, 1986. Diel patterns of active vertical migration in seagrass meiofauna. Mar. Ecol. Prog. Ser. 61: 111–124.

Walters, K. & S. S. Bell, 1994. Significance of copepod emergence to benthic, pelagic, and phytal linkages in a subtidal seagrass bed. Mar. Ecol. Prog. Ser. 108: 237–249.

Yingst, J. Y., 1978. Patterns of micro and meiofauna abundance in marine sediments, measured with the adenosine triphosphate assay. Mar. Biol. 47: 41–54.

*Hydrobiologia* **329**: 79–89, 1996.
*P. Caumette, J. Castel & R. Herbert (eds), Coastal Lagoon Eutrophication and ANaerobic Processes (C.L.E.AN.).*




# The role of phototrophic sulfur bacteria as food for meiobenthic harpacticoid copepods inhabiting eutrophic coastal lagoons

L. P. Souza-Santos, J. Castel & P. J. P. Santos
*Laboratoire d'Océanographie Biologique, Université Bordeaux I, 2 rue du Professeur Jolyet, F-33120 Arcachon, France*

*Key words:* Feeding ecology, meiofauna, harpacticoid copepods, coastal lagoons, phototrophic bacteria, microalgae

## Abstract

Laboratory experiments were undertaken using *Amonardia normani* and *Schizopera cf. compacta*, two meiobenthic harpacticoid copepods commonly found in coastal lagoons. The first experiments were designed to determine if the phototrophic sulfur bacteria *Chromatium gracile* can be ingested by these copepods and at what concentrations. Egestion rate was used as an index of feeding rate. The response of the egestion rate, expressed in numbers of faecal pellets produced by copepod per day, as a function of bacterial concentration followed the functional model. *A. normani* attained constant feeding rates from the bacterial concentration of $1 \times 10^7$ cells $ml^{-1}$ (5 $\mu g$ C $ml^{-1}$) onwards, *S. cf. compacta* attained constant feeding rates from $2.6 \times 10^7$ cells $ml^{-1}$ (13 $\mu g$ C $ml^{-1}$) onwards. The faecal pellet volume changed significantly ($p<0.05$) between food concentrations for *A. normani* but not for *S. cf. compacta* ($p>0.05$). In order to investigate the effect of the phototrophic bacterial diet on the population dynamics of *A. normani* three groups of nauplii were maintained at $2 \times 10^7$ cells $ml^{-1}$ and observed every day. The mortality of these nauplii was very high compared to those maintained on a diatom diet (*Nitzschia constricta*); only in one of the groups did some copepodites develop but no adults were ever observed. Adults fed on bacteria did not have different ($p>0.05$) survival rates compared to those fed on diatoms, nevertheless, the number of nauplii produced was significantly less ($p<0.05$) on the bacterial diet. These results lead us to suggest that although the phototrophic sulfur bacteria (*Chromatium gracile*) can be ingested by both copepod species it cannot sustain the full development of the *A. normani* population. Thus, a bloom of phototrophic sulfur bacteria does not seem to be a favourable situation for opportunistic benthic copepods to colonize eutrophic coastal lagoons after a dystrophic crisis.

## Introduction

The important role played by copepods in the food chains of aquatic systems is now well established. By feeding on microorganisms and detritus they transfer primary production to higher trophic levels and participate in the recycling of organic matter, in both pelagic (Bougis, 1974) and benthic systems (Castel, 1992; Giere, 1993). Literature dealing with feeding activity of pelagic copepods is extensive and several laboratory and *in situ* methods have been developed and employed to quantify feeding rates and to estimate food quality. Surprisingly, very few data on feeding biology of meiobenthic copepods are available. Montagna (1995) recently reviewed the current literature on meiofauna feeding rate studies. Focusing on the utilization of microflora (bacteria and microalgae), this review clearly showed the importance of bacteria as food for several benthic copepods. Rieper (1978; 1982) demonstrated that *Tisbe holothuriae* and *Paramphiascella vararensis* could develop with bacteria as the sole food source and detect differences between bacterial strains, thus showing specific food preferences. Other copepod species, such as *Dactylopodia vulgaris* (Rieper, 1985), seem to depend on algal carbon to reproduce. These examples show how complex and specific are the relationships between meiobenthic copepods and their food items.

Feeding rates of pelagic copepods are known to be affected not only by food quality but also, and strongly, by food concentration (Frost, 1972), therefore, a functional model of ingestion rate as a function of food concentration is usually accepted. For meiobenthic copepods the study of the effect of food concentration on feeding rates is still at an early stage. Rieper (1978; 1985) showed that the increase of food concentration resulted in an increase of feeding rate. The work by Montagna et al. (1995) was the first which tried to model the relationship between meiofauna feeding rate and food concentration.

In this study, the role of a specific bacterial group (the phototrophic sulfur bacteria) as food for meiobenthic copepods is investigated. The choice of these bacteria is related to the possible occurrence of 'red waters' in eutrophic coastal lagoons. Normally, phototrophic sulfur bacteria occupy a well-defined layer in the microbial mat of shallow lagoon sediments, between the oxygenic layers – where diatoms and cyanobacteria grow – and the anoxygenic layers – where sulfate reduction occurs (Caumette, 1992). However, in some coastal lagoons, the proliferation of macroalgae in spring and their decay and degradation in summer may result in anoxic, sulfide-rich waters. During this period, plants and animals die out. Following this dystrophic crisis, waters can turn pink to purple-red ('red-waters') due to the bloom of phototrophic sulfur bacteria which oxidize reduced sulfur compounds, thus removing toxic products from the lagoon waters (Caumette, 1986).

Copepods (*Acartia clausi* and *Tisbe* sp.) feeding on phototrophic sulfur bacteria have been previously observed by Caumette et al. (1983) and Caumette (1986). The purpose of this study was to investigate (1) whether meiobenthic copepods feed on phototrophic sulfur bacteria and (2) whether their populations could be maintained with this sole source of carbon. Two copepod species were tested: *Amonardia normani* (Brady, 1872), an epibenthic opportunist and common inhabitant of coastal lagoons (Castel, 1979), and *Schizopera cf. compacta* Lint, 1922, an endobenthic copepod. The two species are highly abundant in both phytal and sediment environments of the fish ponds of Certes in Arcachon Bay (Castel & Lasserre, 1977). *Chromatium gracile* Strzeszewski, 1913 was the tested phototrophic sulfur bacteria since it is abundant in the same fish ponds (Guyoneaud et al, 1996).

## Materials and methods

### *Collection and cultivation of copepods*

Cultures of meiobenthic copepods were initiated with ovigerous *Amonardia normani* and *Schizopera cf. compacta* females collected in the fish ponds of Certes in Arcachon Bay (France, 44° N, 1°10′ W). The offsprings were cultivated for several generations in glass dishes with 0.7 $\mu$m filtered seawater (salinity: 30‰) and *Nitzschia* sp (diatom isolated from the same fish ponds) or *Nitzschia constricta* (Greg.) Grun. (CCMP 576 strain from Bigelow Laboratory for Ocean Sciences, Maine, USA) as food. The cultures were maintained at 20 °C in a 12 h dark/12 h light photoperiod.

### *Bacterial culture*

The purple sulfur bacteria tested, *Chromatium gracile* strain CE2201, isolated from the fish ponds sediments (Guyoneaud et al., 1996), was originating from the culture collection of the laboratory. It was cultivated in the Pfennig & Trüper (1981) medium with 5 mM of acetate and 5 mM of thiosulphate at pH = 7 and salinity = 30‰. Before being offered to the animals, the bacterial culture was observed under a microscope to verify the absence of sulfur globules inside cells, since the presence of these was shown to be prejudicial for copepods (Caumette, 1985). Afterwards, the bacterial culture was rinsed three times by centrifugation (3000 rpm) with sterile, 0.22 $\mu$m filtered seawater, in order to eliminate toxic compounds present in the medium. Bacterial concentrations were estimated with a Malassez haemocytometer counting-cell, taking care of both sedimentation time of cells and plate desiccation. Dilutions were made with sterile, 0.22 $\mu$m filtered seawater.

Two estimates of the carbon content of bacterial cells were made. Firstly, width and length of 10 bacterial cells were measured by phase contrast microscopy using a micrometric ocular. The carbon content was estimated assuming a cylindrical shape of cells, a density of 1.1 g cm$^{-3}$ and a carbon:fresh weight ratio of 0.1. Secondly, the bacterial culture was filtered in preweighed, 400 °C-burned GF/C filters. These filters were washed with 3% w/v ammonium formate to eliminate salts, dried at 60 °C for 24 h and burned at 400 °C for 24 h, then weighed on a Mettler ME22 microbalance. The carbon:organic matter ratio was assumed to be 0.5 (Karl, 1986).

*Relationship between bacterial concentration and egestion rate of copepods*

A single copepod female was placed in a 23.7 $cm^2$ glass dish containing 10 ml of bacterial cell suspension of known concentration. Five replicate dishes were used for each bacterial concentration and for controls (considered here as copepods in sterile, 0.22 $\mu$m filtered seawater, or without food). These dishes were incubated for 24 h, accounting for the possibility of endogenous diel cycles of feeding rate (Souza-Santos et al., 1995), at the same conditions as copepod cultures. At the end of the incubation period and after checking for copepod survival, the content of each dish was preserved in 4% v/v formalin for further analysis.

The number of faecal pellets produced by each copepod during 24 h was counted under a stereomicroscope. To estimate the total volume of faecal pellets produced, the length and width of 20 to 30 pellets for each food concentration were measured using a camera lucida. The volume of pellets was estimated assuming a cylindrical shape.

Experiments were carried out twice with *A. normani*. At the first experiment the feeding rates increased continually with food concentrations. Regarding the functional model, it means that the food concentrations were lower than the 'critical' food concentration, defined here as the concentration from which no significant increase in feeding rates is observed. In the second experiment higher food concentrations were tested and the feeding rates attained the expected plateau. This resulted in different ranges of tested bacterial concentrations between the two copepod species (Table 1).

*Effect of bacterial diet on the population dynamics of* Amonardia normani

*First experiment*

A new culture of *A. normani* was initiated just before the experiment to avoid the possible negative effect of inbreeding on the population estimates. The results obtained for *A. normani* fed on *C. gracile* strain CE2201 were compared to those obtained for *A. normani* fed on the diatom *N. constricta*, to detect any experimental artifact. These latter results will be completely reported in a subsequent paper (Souza-Santos, in preparation).

At the start of the experiment two groups of 30 pregnant females of the first generation that were born in the laboratory were separated from the maintenance culture. Each group received a bacterial or a diatom suspension at concentrations which resulted in maximum egestion rates, i e., $2 \times 10^7$ cells $ml^{-1}$ for *C. gracile* (see Results) and 0.13 $\mu$g chlorophyll *a* $ml^{-1}$ (about $7 \times 10^5$ cells $ml^{-1}$) for *N. constricta* (Souza-Santos, in preparation). One day later, the hatched nauplii were placed individually in the cavities of sterile microplates, containing about 0.3 ml of the food suspension. To reduce evaporation, the cavities situated at the border of the microplates were filled with distilled water and the micro-plates were placed inside aquaria with a humid atmosphere. Two groups of 45 and 43 nauplii were fed on *N. constricta* and three groups of 76, 49 and 50 nauplii were fed on *C. gracile*.

All copepod stages were observed every day until they died. The criterion used to identify the dead animals was the absence of mobility after mechanical stimulation. Once and if the copepodite stage was attained, the animals were put together in a sterile glass dish containing 10 ml of the food suspension. After laying the first egg sac, each female was isolated with a male in a sterile plastic flask containing 10 ml of the food suspension. As soon as a new brood hatched, the adult couple was transferred to a new flask and the nauplii released were preserved in 4% v/v formalin and stained with Rose Bengal to be counted.

Food in microplates (nauplius development) was not renewed. During the copepodite stages the food suspension was renewed every two days and the living animals counted. During the reproductive period, the food suspension was changed every time a new brood was produced (between two and three days) or each five days for the post-reproductive period.

*Second experiment*

This short experiment was made in order to investigate the effect of the diet on the survival and fecundity (expressed in hatched nauplii) of adult *A. normani*, since the first experiment showed that the nauplii could not develop to adults when fed on *C. gracile*.

Six groups of about 30 females and 5 males were sampled in the same maintenance culture (fed on *N. constricta* at this time). Three groups were fed on *C. gracile* ($2 \times 10^7$ cells $ml^{-1}$) and the other ones fed on axenic *N. constricta* (0.13 $\mu$g chlorophyll-*a* $ml^{-1}$). The copepods fed on bacteria were acclimated on this food one day before the experiment.

Every day during one week, the living animals were counted and transferred to a new food suspension. The

*Table 1.* Results of the experiments dealing with the relationship between bacterial concentration (cells $ml^{-1}$) and egestion rate of copepods (No faecal pellets $cop^{-1}$ $day^{-1}$).

| *Amonardia normani* | | *Schizopera cf. compacta* | |
|---|---|---|---|
| Bacterial concentration (cells· $ml^{-1}$) | No. faecal pellets·$cop^{-1}$·$day^{-1}$ ± confidence intervals | Bacterial concentration (cells· $ml^{-1}$) | No. faecal pellets·$cop^{-1}$·$day^{-1}$ ± confidence intervals |
| 0 (control) | 0.75 ± 0.78 | 0 (control) | 1 ± 1.2 |
| $3 \times 10$ | 1 ± 0.71 | | |
| $3 \times 10^2$ | 0.8 ± 1.1 | | |
| $3 \times 10^3$ | 1.6 ± 1.5 | | |
| $3 \times 10^4$ | 3.8 ± 4.15 | | |
| $3 \times 10^5$ | 6.6 ± 7.06 | $2.6 \times 10^5$ | 1.2 ± 1.10 |
| $2 \times 10^6$ | 18.3 ± 9.4 | | |
| $3 \times 10^6$ | 25.4 ± 16.0 | $2.6 \times 10^6$ | 0.4 ± 0.55 |
| $1 \times 10^7$ | 96 ± 13.7 | | |
| $3 \times 10^7$ | 82.8 ± 22.3 | $2.6 \times 10^7$ | 12.4 ± 8.6 |
| $4 \times 10^7$ | 102.3 ± 42.6 | $5.3 \times 10^7$ | 12 ± 11.2 |
| $1 \times 10^8$ | 101.8 ± 42.7 | $1.3 \times 10^8$ | 20.4 ± 13.2 |
| $2 \times 10^8$ | 56.4 ± 19.4 | $2.6 \times 10^8$ | 22.2 ± 12.0 |

old suspensions were preserved in 4% v/v formalin and stained with Rose Bengal for further counting of hatched nauplii.

## Results

*Relationship between bacterial concentration and egestion rate of copepods*

The carbon content of *C. gracile* strain CE2201 cell was estimated as $5.19 \times 10^{-1}$ pg from the measure of cell volume (4.72 $\mu m^3$) and as $5.27 \times 10^{-1}$ pg by weighing the bacterial culture. The mean value of 0.52 pg C $cell^{-1}$ was thus used for the transformation of cell concentration in biomass.

The tested food items sedimented on the bottom of the flasks; therefore the copepods could feed on them without another kind of substrate.

Under the control conditions, without food, copepods normally produce from 0 to 3 typical faecal pellets per copepod per day (Table 1) or seldom many (about 20) very small, transparent pellets (not considered here). This faecal pellet production should be a result of the feeding activity in the maintenance cultures.

*A. normani* egestion rates were different from the control values from $3 \times 10^4$ cells $ml^{-1}$ onwards (Table 1). The number of faecal pellets produced and the mean pellet volume for *A. normani* increased significantly up to a concentration of $1 \times 10^7$ cells $ml^{-1}$ or 5.2 $\mu$g C $ml^{-1}$ (Table 1, Figure 1), from this concentration onwards the number of faecal pellets did not change significantly (ANOVA, F=2.2, d.f.=4/19, $p$=0.1095). The pellet volume, although, decreased significantly (ANOVA, F=4.43, d.f.=4/104, $p$=0.0024) from $4 \times 10^7$ cells $ml^{-1}$ to $10 \times 10^7$ cell $ml^{-1}$. Only from $2.6 \times 10^7$ cells $ml^{-1}$ and above did the number of faecal pellets produced by *S. cf. compacta* differ from control values (Table 1). The number of faecal pellets produced from $2.6 \times 10^7$ cells $ml^{-1}$ or 13.6 $\mu$g C $ml^{-1}$ onwards did not change significantly (Figure 2) (ANOVA, F= 1.0, d.f.=2/12, $p$=0.3961). The mean pellet volume did not change significantly (ANOVA, F=0.54, d.f.=3/98, $p$=0.66) between bacterial concentrations at least between $2.6 \times 10^7$ cells $ml^{-1}$ and $26 \times 10^7$ cells $ml^{-1}$.

*Effect of bacterial diet on the population dynamics of* Amonardia normani

*First experiment*

The nauplii of *A. normani* are strictly benthic and could feed on both *C. gracile* and *N. constricta*, based on the presence of many faecal pellets in cavities of the micro-plate.

The survival of *A. normani* was very low in the three groups of nauplii fed on *C. gracile* (Figure 3) compared with those fed on the diatom *N. constricta* (Figure 4). The lethal time for 50% of the population

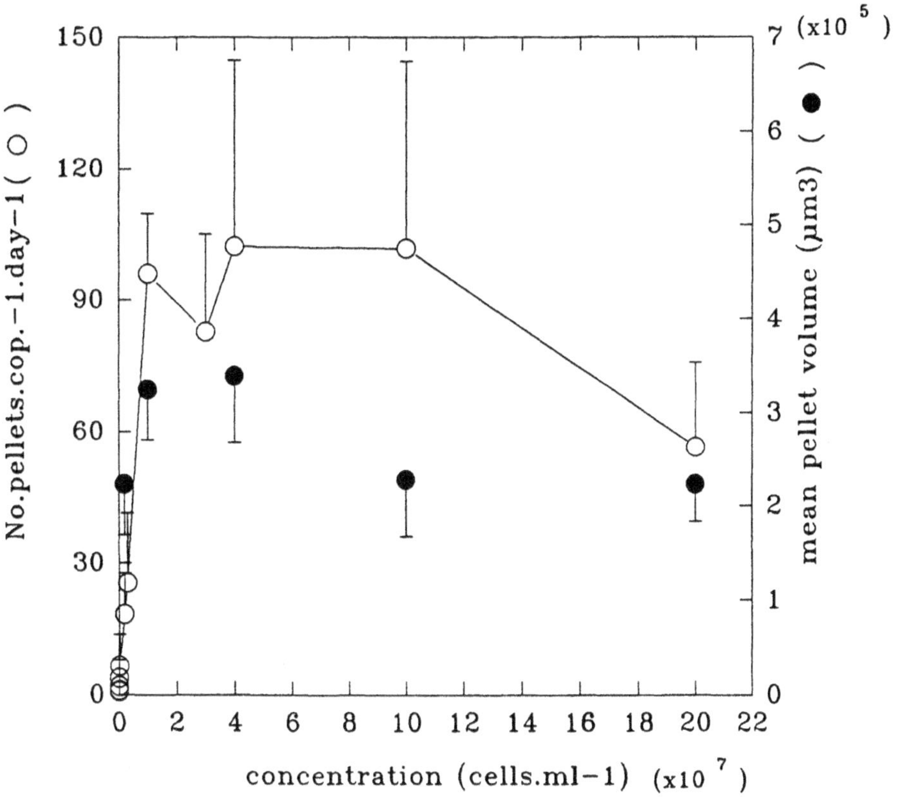


*Figure 1.* Variation of the number of faecal pellets produced (No. pellets. $cop^{-1}$ $day^{-1}$) and mean faecal pellet volume ($\mu m^3$) of *Amonardia normani* in relation to *Chromatium gracile* concentration (cells $ml^{-1}$). Vertical bars denote the confidence intervals of the means ($p = 0.05$).

ranged from 7 to 10 days for the bacterial diet and was about 37 days for the diatom diet. Only in one group fed on bacteria, nauplii developed to copepodites (C 5 stage) but no adults were observed. *A. normani* presented a complete life-cycle on the diatom diet with an intrinsic rate of increase ($r_o$) of 0.3 (Souza-Santos, in preparation).

*Second experiment*

The survival of adult *A. normani* fed during one week on *N. constricta* (95%) or on *C. gracile* (87%) was not significantly different (Student's *t*-test, $t = 1.86$, d.f. = 4, $p = 0.14$) at the end of the experiment (Figure 5). On the other hand, fecundity showed different variations through time. In the *N. constricta* diet the fecundity increased significantly with time (regression analysis – $R^2 = 76\%$, $p = 0.01$). On the *C. gracile* diet the fecundity decreased significantly with time (regression analysis – $R^2 = 60\%$, $p = 0.04$). At the end of the experiment the fecundity was significantly different (Student's *t*-test, $t = 22.4$, d.f. = 4, $p < 0.0001$) between the diatom diet and the bacterial diet: 13.8 and 0.41 nauplii · $female^{-1}$ · $day^{-1}$, respectively.

## Discussion

Data on the effect of food concentration on the physiological rates of meiobenthic copepods are scarce. The results of Heinle et al. (1977) indicated that *Coullana canadensis* produced more eggs at higher algal concentration. Rieper (1978, 1985) and Rieper & Flotow (1981) showed that the feeding rates of both *Tisbe holothuriae* and *Paramphiascella vararensis* increased with food (bacteria, algae and ciliate) concentrations. *Heteropsyllus pseudonunni* had higher egg production at increasing algal concentration but not at increasing detritus concentration (Ustach, 1982). However, these studies did not test a large range of food concentration and did not discuss the existence of a 'critical' food concentration above which the physiological rates (such as feeding and egg production) are independent of the food concentration. Only for *Coullana*

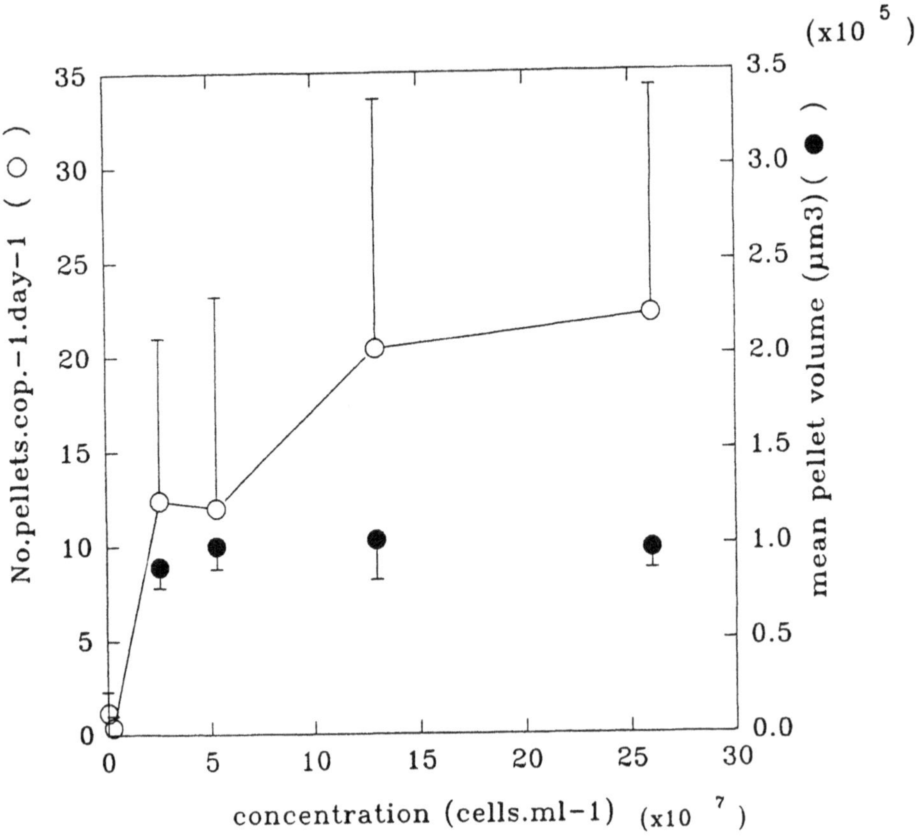


*Figure 2.* Variation of the number of faecal pellets produced (No. pellets. $cop^{-1}$ $day^{-1}$) and mean faecal pellet volume ($\mu m^3$) of *Schizopera cf. compacta* in relation to *Chromatium gracile* concentration (cells $ml^{-1}$). Vertical bars denote the confidence intervals of the means ($p = 0.05$).

*canadensis* it has been described a saturation of fertility (Harris, 1977) and feeding rates (Lonsdale & Levinton, 1989) above $1 \times 10^5$ cells $ml^{-1}$ and $2.5 \times 10^5$ cells $ml^{-1}$ of *Isochrysis galbana*, respectively. More recently, Shaw et al. (1994) presented the effect of *Thalassiosira pseudonana* concentration on the feeding rate, expressed as the number of faecal pellets produced per hour, of *Tigriopus californicus*. In this work, the feeding response was well-described by a Michaelis-Menten equation ($F = Fmax \times C\ /(K + C)$) and the 'critical' food concentration was $1 \times 10^5$ cells $ml^{-1}$. *Tisbe battagliai* showed a constant development time above the concentration of 1.3 $\mu g$ C $ml^{-1}$ or $2 \times 10^5$ cells $ml^{-1}$ of *I. galbana* (Williams & Jones, 1994). Montagna et al. (1995) could fit a functional model ($F = Fmax\ (1 - e^{-kC})$) to the meiofauna feeding response to algal concentration in sediments.

The present study confirms that the feeding rates, expressed in number of faecal pellets produced per day, of *Amonardia normani* and *Schizopera cf. compacta* followed the expected functional response as a function of bacteria concentration, with the 'critical' bacterial

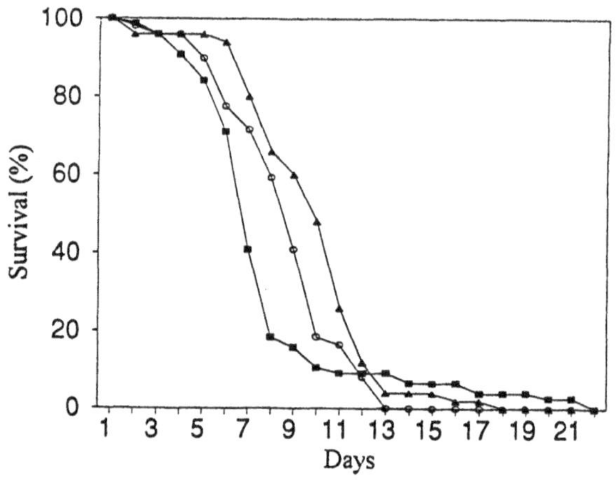


*Figure 3.* Survival curves of the three groups of *Amonardia normani* fed on *Chromatium gracile*, a phototrophic sulfur bacteria, at the concentration of $2 \times 10^7$ cells $ml^{-1}$ (n = 76, 49 and 50).

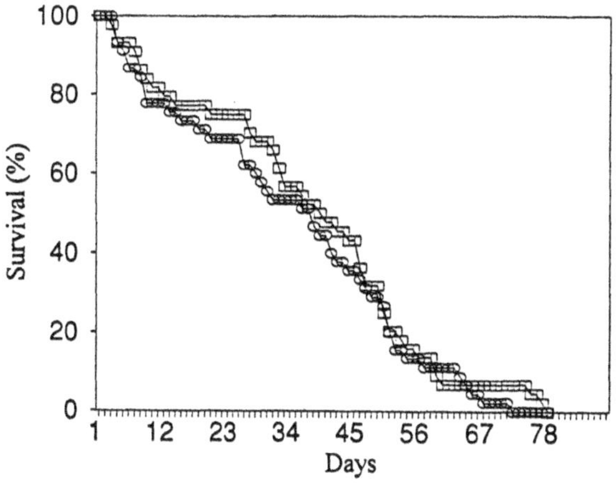


*Figure 4.* Survival curves of the two groups of *Amonardia normani* fed on *Nitzschia constricta*, a diatom, at the concentration of 0.13 μg Chla $ml^{-1}$ (n = 45 and 43).

concentrations of $1 \times 10^7$ cells $ml^{-1}$ (5.2 μg C $ml^{-1}$) and $2.6 \times 10^7$ cells $ml^{-1}$ (13.6 μg C $ml^{-1}$), for each species respectively.

As proposed by Frost (1972) and Mullin et al. (1975) three models were adjusted to the results obtained for *A. normani* the Ivlev equation as modified by Parsons et al. (1967): $F = Fmax\ [1 - e^{-k(C-C')}]$, where F is the feeding rate, Fmax the maximum rate, K a constant, C the food concentration and C′ the food concentration (if any) at which feeding ceases, the analogous Michaelis-Menten equation: $F = Fmax\ (C-C')/[K+(C-C')]$, where K is the half-saturation constant, and two rectilinear models, one up to the 'critical' concentration but not including it and another from this one onwards. The three models fitted significantly the data and can be observed in Figure 6. The rectilinear models explained more of the data variability, especially in the bacterial concentrations below the 'critical' one (n = 5, $R^2 = 98\%$, $p = 0.0009$) (Figure 6b) and showed the independency of feeding rates in relation to the bacterial concentration from the 'critical' concentration onwards (n = 5, $R^2 = 55\%$, $p = 0.15$). The Ivlev model (n = 10, $R^2 = 87\%$, $p = 0.00008$) and Michaelis-Menten model (n = 10, $R^2 = 83\%$, p = 0.0002) also explained a great part of the data variability. The same tendency on data explicability was observed by Mullin et al. (1975) using Frost (1972)'s results for *Calanus*. We did not attempt to fit any model to the *S. cf. compacta* results since only a few number of different bacterial concentrations were tested and almost all of them were higher than the 'critical' one (Figure 2).

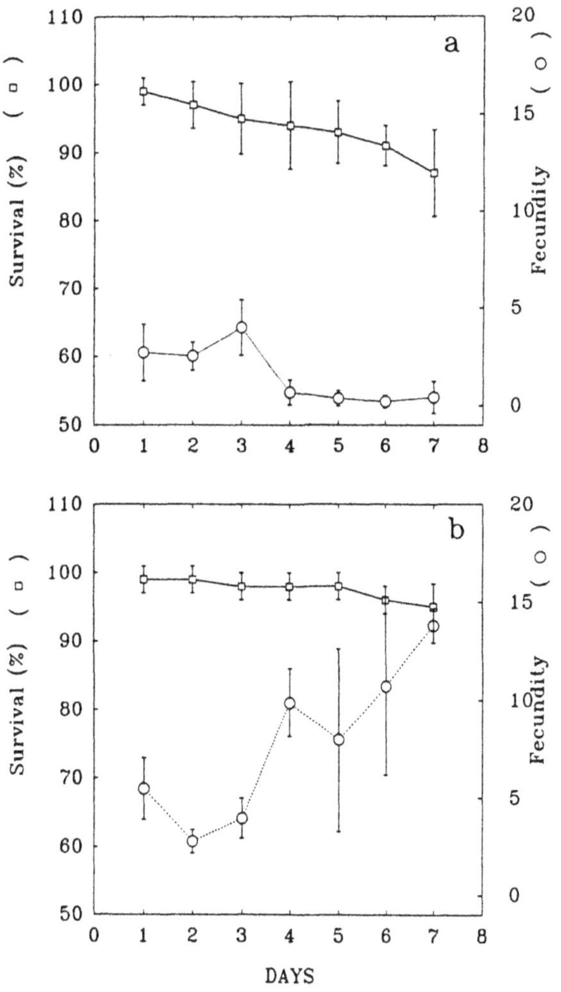


*Figure 5.* Percentage of survival and fecundity, expressed in hatched nauplii. $female^{-1}$ $day^{-1}$, of three groups (each one composed of 30 females and 5 males) of *Amonardia normani* adults fed on *Chromatium gracile* ($2 \times 10^7$ cells $ml^{-1}$) (a) and on *Nitzschia constricta* (0.13 μg Chla $ml^{-1}$) (b).

In view of practical utilization, it is of utmost importance to estimate the 'critical' food concentration. Studies on food preference, effect of the diet on the physiological rates and measures of carbon utilization should use food concentrations higher than the 'critical' one in order to avoid misinterpretation of the results. Montagna (1995) re-analysed the data of Vanden Berghe & Bergmans (1981) and proposed that the intrageneric differences on feeding rate and food preference of *Tisbe* spp., as suggested by Vanden Berghe & Bergmans, could be simply explained by the differences in the food concentrations tested for each

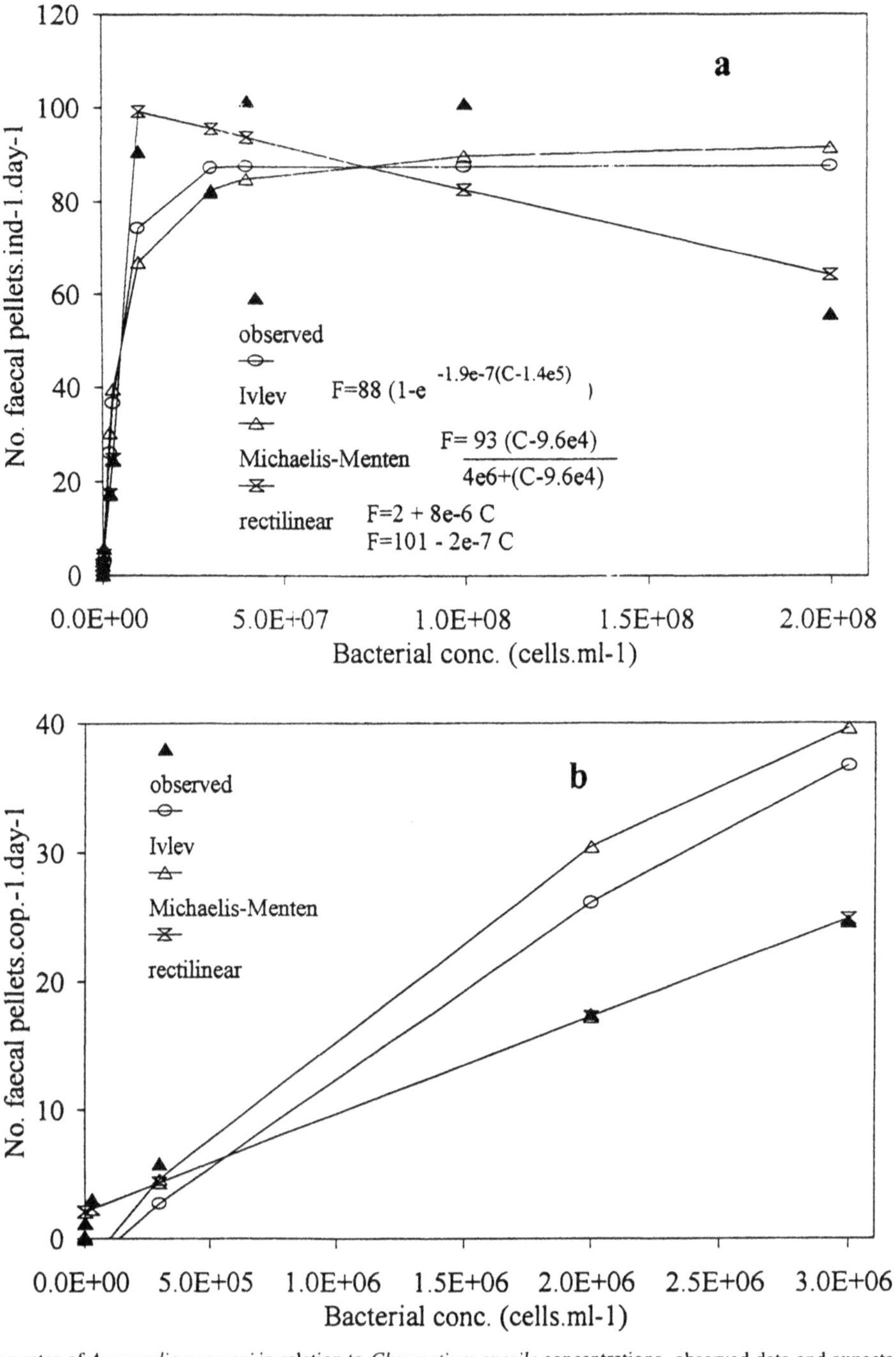


*Figure 6.* Feeding rates of *Amonardia normani* in relation to *Chromatium gracile* concentrations, observed data and expected curves estimated by the Ivlev model, Michaelis-Menten model and the rectilinear model. a) all data, b) bacterial concentrations below the 'critical' one.

species and a functional response to these concentrations. However, Montagna plotted a relative algal feeding rate ($\times 10^{-4}$ h$^{-1}$) versus the total carbon added (see his Figure 1). This relative rate, although interesting to measure meiofauna grazing impact on food stocks, is worthless in physiological studies where it is preferable to express the feeding rate by individual or by body biomass per unit of time. Thus, we have replotted the results of Vanden Berghe & Bergmans (1981), now with the algal feeding rate expressed in $\mu$g C. ind$^{-1}$ day$^{-1}$ versus the total carbon added or the algal carbon added (Figure 7a and b). The linear regression

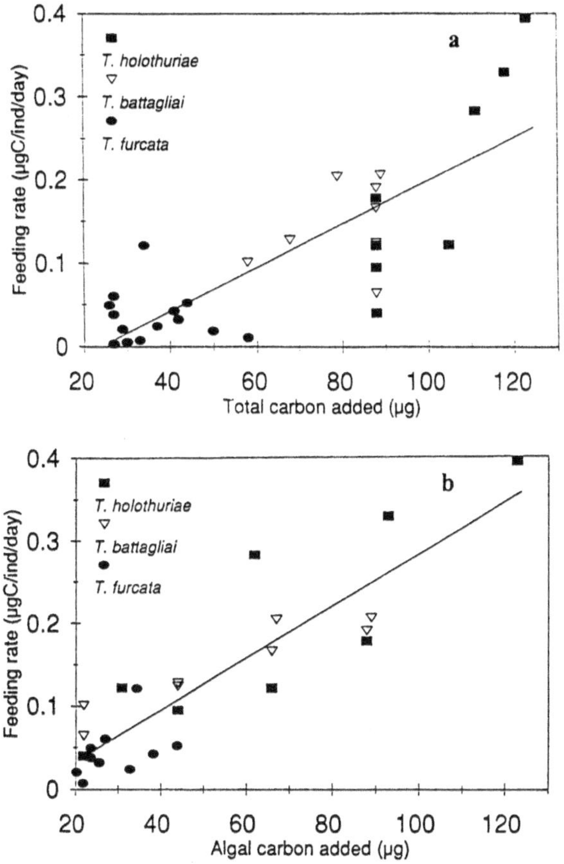


*Figure 7.* Vanden Berghe & Bergmans (1981) results of algal feeding rates (expressed in $\mu g$ C. $ind^{-1}$ $day^{-1}$) of three species of the genus *Tisbe* versus: (a) total carbon added and (b) algal carbon added.

between these two variables was significant (total carbon – $R^2 = 65\%$, $p<0.0001$, algal carbon – $R^2 = 82\%$, $p<0.0001$) although the algal carbon explained better the algal feeding rate, as expected.

The 'critical' bacterial concentrations, even expressed in carbon, observed for the two tested copepod species were quite similar to each other although much higher than critical algal concentrations estimated for other studied meiobenthic copepod species. This result can be explained by the differences of volume between the bacteria tested (4.72 $\mu m^3$) and the algae (about 40 $\mu m^3$ for *I. galbana* and 50 $\mu m^3$ for *T. pseudonana*) and the inverse relationship between the cell volume of the food item and its critical carbon concentration (Frost, 1972). Schiemer et al. (1980) and Schiemer (1982) found that growth of two nematode species fed on *Escherichia coli* could be effective with at least bacterial concentrations of about $10^8$ cells $ml^{-1}$, this higher bacterial concentration can be explained once again by the small size of *E. coli* cells. The phototrophic bacterial densities in the sediments of the fish ponds of Certes were higher than $10^7$ cells $ml^{-1}$ during the three sampling periods (March, June and September 1993) (Guyoneaud et al., 1996). Caumette (1986) observed phototrophic bacteria densities of $10^7$ cells $ml^{-1}$ in the sediment of Prevost Lagoon during the summer and in the water column the densities attained $10^8$ cells $ml^{-1}$ during 'red waters'. These results indicate that the 'critical' phototrophic bacterial concentrations estimated in the present study can be found by meiobenthic copepods in the field, making possible the exploitation of this food at maximum rates in natural conditions.

Variation of the faecal pellet volume as a function of bacterial concentration differed for each copepod species. *S. cf. compacta* faecal pellet volume did not change significantly between the tested concentrations (Figure 2), for *A. normani* the faecal pellet volume increased up to the 'critical' concentration, did not change significantly up to $5.3 \times 10^7$ cells $ml^{-1}$ and decreased at $10 \times 10^7$ cells $ml^{-1}$ (Figure 1). This faecal pellet volume decrease together with the small (non-significant) decrease of the number of faecal pellets produced by copepod per day (Figure 1) lead us to suggest that the real egestion rate of *A. normani* tends to decrease at the highest bacterial concentration, a non-expected result in view of the theoretical models proposed above. The decrease of feeding rate of copepods above 'critical' concentrations was already observed (Mullin, 1963; Conover, 1966) but not well understood. A hypothesis may be the production of chemical feeding deterrents by the food organisms that can inhibit copepod feeding above a specific food concentration. The diatom *Phaeodactylum tricornutum* deterred feeding of *Tigriopus californicus* at concentrations of $6 \times 10^5$ cells $ml^{-1}$, a characteristic already known for some dinoflagellate species (Shaw et al., 1994). This decrease may be also an experimental artifact and more experiments are needed to confirm it. For this reason, however, the experiments on *A. normani* population dynamics used the concentration of $2 \times 10^7$ cells $ml^{-1}$ at which the total egestion rate was almost constantly high.

The results of the experiment on population dynamics clearly showed that *A. normani* population could not grow with *C. gracile* strain CE2201 as the single food source, since the nauplii did not develop to adults. To make sure that these results were not a methodological artifact related to the culture of nauplii in micro-plates (restricted volume), six other groups of nauplii (n = 30) were cultivated in glass dishes containing 10 ml of the bacterial suspension. The nauplii were observed (though not counted) every day until they died, but development to copepodites was not observed.

The adults that fed on *C. gracile* had similar survival rates to those fed on diatoms and were able to produce nauplii, although in smaller number than in the diatom diet (Figure 5). The increase of fecundity observed in the diatom diet during the experiment can be a result of the lower density of copepods in the experimental dishes as compared to the maintenance dishes (Fava & Crotti, 1979) or can be due to a higher food supply (in the maintenance dishes the food concentration was not controlled and was renewed only once a week). The decrease of fecundity observed in the bacterial diet can represent a delay of the response to diet change, meaning that the first broods could still be related to the diatom diet in the maintenance cultures.

Caumette et al. (1983) demonstrated, by means of *in situ* gut-content analyses, that *Acartia clausi* fed on phototrophic bacteria (*Rhodopseudomonas* sp. and *Chromatium* sp.) present in the water column of a tropical lagoon. Laboratory experiments also showed that this copepod could lay eggs when feeding on three different strains of phototrophic bacteria, indicating assimilation. Nevertheless, egg production was lower than when the copepods were fed on algae (Caumette, 1985). Rieper (1984) tested *Chromatium vinosum* and *Rhodopseudomonas palustris* as food for *Tisbe holothuriae* and *Paramphiascella vararensis*. For *P. vararensis* both phototrophic bacteria represented adequate food resource for its complete life-cycle. In the case of *T. holothuriae*, a species known to eat on several bacteria strains (Rieper, 1978; Guérin & Rieper-Kirchner, 1991), the *C. vinosum* diet resulted in no egg production and the *R. palustris* diet in an incomplete life-cycle. Decho & Castenholz (1986) proposed that *Thiocapsa* sp (even with sulfur globules) is an important food resource for *Mesochra lilljeborgi* based on distributional data and laboratory feeding experiments.

In conclusion, the meiobenthic copepods *Amonardia normani* and *Schizopera cf. compacta* can ingest the phototrophic bacteria *Chromatium gracile*. Their higher feeding rates were observed from the bacterial concentration of $10^7$ cells $ml^{-1}$ onwards. Nevertheless, it can be suggested that adults of the opportunistic species *Amonardia normani* entering shallow coastal lagoon from the sea during blooms of phototrophic sulfur bacteria will produce a small number of nauplii that will not develop, making it difficult or even impossible the population growth and colonization of this lagoon, unless another food resource is available. Other meiobenthic copepods, such as *P. vararensis* (Rieper, 1984), can grow using phototrophic bacteria as sole food source and may colonize lagoons during 'red waters'. On the other hand, the utilization of the phototrophic sulfur bacteria as a food complement for both tested copepods cannot be discarded.

## Acknowledgments

The authors are grateful to P. Caumette and R. Guyoneaud for providing the original culture of *Chromatium gracile* strain CE2201 isolated from Certes sediments and their suggestions during all the work. Sincere thanks to all staff of Prof. P. Caumette for practical help during the preparation of bacterial culture media. The authors thank G. Bachelet, M. Kirchner and V. U. Ceccherelli for critical reading of the manuscript. LPSS and PJPS acknowledge a CAPES postgraduate studentship (Brazil). This study was partly supported by a CEC Environment programme (CLEAN contract EV5V – CT92 – 0080).

## References

Bougis, P., 1974. Ecologie du plancton marin. Tome II – Le zooplancton. Masson, Paris, 200 pp.

Castel, J., 1979. Adaptation and reproductive cycle of the harpacticoid copepod *Amonardia normani* (Brady, 1872) in semi-enclosed lagoons of Arcachon Bay, France. In E. Naylor and G. Hartnoll (eds), Cyclic phenomena in marine plants and animals. Pergamon Press, Oxford and New York: 131–138.

Castel, J., 1992. The meiofauna of coastal lagoon ecosystems and their importance in the food web. Vie Milieu 42: 125–135.

Castel, J. & P. Lasserre, 1977. Colonisation et distribution spatiale des copépodes dans des lagunes semi-artificielles. In B. F. Keegan, P. O. Ceidigh & P. J. S. Boaden (eds), Biology of benthic organisms. Pergamon Press Oxford and New York: 129–146.

Caumette, P., 1985. Développement des bactéries phototrophes et sulfato-rédutrices dans des lagunes peu profondes et des lagunes stratifiées. Etude de leur rôle dans le cycle du soufre et dans la production de biomasse. Doctoral Thesis. University of Aix-Marseille III. France, 325 pp.

Caumette, P., 1986. Phototrophic sulfur bacteria and sulfate-reducing bacteria causing red waters in a shallow brackish coastal lagoon (Prévost Lagoon, France). FEMS Microbiol. Ecol. 38: 113–124.

Caumette, P., 1992. Bacterial communities in coastal lagoons. An overview. Vie Milieu 42: 111–123.

Caumette, P., M. Pagano & L. Saint-Jean, 1983. Répartition verticale du phytoplancton, des bactéries et du zooplancton dans un milieu stratifié en Baie de Biétri (Lagune Ebrié, Côte d'Ivoire). Relations trophiques. Hydrobiologia 106: 135–148.

Conover, R. J., 1966. Factors affecting the assimilation of organic matter by zooplankton and the question of superfluous feeding. Limnol. Oceanogr 11: 1066–1071.

Decho, A. W. & R. W. Castenholz, 1986. Spatial patterns and feeding of meiobenthic harpacticoid copepods in relation to resident microbial flora. Hydrobiologia 131: 87–96.

Fava, G. & E. Crotti, 1979. Effect of crowding on nauplii production during mating time in *Tisbe clodiensis* and *T. holothuriae* (Copepoda, Harpacticoida). Helgoländer wiss. Meeresunters. 32: 466–475.

Frost, B. W., 1972. Effects of size and concentration of food particles on the feeding behavior of the marine planktonic copepod *Calanus pacificus* Limnol. Oceanogr. 17: 805–815.

Giere, O., 1993. Meiobenthology. The microscope fauna in aquatic sediments. Springer-Verlag, Berlin, 328 pp.

Guérin, J.-P. & M. Rieper-Kirchner, 1991. Influence of three bacteria strains on the population dynamics of *Tisbe holothuriae* (Copepoda, Harpacticoida). Helgoländer. wiss. Meeresunters. 45: 493–511.

Guyoneaud, R., R. Matheron, R. Baulaigue, K. Podeur, A. Hirschler & P. Caumette, 1996. Anoxygenic phototrophic bacteria in eutrophic coastal lagoons of the french mediterranean and atlantic coasts (Prévost Lagoon, Arcachon Bay). Hydrobiologia 329 (Dev. Hydrobiol. 117): 33–43.

Harris, R. P., 1977. Some aspects of the biology of the Harpacticoid Copepod, *Scottolana canadensis* (Willey), maintained in laboratory culture. Chesapeake Sci. 18: 245–252.

Heinle, D. R., R. P. Harris, J. F. Ustach & D. A. Flemer, 1977. Detritus as food for estuarine copepods. Mar. Biol. 40: 341–353.

Karl, D. M., 1986. Determination of in situ microbial biomass, viability, metabolism, and growth. In J. S. Pointdexter & E. R. Leadbetter (eds), Bacteria in nature (Vol. 2) Methods and special applications in bacterial ecology. Plenum Press, New York – London: 85–176.

Lonsdale, D. J. & J. S. Levinton, 1989. Energy budgets of latitudinally separated *Scottolana canadensis* (Copepoda: Harpacticoida). Limnol. Oceanogr. 34: 324–331.

Montagna, P. A., 1995. Rates of metazoan meiofaunal microbivory: a review. Vie Milieu 45: 1–9.

Montagna, P. A. & G. F. Blanchard & A. Dinet, 1995. Effect of production and biomass of intertidal microphytobenthos on meiofaunal grazing rates. J. exp. mar. Biol. Ecol. 185: 149–165.

Mullin, M. M., 1963. Some factors affecting the feeding of marine copepods of the genus *Calanus*. Limnol. Oceanogr. 8: 239–250.

Mullin, M. M., E. F. Stewart & F. J. Fuglister, 1975. Ingestion by planktonic grazers as a function of concentration of food. Limnol. Oceanogr. 20: 259–262.

Parsons, T. R., R. J. LeBrasseur & J. D. Fulton, 1967. Some observations on the dependence of zooplankton grazing on the cell size and concentration of phytoplankton blooms. J. Oceanogr. Soc. Jap. 23: 10–17.

Pfennig, N. & H. G. Trüper, 1981. Isolation of members of the families Chromatiaceae and Chlorobiaceae. In Starr M. P., Stolp H., Trüper H. G., Balows A., Schlegel H. G. (eds), The Prokaryotes. Springer Verlag, Berlin: 279–289.

Rieper, M., 1978. Bacteria as food for marine harpacticoid copepods. Mar. Biol. 45: 337–345.

Rieper, M., 1982. Feeding preferences of marine harpacticoid copepods for various species of bacteria. Mar. Ecol. Prog. Ser. 7: 303–307.

Rieper, M., 1984. Relationships between bacteria and marine copepods. In: Bactériologie Marine. CNRS, Paris: 169–172.

Rieper, M., 1985. Some lower food web organisms in the nutrition of marine harpacticoid copepods: an experimental study. Helgoländer. wiss. Meeresunters. 39: 357–366.

Rieper, M. & C. Flotow, 1981. Feeding experiments with bacteria, ciliates and harpacticoid copepods. Kieler Meeresforsch., Sonderh. 5: 370–375.

Schiemer, F., 1982. Food dependence and energetics of free living nematodes. I. Respiration, growth and reproduction of *Caenorhabditis briggsae* (Nematoda) at different levels of food supply. Oecologia (Berl.) 54: 108–121.

Schiemer, F., A. Duncan & R. Z. Klekowski, 1980. A bioenergetic study of a benthic nematode, *Plectus palustris* de Man 1881, throughout its life cycle. II. Growth, fecundity and energy budgets at different densities of bacterial food and general ecological considerations. Oecologia (Berl.) 44: 205–212.

Shaw, B. A., P. J. Harrison & R. J. Andersen, 1994. Evaluation of the copepod *Tigriopus californicus* as a bioassay organism for the detection of chemical feeding deterrents produced by marine phytoplankton. Mar. Biol. 121: 89–95.

Souza-Santos, L. P., J. Castel & P. J. P. dos Santos, 1995. Feeding rate cycle of the epibenthic harpacticoid copepod *Harpacticus flexus*: Laboratory experiments using faecal pellets counts. Vie Milieu 45: 75–83.

Ustach, J. F., 1982. Algae, bacteria and detritus as food for the harpacticoid copepod, *Heteropsyllus pseudonunni* Coull and Palmer. J. exp. mar. Biol. Ecol. 64: 203–214.

Vanden Berghe, W. & M. Bergmans, 1981. Differential food preferences in three co-occurring species of *Tisbe* (Copepoda, Harpacticoida). Mar. Ecol. Prog. Ser. 4: 213–219.

Williams, T. D. & M. B. Jones, 1994. Effects of temperature and food quantity on postembryonic development of *Tisbe battagliai* (Copepoda: Harpacticoida). J. exp. mar. Biol. Ecol. 183: 283–298.

# PART II

## Eutrophication effects on biogeochemistry of nitrogen and sulfur in coastal lagoons

*Hydrobiologia* **329**: 93–103, 1996.
*P. Caumette, J. Castel & R. Herbert (eds), Coastal Lagoon Eutrophication and ANaerobic Processes (C.L.E.AN.).*



# Growth of the seaweed *Ulva rigida* C. Agardh in relation to biomass densities, internal nutrient pools and external nutrient supply in the Sacca di Goro lagoon (Northern Italy)

Pierluigi Viaroli[1], Mariachiara Naldi[1], Cristina Bondavalli[1] & Silvano Bencivelli[2]
[1]*Dipartimento di Scienze Ambientali, Università di Parma, viale delle Scienze, 43100 Parma, Italy*
[2]*Amministrazione Provinciale di Ferrara, Servizio Ambiente, 44100 Ferrara, Italy*

*Key words: Ulva rigida*, growth rates, nitrogen, phosphorus

## Abstract

Growth of the seaweed *Ulva rigida* C. Agardh was investigated in relation to biomass densities, internal nutrient pools and external nutrient supply. Research was carried out from 23 March to 5 July 1994 in the Sacca di Goro (Po Delta, Northern Italy), whose south-eastern part was covered by extensive mats of *Ulva rigida*. Two types of field experiments were conducted by incubating *Ulva* thalli inside large cages. In the first experiment, beginning on 23 March, 100 g of wet thalli were placed into the cages, allowed to grow for two weeks, then collected and replaced. This procedure was repeated 8 times over the study period. In the second experiment, *Ulva* thalli were left inside the cages and collected at selected time intervals (14, 27, 41, 64 and 76 days) in order to simulate the effects of increased density on growth and nutrient storage.

We recorded specific growth rates (NGR) ranging from 0.025 to 0.081 $d^{-1}$ for a period up to two months in the repeated short-term experiments performed at relatively low initial algal densities (300–500 g AFDW $m^{-3}$). These NGR resulted significantly related to dissolved inorganic nitrogen (DIN) in the water column. Tissue concentrations of total Kjeldahl nitrogen (TN) were almost constant, while extractable nitrate decreased in a similar manner to DIN in the water column. Total phosphorus showed considerable variation, probably linked to pulsed freshwater inflow.

In the long-term incubation experiment, NGR of *Ulva* was inversely related to density. Internal concentrations of both total P and TN reached maximum values after one month; thereafter P concentration remained almost constant, while TN decreased below 2% w/w (by dry weight). The TN decrease was also accompanied by an abrupt decrease in nitrate tissue concentration. The biomass incubated over the two month period suffered a progressive N limitation as shown by a decreasing N:P ratio (49.4 to 14.6). The reciprocal control of *Ulva* against biogeochemical environment and viceversa is a key factor in explaining both resource competition and successional stages in primary producer communities dominated by *Ulva*. However, when the biomass exceeds a critical threshold level, approximately 1 kg AFDW $m^{-3}$, the macroalgal community switches from active production to rapid decomposition, probably as a result of selfshading, biomass density and development of anaerobic conditions within the macroalgal beds.

## Introduction

In recent years there has been an increased production of large seaweeds in shallow coastal waters. These macroalgae are mostly ephemeral Chlorophyceans, able to survive in fluctuating environments because of their ability to rapidly take-up and store nutrients. The wide spread distribution of large opportunistic seaweeds has been recognized to depend on their growth strategy, *viz.* the species-specific growth rate and its seasonal timing (Rosenberg & Ramus, 1981; Luning, 1993). Moreover, macroalgal growth is modulated by factors such as light, nutrient availability, and temperature, although in subtidal environments the period of maximum growth appears to be determined mostly by nitrogen availability (Lapointe & Tenore, 1981; Duke et al., 1989a,b; Fujita et al., 1989; Sand-Jensen, 1991; Couthino & Zingmark, 1993). Compared with ammonium, little work has focused on the effects of nitrate upon growth and biochemical composition of Chlorophyceans (Lapointe & Tenore, 1981; O'Brien & Wheeler, 1987; Corzo & Niell, 1992). However, it has been proposed that nitrate is the main macronutrient affecting the growth cycle of other macroalgae (Chapman & Craigie, 1977; Da Costa Braga & Yoneshigue-Valentin, 1994). Several Phaeophyceans may accumulate nitrate in vacuoles to very high concentrations thereby gaining an additional nitrogen pool in addition to amino acids and proteins (Chapman & Craigie, 1977; DeBoer, 1981).

The excessive growth of green macroalgae (green tides) is causing severe dystrophic crises in a number of lagoons located along the northern Mediterranean coast (Amanieu et al., 1975; Izzo & Hull, 1991; Sfriso et al., 1992). Amongst these, the Sacca di Goro has been experiencing summer dystrophic episodes due to *Ulva rigida* C. Ag., whose excessive growth and decomposition has been related to nitrate availability (Viaroli et al., 1992; 1993; 1995).

The data presented in this paper are part of a long-term research project on the eutrophication of the Sacca di Goro. We report field experiments which simulate natural growth conditions of *Ulva*. The experiments aimed (1) to analyze biomass parameters of *Ulva* as related to biomass densities, internal nutrient pools and external nutrient supply and (2) to evaluate the ability of *Ulva* to store nutrients over short and long time periods. The latter objective is especially relevant in order to assess to what degree *Ulva* may act as a nutrient sink.

## Study area

A typical estuarine environment of the Po Delta is represented by shallow-water embayments locally called 'sacca'. The Sacca di Goro lies south of the Po di Goro, the southernmost deltaic branch, at approximately 44°47′–44°50′ N and 12 °15′–12°20′ E. The lagoon has an area of 26 km$^2$ and an average depth of about 1.5 m; it is connected to the sea by a 1.5 km-wide mouth (Figure 1). The bottom is flat and the sediment is composed of typical alluvial mud with a high clay and silt content in the northern and central zones; sand is more abundant near the southern shoreline, while sandy-mud prevails in the eastern area (Dal Cin & Pambianchi, 1991). Freshwater flows into the northeastern part of the lagoon from the Po di Goro, while the western and central zones are affected mainly by the Po di Volano canal. In 1992, the Po di Volano canal discharged about 3100 tons of nitrate, 560 tons of ammonium, and 46 tons of total phosphorus into the Sacca di Goro (Alvisi et al., 1993).

## Materials and methods

Research was carried out from 23rd March to 5th July 1994 in the southeastern part of the lagoon where the mean water depth is 1.2 m (station M, Figure 1). Two types of experiments were conducted incubating *Ulva* thalli inside cylindrical enclosures positioned at a depth of 0.5 – 0.7 m in the lagoon. Enclosures were built using a plastic fencing net (mesh = 10 mm) to obtain cages with no limits to water exchanges.

### *Field observations*

In addition to the experiments, three to six macroalgal samples were collected around the experimental site within a 0.5 km$^2$ area. The macroalgal biomass was harvested with a rake-like device over an area ranging from 2 to 10 m$^2$.

Commencing on the 23rd March, water samples were collected approximately twice a week at site M at a depth of about 0.5 m with a Ruttner bottle. Nitrite was determined spectrophotometrically after diazotation; nitrate was measured as nitrite after cadmium reduction (APHA, 1975); ammonium was determined by the indophenol-blue method (Koroleff, 1970) and soluble reactive phosphorus (SRP) by the ascorbic acid method (Valderrama, 1977).

Water temperature, salinity, and dissolved oxygen were recorded approximately every four hours from an *in situ* buoy equipped with specific probes (Hydronaut 601). Total daily irradiance was measured by hemispherical pyroheliometer at the meteorological station in Volano, near to the mouth of the Po di Volano (courtesy of the Ufficio Meteorologico Regionale, Regione Emilia Romagna).

### *Long-term experiment*

From 23rd March to 7th June macroalgal thalli were incubated in cylindrical enclosures in the field and collected at selected time intervals to simulate the effects

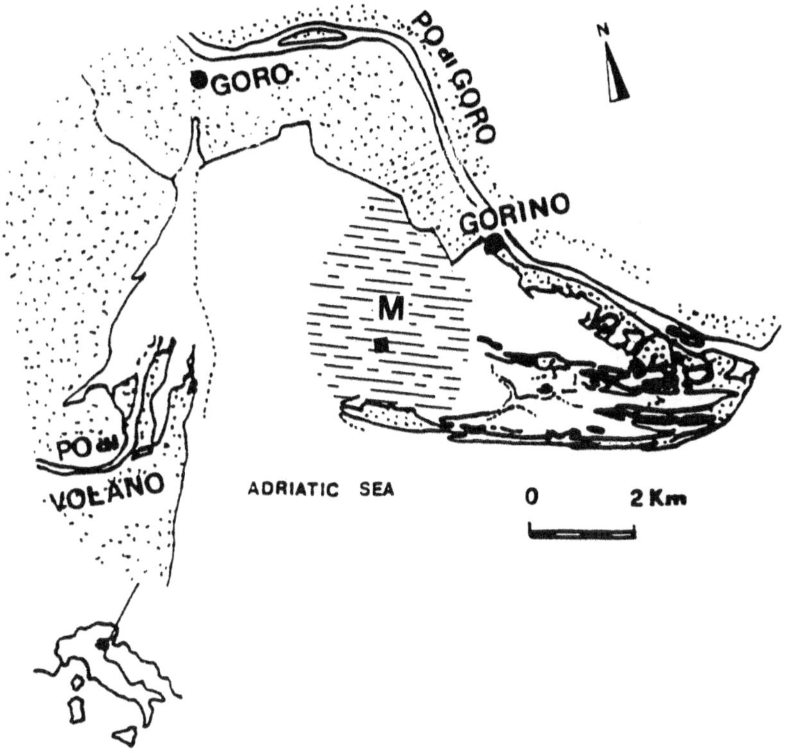


*Figure 1.* General layout of the Sacca di Goro, and position of the sampling site (st. M.). Shaded area indicates the maximum spreading of *Ulva* in June 1994.

of increased density upon growth and nutrient storage. On 23rd March, 25 cylindrical enclosures (mean ⌀ = 30 cm; height = 50 cm, mesh = 10 mm) were positioned at station M. *Ulva* thalli were collected around the experimental site, washed with sea water, chopped into small pieces (approximately 20 × 20 cm) and weighed. Five subsamples of the biomass were collected and stored in an ice box and transferred to the laboratory. Each cage was then filled with 100 g of wet *Ulva* thalli, corresponding to $10.8 \pm 1.1$ g dry weight. Afterwards, the cages were sampled on the 6th (the 14th day of incubation) and 19th April (the 27th day), on 3rd (the 41st day) and 26th (the 64th day) May and 7th June (the 76th day). Each time, five cages were sampled and not replaced. The harvested biomass was washed with sea water and kept in an ice box until transferred to the laboratory.

*Short-term experiments*

*Ulva* thalli were incubated over the short-term periods (approximately two weeks) in order to simulate the ephemeral behaviour of this seaweed. From 23rd March to 5th July, eight subsequent experiments were carried out incubating the macroalga inside cylindrical plastic enclosures (mean ⌀ = 30 cm; height = 35 cm, mesh = 10 mm). On 23rd March, five enclosures were positioned at the experimental site (station M, Figure 1). Each cage was filled with 100 g of wet *Ulva* thalli, as described for the long-term experiment. From then onwards, the same five cages were sampled and refilled approximately every two weeks: sampling dates (end of an incubation and beginning of the next one) were 6th and 19th April, 3rd, 10th and 26th May, 7th and 23rd June, and 5th July. Sampling of the cages and refilling were performed in a similar manner to that described for the long-term experiment.

*Analytical procedure for macroalgal samples*

In the laboratory *Ulva* thalli were cleaned to remove the epiphytes, rinsed with tap water to remove salt, and oven- dried at 70 °C to determine the dry weight (DW).

Subsamples ranging from 20 to 50 mg were analyzed for the determination of total nitrogen content using the Kjeldahl method (APHA, 1975) modified according to the TECATOR (LKB) user's manual.

Soluble inorganic nitrogen (nitrate, nitrite, and ammonium) content of the thalli was determined after extraction. Approximately 100 mg DW sample was placed into flask containing 50 ml of distilled water and shaken (Corzo & Niell, 1992). After two hours the previously described spectrophotometric measurements were performed on the filtered extract.

Approximately 1 g of dried thalli was ignited over 24 hours at 550 °C in a muffle furnace to determine the Ash Free Dry Weight (AFDW). Total phosphorus was determined by spectrophotometry after acid extraction of the ash (Aspila et al., 1976).

Net growth rates were estimated considering an exponential growth, according to the following equation: $B_t = B_o \cdot e^{kt}$ where $B_o$ = initial biomass (g AFDW), $B_t$ = biomass (g AFDW) after t days of incubation and k ($d^{-1}$) = net growth rate (DeBoer et al., 1978).

## Results

### *Field observations*

In the experimental area, *Ulva* growth started in early spring when temperature and light conditions became favourable (Figure 2). Afterwards, the biomass increased rapidly until June, reaching the highest net growth rate at the end of May (Figure 3). The growth period ceased at the end of June when a nearly continuous bed of *Ulva* extended as far as the centre of the lagoon. In the more sheltered areas, macroalgal biomass reached values of around 500 g AFDW $m^{-2}$, corresponding to a wet weight greater than 5 kg $m^{-2}$. If the actual thickness of the macroalgal bed (50–70 cm) is taken into account the density of *Ulva* is approximately 700–1000 g AFDW $m^{-3}$.

At the end of June, the *Ulva* biomass underwent a rapid decay which was coincident with a progressive decrease in dissolved oxygen concentration in the water column (Figure 2).

In the water column at station M, the most important form of dissolved inorganic nitrogen was nitrate, especially in April and to a lesser extent in May. Ammonium was almost always undetectable (Figure 2). Nitrate was in excess in April, and at the end of May. Soluble reactive phosphorus was undetectable at all sampling dates, as is usual in the spring phase of macroalgal growth (Viaroli et al., 1993).

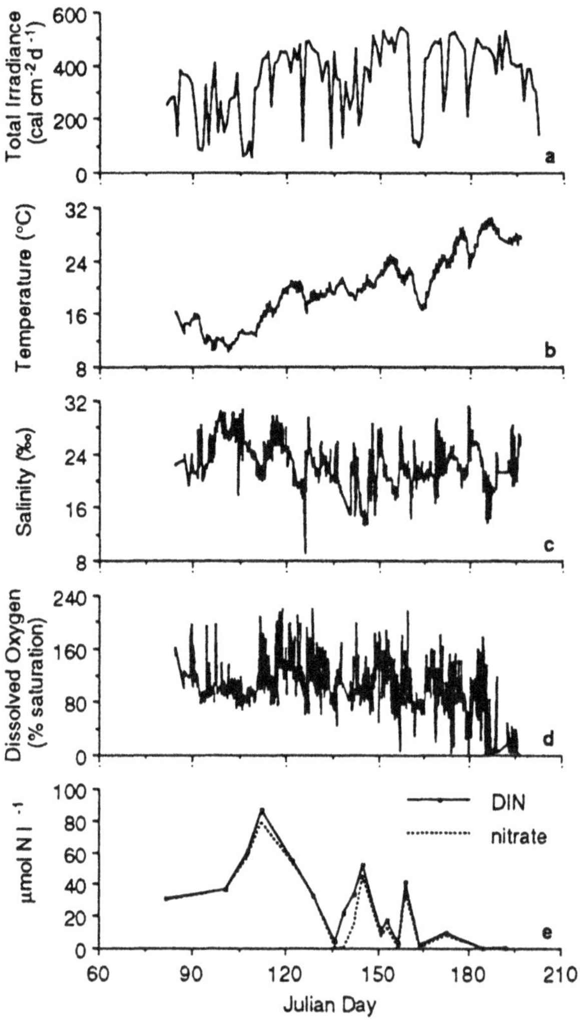


*Figure 2.* Total irradiance measured at the meteorological station in Volano (a), water temperature (b) and salinity (c), dissolved oxygen (d), dissolved inorganic nitrogen (DIN) and nitrate (e) concentrations in the water column at station M (23 March–15 July 1994).

Total nitrogen content in the *Ulva* thalli ranged from 2.5 to 3.0% DW, and was similar to the levels measured in previous years; minima of about 2.0% DW were also typical of the biomass immediately prior the decomposition phase (Viaroli et al., 1992, 1993). The increase in tissue-N which accompanied the decomposition of *Ulva* is probably dependent on the nitrogen enrichment of the decomposing biomass due to leaching of nitrogen-free compounds (Pugnetti et al., 1992) as well as growth of the decomposers. Total phosphorus content in *Ulva* biomass ranged from 0.11 to 0.24% DW and showed no well defined seasonal trend. However, the atomic N:P ratio in the macroalgal tissue (30.9 to

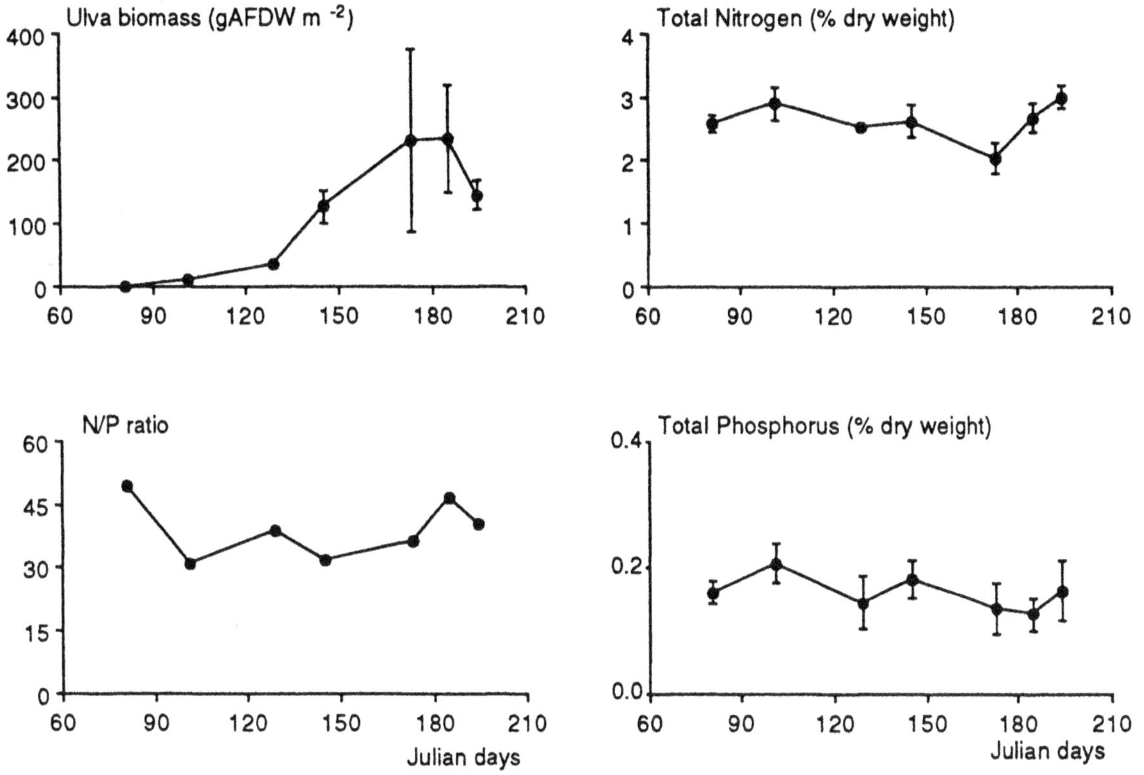


*Figure 3.* Free-floating *Ulva* biomass at site M and total Kjeldahl nitrogen, total phosphorus and N:P ratios (by atoms) in macroalgal tissue (23 March–15 July 1994). Bars indicate one standard deviation.

49.4) suggested that free-floating *Ulva* had a nutritional status which tended toward P-limitation rather than N- limitation (Atkinson & Smith, 1983; Lapointe et al., 1992).

*Long-term experiment*

Figure 4 shows the main results of the long-term experiment which simulated the natural growth of *Ulva* subjected to increased macroalgal density. From the 23rd March, the biomass incubated inside the enclosures underwent an almost linear increase that lasted approximately 60 days. During the growth phase the macroalgal biomass stored a considerable quantity of nitrogen (from $8.0 \pm 0.8$ to $68.2 \pm 15.3$ g N $m^{-3}$) and phosphorus (from 0.4 to $7.3 \pm 3.2$ g P $m^{-3}$), thus functioning as a nutrient sink.

The maximum biomass value ($1351 \pm 212$ g AFDW $m^{-3}$) reached on 26th May was about six times higher than that recorded at the commencement of the experiment ($231 \pm 23$ g AFDW $m^{-3}$). Similarly, such a major increase of biomass within a single enclosure caused a large proportion of the *Ulva* tissue to suffer physiological stress. Subsequently, *Ulva* decomposed rapidly losing approximately 50% biomass within two weeks.

The data suggest that the observed growth pattern may be related to the elemental composition of the *Ulva* tissue. From 23rd March to 19th April, total phosphorus content in the *Ulva* thalli increased from $0.12 \pm 0.02$ to $0.31 \pm 0.01$% DW. Concurrently, the total nitrogen reserve grew from $2.59 \pm 0.14$ to $3.04 \pm 0.13$% DW. Afterwards, tissue-P continued to remain almost constant in the range 0.2–0.3% DW, while tissue-N decreased to $1.76 \pm 0.39$% DW. The latter value is significantly lower than that required for active growth and might indicate that throughout the incubation period the macroalga was nitrogen limited (Fujita et al., 1989). A further indication that nitrogen was limiting was that atomic N:P ratio was less than 30, well below the threshold proposed by Atkinson & Smith (1983) for balanced growth.

The dilution of total nitrogen due to the biomass increase coincided with an almost linear decrease of nitrate in the macroalgal tissue. Nevertheless, during the early incubation phase *Ulva* took up nitrate active-

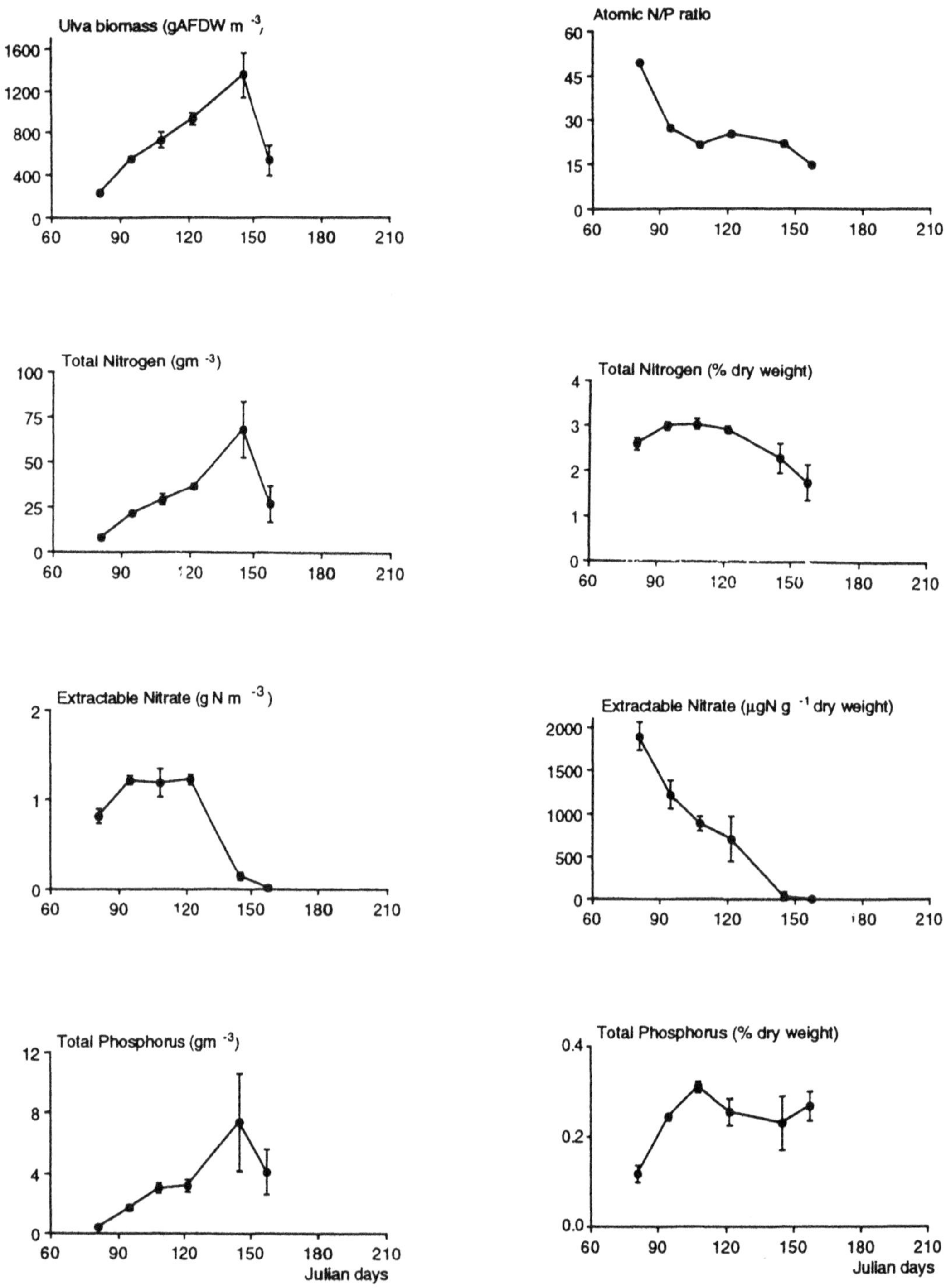


*Figure 4.* Long-term experiment (23 March–7 June). Changes of *Ulva* biomass, atomic N:P ratio, and nitrogen and phosphorus content in macroalgal tissue during the long-term incubation at station M. *Left*: units as g $m^{-3}$; *right*: units as % dry weight. Bars indicate one standard deviation.

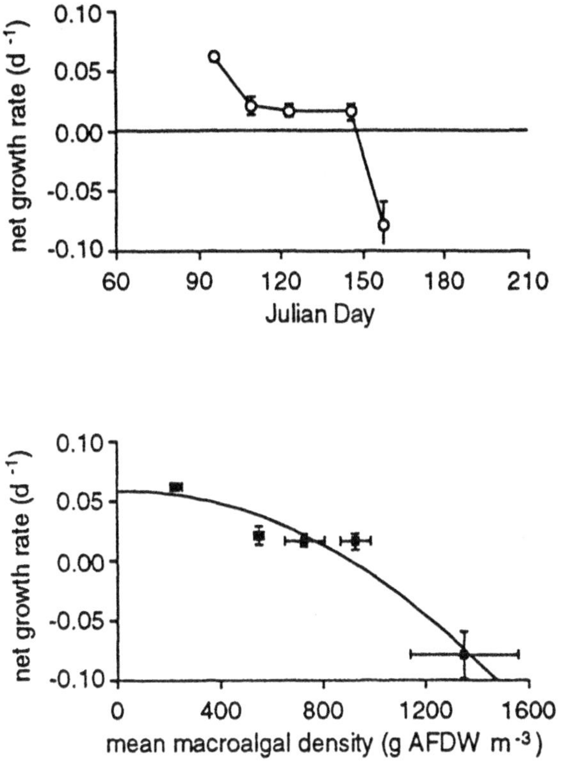


*Figure 5.* Long-term experiment. *Upper panel*: net growth rates of *Ulva* incubated over the long-term period (23 March–7 June). *Lower panel*: relationship between net growth rates and mean densities of *Ulva* incubated over the long-term period. Bars indicate one standard deviation.

ly storing it at an almost constant level (0.9 to 1.2 g $m^{-3}$). Nitrate accumulation into the *Ulva* biomass was coincident with high nitrate concentration in the water column as well as with macroalgal density less than 850 g AFDW $m^{-3}$. Thereafter, depletion of the internal nitrate reserve in the algal tissue is indicative that the nitrogen demand of *Ulva* exceeded the external supply.

Since the beginning of the experiment, net growth rates began to decrease with a sudden collapse during the final incubation phase. At lower biomass densities, rates decreased as density increased; at densities ranging from 500 to 1000 g AFDW $m^{-3}$ rates continued to remain almost constant, while at densities gretaer than 1000 g AFDW $m^{-3}$ they became negative, and rapid decomposition occurred (Figure 5).

*Short-term experiments*

In the short-term experiments, the macroalgal density inside the cages at the beginning of each fortnightly incubation period was nearly 400 g AFDW $m^{-3}$. As this quantity had been observed during exponential growth of the macroalga, we expected that *Ulva* growth inside the cages was unlikely to be density-dependent. The biomass increase measured inside cages should therefore represent the specific growth rate and would be dependent only on the physical and biogeochemical environment.

The main results obtained with the eight short-term incubations are shown in Figure 6. On average, biomass inside the enclosures underwent significant increases, with a maximum from 19th April to 3rd May (from 488 to 1024 ± 44 g AFDW $m^{-3}$ and a minimum from 10th to 26th May (from 488 to 548 ± 72 g AFDW $m^{-3}$). Macroalgal growth was accompanied by nutrient uptake and luxury storage in the biomass. The maximum nitrogen uptake was observed from 19th April to 3rd May (from 15.2 to 36.4 ± 3.0 g N $m^{-3}$). Phosphorus uptake showed a greater variability, with very high values recorded in April and June. During these periods, phosphorus uptake was also accompanied by an increase in tissue reserve, with maximum values recorded during the period from 23rd March to 6th April (from 0.11 to 0.24% DW). Significant increases were also observed on 23rd June and on 5th July, when incubated thalli grew actively although in the lagoon *Ulva* was undergoing decomposition. In the latter situation, early decomposition of the macroalga outside the enclosures might provide a source of phosphorus due to leaching of the more soluble and labile cell constituents. From 22nd April to 3rd May, from 3rd to 10th May, and to a lesser degree from 26th May to 7th June incubated macroalgae were probably phosphorus limited as shown by the decrese in tissue-P at the end of each incubation period.

From 23rd March to 7th June, total nitrogen content in algal tissue ranged from 2.59 to 3.02% DW and did not undergo significant variations during each incubation period. In contrast, in the final phase, *Ulva* tended to accumulate nitrogen: from 1.62 to 2.73% from 7th to 23rd June and from 2.30 to 3.28% in the subsequent period (23rd June–5th July). In the latter case, one might expect that the observed nitrogen increase may be due both to macroalgal assimilation and the development of epiphytes on the *Ulva* thalli.

Changes in nitrogen content in algal tissue were different from what had been anticipated by analysis of the inorganic nitrogen concentrations in the water column and indicate that, over short-term periods, incubated *Ulva* thalli are able to maintain a nearly constant nutritional status independent of the external nitrogen

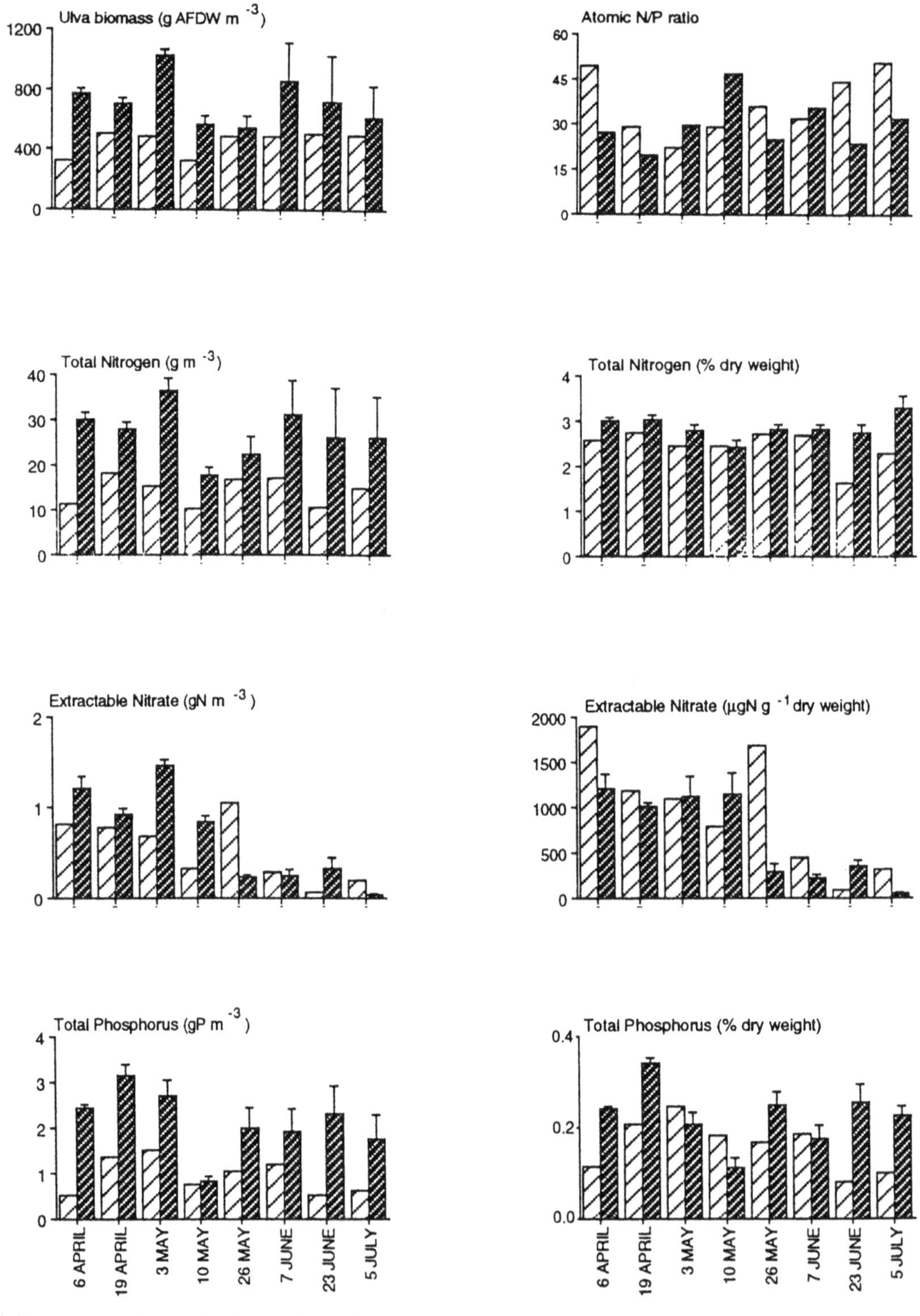


*Figure 6.* Short-term experiments (23 March–5 July). Changes of *Ulva* biomass, atomic N:P ratio, and nitrogen and phosphorus content in macroalgal tissue during the long-term incubation at station M. *Left*: units as g $m^{-3}$; *right*: units as % dry weight. For each experiment initial (light) and final (dark) values of each parameter considered are shown. Dates indicated in the lower panels represent the end of an incubation period and the beginning of the subsequent one. Bars indicate one standard deviation.

concentration. However, pulses of dissolved nitrate in the external environment clearly affected nitrate metabolism of the *Ulva* population. It was evident that in April and early May, when nitrate concentration in the water was high, the macroalga was able to store temporarily nitrate. Thereafter, when nitrate in the water decreased *Ulva* was still maintaining relatively high growth rates, by using internal nitrate reserves. This is unequivocally demonstrated by the fact that from 10th to 26th May the tissue nitrate content declined from $1638 \pm 239$ to $291 \pm 95$ $\mu$g N $g^{-1}$ DW.

*Ulva* incubated over short-term periods underwent maximum net growth rate in the first ten days of May ($0.081 \pm 0.012$ $d^{-1}$), whereas minimum growth ($0.007 \pm 0.009$ $d^{-1}$) was recorded in the subsequent period (Figure 7). Excluding net growth rates estimated on 19th April as well as the 26th May, the remaining growth rate estimates correlated significantly with mean DIN concentration in the water column ($r = 0.966$, $p < 0.007$), with the linear model explaining 93.2% of the total variance. The lower values observed on 19th April and 26th May coincided, to a large extent, with a decrease in total light irradiance and water temperature (see Figure 2).

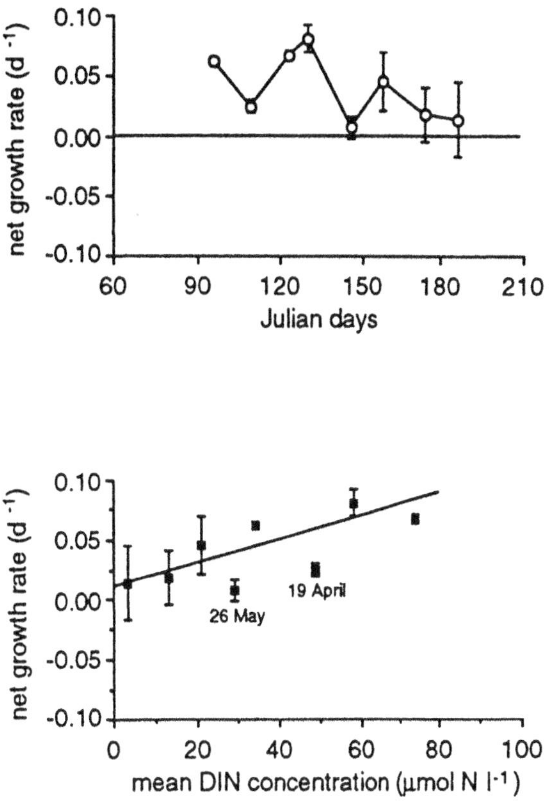


*Figure 7.* Short-term experiments (23 March–5 July). *Upper panel*: net growth rates of *Ulva* incubated over the short-term periods. *Lower panel*: relationship between net growth rates and the corresponding mean dissolved inorganic nitrogen (DIN) concentration in the water column. Bars indicate one standard deviation.

## Discussion

Light, temperature, and nutrient availability in the water are thought to exert direct control on macroalgal productivity, although the seasonal growth cycle can also be conditioned by internal factors, such as circannual rhythms (Luning, 1993). Moreover, many chlorophytes have been found to integrate short-term environmental variability, by virtue of either light or nutrient saturation characteristics. Thus, at a relatively low growth rate they can control a fluctuating environment (Ramus & Venable, 1987; Duke et al., 1989a). However, when macroalgal densities exceed a certain critical threshold, growth becomes more controlled by internal factors rather than by the biogeochemical environment.

Despite a fluactuating and sometimes very stressful environment, Ulvaceans are capable of utilizing ammonium when available in transiently large quantities and convert it rapidly into new biomass; furthermore, it seems that these macroalgae have evolved mechanisms of coping with variable supply (Duke et al., 1989a). In fluctuating environment dominated by nitrate inputs, many seaweeds appear to control the transient nitrate availability by means of active transport system which leads to a luxury uptake saturating the nitrate and nitrite reductase systems and leading to nitrate accumulation within vacuoles (DeBoer, 1981).

Our experimental site is a similarly variable environment, with nitrate concentrations varying on both a tidal and seasonal basis. Nitrate can be also supplied by episodic freshwater discharge from the Po di Volano canal. Thus the control of nitrate availability by nitrate storage in macroalgal tissue is of paramount importance in regulating the growth of *Ulva*. In the long-term incubation experiments, *Ulva* appears to control nitrate availability for approximately 40 days. This is coincident with macroalgal densities less than 850 g AFDW $m^{-3}$, as well as with high nitrate concentrations in the water column. Thereafter, the progressive decrease of tissue nitrate can be considered as an indication of the onset of nitrogen limitation. As a matter of fact, the decrease in internal nitrate reserve may indicate that the nutritional requirements has outstripped the external nitrate availability. The net result is that the storage capacity allows this macroalga to maintain an active growth for at least three weeks after nitrate depletion.

We would also consider that *Ulva* is a ruderal species specialised in rapid colonisation, with large individuals suffering frequent extinctions due to the removal by waves and tidal currents (Littler & Littler, 1980). In the short-term experiments, resembling this ephemeral behaviour of *Ulva*, net growth rates were directly related to the dissolved nitrate concentrations in the water column since there was a significant decrease in net growth rates when nitrate became limiting in the water. Thus we can postulate that in the Sacca di Goro nitrate is one of the key factors determining *Ulva* biomass production, and it is in agreement with previous studies by Viaroli et al. (1992; 1993).

Temporary luxury uptake and storage of nutrients may confer selective advantages to Ulvaceans by buffering against transient nutrient availability. Furthermore, several Ulvaceans are able to control prolonged nitrogen depletion via storage of nitrogen in photosynthetic pigments and rubisco molecules (Duke et al., 1989a). Phosphorus is a different matter due to its rapid turn-over. However, in both long and short-term experiments we recorded considerable accumulation of P within the *Ulva* biomass. In the short-term experiments, the potential importance of N relative to P in the incubated biomass is difficult to assess due to the high degree of variability. On 23rd March, at the beginning of the experiments, the N:P ratio was 49.4:1 indicating a strong P- limitation, whereas in the subsequent period *Ulva* tended to become N-limited (N:P ratio = 19.6:1). In early May, after the increase of nitrate concentration in the water column, availability of P, rather than N, seemed of paramount importance in limiting macroalgal growth (N:P ratio = 47.0:1). Considering that N:P ratios less than 10 indicate N-limitation, while N:P ratios greater than 30 indicate P-limitation (Atkinson & Smith, 1983; Smith, 1984), we can therefore conclude that the nutritional status of macroalgae incubated over short periods would seem to depend mainly on nutrients supplied exogenously from the surrounding water. In contrast, macroalgae incubated in the long-term experiment underwent a clear decrease in N:P ratios from 49.4:1 to 14.6:1 indicating a progression to N-limitation. In this case, the nutritional status was controlled by nutrient dilution due to the increased macroalgal density rather than by the biogeochemical environment.

The ability of *Ulva* to store P and N may have relevant effects not only upon the macroalgal growth, but also on nutrient cycling in the lagoon ecosystem. In the long-term experiments nutrients were accumulated within macroalgal biomass for about two months, with mean rates of 3.3% $d^{-1}$ for N and 4.6% $d^{-1}$ for P. In the short-term experiments, beside a greater variability, we recorded accumulation rates of approximately 8% $d^{-1}$ for N and 10% $d^{-1}$ for P. Thus *Ulva* can be considered as a nutrient sink over very long periods of time. Furthermore, high uptake rates coupled with the removal of floating macroalgae by currents and storms would likely favour a considerable nutrient loss.

The reciprocal control of *Ulva* against biogeochemical environment and viceversa is a key factor in explaining both resource competition and successional stages in primary producer communities dominated by *Ulva* (Smith & Horne, 1988; Pugnetti et al., 1992). However, it is evident that when biomass exceeds certain critical tresholds, approximately a thousand g AFDW $m^{-3}$, the macroalgal community shifts from active production to rapid decomposition, probably due to strong selfshading, biomass density and anaerobic processes starting within the macroalgal beds. This is likely the case for *Ulva rigida* which accumulates in the sheltered lagoons of the northern Adriatic coast: at low biomass, growth appears controlled by nutrient availability, whereas when biomass exceeds 2–3 kg wet weigth, the density-dependent factors become dominant in driving the entire system towards dystrophic crisis.

## Acknowledgments

This research was supported by the joint EU project 'CLEAN': Coastal Lagoon Eutrophication and ANaerobic processes (Contract No EV5V-CT92-0080). Part of our work was made possible thanks to Mr Vittorio Gaiani (Dipartimento di Biologia, Università di Ferrara), Mr Tony Zappata (Battello Hydra, Amministrazione Provinciale di Ferrara), Mr Mimmo Cavalca (Istituto di Ecologia, Università di Parma) and Dr Matteo Cattadori (ITIS 'G. Galilei', San Secondo, Parma). We are indebted with Prof. Paolo Menozzi (Istituto di Ecologia, Università di Parma) and Prof. R. A. Herbert (Department of Biology, Dundee University) who revised the manuscript.

## References


Alvisi, A., D. Marzocchi & S. Edelwais, 1993. Indagine quali-quantitativa delle acque superficiali dei bacini Burana-Volano e Canal Bianco. Dimensione Ambiente, Provincia di Ferrara, Servizio Ambiente. 207 pp.

Amanieu, M., B. Baleux, O. Guélorget & P. Michel, 1975. Etude biologique et hydrologique d'une crise dystrophique (malaigue) dans l'étang du Prévost à Palavas (Hérault). Vie Milieu 25: 175–204.

A.P.H.A., A.W.W.A., W.P.C.F., 1975. Standard methods for the examination of water and wastewaters. 14th ed. A.P.H.A. Washington. 1193 pp.

Aspila, K. I., H. Agemian & A. S. Y. Chau, 1976. A semiautomated method for the determination of inorganic, organic and total phosphate in sediments. Analyst 101: 187–197.

Atkinson, M. J. & S. V. Smith, 1983. C:N:P ratios of benthic marine plants. Limnol. Oceanogr. 28: 568–574.

Chapman, A. R. O. & J. S. Craigie, 1977. Seasonal growth in *Laminaria longicruris*: relation with dissolved inorganic nutrients and internal reserves of nitrogen. Mar. Biol. 40: 197–205.

Corzo, A. & F. X. Niell, 1992. Inorganic nitrogen metabolism in *Ulva rigida* illuminated with blue light. Mar. Biol. 112: 223–228.

Couthino, R. & R. Zingmark, 1993. Interactions of light and nitrogen on photosynthesis and growth of the marine macroalga *Ulva curvata* (Kützing) De Toni. J. exp. mar. Biol. Ecol. 167: 11–19.

Dal Cin, R. & P. Pambianchi, 1991. I sedimenti della Sacca di Goro. In S. Bencivelli & N. Castaldi (eds), Studio integrato sull'ecologia della Sacca di Goro. F. Angeli, Milano: 253–263.

Da Costa Braga, A. & Y. Yoneshigue-Valentin, 1994. Growth of *Laminaria abissalis* (Phaeophyta) at different nitrate concentrations. Phycologia 33: 271–274.

DeBoer, J. A., 1981. Nutrients. In C. S. Lobban & M. J. Wynne (eds), The Biology of Seaweeds. Botanical Monographs, Vol. 17. Univ. California Press, Berkeley and Los Angeles: 356–392.

DeBoer, J. A., H. J. Guigli, T. L. Israel & C. F. D'Elia, 1978. Nutritional studies of two red algae. I. Growth rate as a function of nitrogen source and concentration. J. Phycol. 14: 216–266.

Duke, C. S., W. Litaker & J. Ramus, 1989a. Effect of temperature on nitrogen-limited growth rate and chemical composition of *Ulva curvata* (Ulvales: Chlorophyta). Mar. Biol. 100: 143–150.

Duke, C. S., W. Litaker & J. Ramus, 1989b. Effects of temperature, nitrogen supply, and tissue nitrogen on ammonium uptake rates of the chlorophyte seaweeds *Ulva curvata* and *Codium decorticatum*. J. Phycol. 25: 113–120.

Fujita, R. M., P. A. Wheeler & R. L. Edwards, 1989. Assessment of macroalgal nitrogen limitation in a seasonal upwelling region. Mar. Ecol. Prog. Ser. 53: 293–303.

Izzo, G. & V. Hull, 1991. The anoxic crises in dystrophic processes of coastal lagoons: an energetic explanation. In C. Rossi and E. Tiezzi (eds), Ecological Physical Chemistry. Elsevier Sci. Pbls., Amsterdam: 559–572.

Koroleff, F., 1970. Direct determination of ammonia in natural waters as indophenol blue. Information on techniques and methods for seawater analysis. I.C.E.S. Interlaboratory Rep. No. 3: 19–22.

Lapointe, B. E. & K. R. Tenore, 1981. Experimental outdoor studies with *Ulva fasciata* Delile. I. Interaction of light and nitrogen on nutrient uptake, growth, and biochemical composition. J. exp. mar. Biol. Ecol. 53: 135–152.

Lapointe, B. E., M. M. Littler & D. S. Littler, 1992. Nutrient availability to marine macroalgae in siliciclastic versus carbonate-rich coastal waters. Estuaries 15: 75–82.

Littler, M. M. & D. S. Littler, 1980. The evolution of thallus form and survival strategies in benthic marine macroalgae: field and laboratory tests of a functional form model. Am. Nat. 116: 25–44.

Luning, K., 1993. Environmental and internal control of seasonal growth in seaweeds. Hydrobiologia 260/261: 1–14.

O'Brien, M. C. & P. A. Wheeler, 1987. Short-term uptake of nutrients by *Enteromorpha prolifera* (Chlorophyceae). J. Phycol. 23: 547–556.

Pugnetti, A., P. Viaroli & I. Ferrari, 1992. Processes leading to dystrophy in a Po River Delta lagoon (Sacca di Goro): phytoplankton-macroalgae interactions. Sci. Total Envir. suppl. 1992: 445–456.

Ramus, J. & M. Venable, 1987. Temporal ammonium patchiness and growth rate in *Codium* and *Ulva* (Ulvophyceae). J. Phycol. 23: 518–523.

Rosenberg, G. & J. Ramus, 1981. Ecological growth strategies in the seaweeds *Gracilaria foliifera* (Rhodophyceae) and *Ulva* sp. (Chlorophyceae): the rate and timing of growth. Bot. mar. 24: 583–589.

Sand-Jensen, K., 1991. Photosynthetic responses of *Ulva lactuca* at very low light. Mar. Ecol. Prog. Ser. 50: 195–201.

Smith, S. V., 1984. Phosphorus versus nitrogen limitation in the marine environment. Limnol. Oceanogr. 29: 1149–1160.

Smith, D. W. & A. J. Horne, 1988. Experimental measurement of resource competition between planktonic microalgae and macroalgae (seaweeds) in mesocosms simulating the San Francisco bay-Estuary, California. Hydrobiologia 159: 259–268.

Sfriso, A., B. Pavoni, A. Marcomini & A. A. Orio, 1992. Macroalgae, nutrient cycles, and pollutants in the lagoon of Venice. Estuaries 15: 517–528.

Valderrama, J. C., 1977. Methods used by the Hydrographic Department of National Board of Fisheries, Sweden. In K. Grasshoff (ed.), Report of the Baltic Intercalibration Workshop. Annex Interim Commission for the Protection of the Environment of the Baltic Sea: 14–34.

Viaroli, P., A. Pugnetti & I. Ferrari, 1992. *Ulva rigida* growth and decomposition processes and related effects on nitrogen and phosphorus cycles in a coastal lagoon (Sacca di Goro, Po River Delta). In G. Colombo, I. Ferrari, V. U. Ceccherelli and R. Rossi (eds), Marine eutrophication and population dynamics. Olsen & Olsen, Fredensborg: 77–84.

Viaroli, P., M. Naldi, R. R. Christian & I. Fumagalli, 1993. The role of macroalgae and detritus in the nutrient cycles in a shallow-water dystrophic lagoon. Verh. Int. Ver. Limnol. 25: 1048–1051.

Viaroli, P., M. Bartoli, C. Bondavalli & M. Naldi, 1995. Oxygen fluxes and dystrophy in a coastal lagoon colonized by *Ulva rigida* (Sacca di Goro, Po River Delta, Northern Italy). Fresenius Envir. Bull. 4: 381–386.

*Hydrobiologia* **329**: 105–119, 1996.
*P. Caumette, J. Castel & R. Herbert (eds), Coastal Lagoon Eutrophication and ANaerobic Processes (C.L.E.AN.).*



# Macrophyte communities and their impact on benthic fluxes of oxygen, sulphide and nutrients in shallow eutrophic environments

Pierluigi Viaroli[1], Marco Bartoli[1], Cristina Bondavalli[1], Robert R. Christian[2], Gianmarco Giordani[1] & Mariachiara Naldi[1]
[1]*Dipartimento di Scienze Ambientali, Università di Parma, 43100 Parma, Italy*
[2]*Department of Biology, East Carolina University, Greenville, NC, USA*

*Key words:* Seaweeds, seagrasses, benthic fluxes, oxygen, sulphide, nutrients

## Abstract

The impact of macrophyte communities on benthic fluxes has been analyzed in three shallow coastal environments: Etang du Prévost (Mediterranean coast of France), characterized by the large floating macro-alga *Ulva rigida*; Certes fishponds (Bassin d'Arcachon), covered by *Ruppia cirrhosa*; and the inner intertidal mud-flat in the Arcachon Bay (French Atlantic coast), which has extensive *Zostera noltii* meadows. In these bodies of water, primary production is dependent primarily on the dominant seagrasses and macroalgae that are also responsible for the large quantity of organic matter deposited on the sediment surface. In 1993 and 1994, fluxes of oxygen, sulphide and nutrients were measured in early and late summer, which were selected in order to represent the production and decomposition phases of the dominant macrophytes. Experimental work was undertaken to measure: (1) standing crop of dominant macroalgae and rooted phanerogams and the elemental and macromolecular composition of plant biomass; (2) benthic fluxes of oxygen, sulphide, nitrogen and phosphorus using incubation of multiple dark and light benthic chambers; (3) water-sediment profiles of free-sulphide in sediment cores with rooted phanerogams (*Ruppia*) as well as with floating seaweeds (*Ulva*).

At the selected sampling sites, in addition to external (tides) and/or internal (sediment reactivity) factors, we observed differences in benthic fluxes which were clearly related to growth patterns and structure of the macrophyte communities. The *Z. noltii* meadows were stable and characterized by slow growth and almost constant biomass. In the more sheltered sampling station in the Certes fishponds, *R. cirrhosa* beds showed a summer decrease due to extensive epiphyte growth. During the decomposition phase, significant fluxes of free-sulphide occurred inside the dark benthic chambers, probably due to the metabolism of the epiphytic layer. In the Etang du Prévost, *U. rigida* achieved high biomass levels, even though the macroalgal beds exhibited a patchy distribution due to wind action and the hydrodynamics of the lagoon. In the decomposition phase, which was coincident with the annual dystrophic crisis the rapid decomposition of *Ulva* led to high fluxes of free sulphide.

The shift from the production to decomposition phase resulted in substantial changes in nutrient recycling only in the macro-algal-dominated system. During the growth period dissolved inorganic nitrogen and phosphorus were kept at low levels due to macrophyte uptake. In contrast during the decomposition phase when the macroalgal biomass was mineralised, nitrogen and phosphorus were rapidly recycled. The same processes did not occur in the Certes fishponds probably because of the greater internal buffering capacity linked either to plant morphology/physiology or to the properties of the sediment.

## Introduction

Shallow intertidal and subtidal environments are considered to have a high primary productivity due to the growth of extensive meadows of rooted angiosperms. In the last twenty- five years, losses of seagrass meadows have been occurred in eutrophic coastal areas, principally in sheltered habitats where the effects of eutrophication are more severe, either due to reduced tidal flushing or to other physical constraints (Stevenson, 1988; Giesen et al., 1990). In general, increased nutrient loading results in enhanced phytoplankton growth, which in turn is accompanied by a decline of attached phanerogam communities due to decreased water transparency, build up of epiphytes and occurrence of reducing conditions in the bottom water and surface sediments (Twilley et al., 1985; Sand-Jensen & Borum, 1991; Nienhuis, 1992; Williams & Ruckelshaus, 1993). Aquatic environments with extensive phytoplankton blooms frequently progress to systems dominated by opportunistic, free-floating macroalgae belonging to the Ulvales and Cladophorales.

The development of such large and fast growing primary producers plays a key role in oxygen production and consumption and regulation of the entire benthic metabolism (for an up-dated review see Sand-Jensen & Borum, 1991). Usually, fast growing macroalgae exibit nutrient uptake and storage greatly in excess of their physiological needs. This is the case for *Ulva rigida* C. Agardh, whose luxury uptake coupled with high growth rates results in nutrient depletion in the water column, which in turn results in the pulsing trends seen especially in the dissolved inorganic nitrogen pool. As a consequence, *Ulva* outcompetes other primary producers due to its ability to store nitrogen, bringing about a shortened food chain (Smith & Horne, 1985; Pugnetti et al., 1992; Fong et al., 1994). There is also experimental evidence of a strict relationship between growth and decomposition rates: fast growing plants tend to have high nutrient concentrations and low fibre content and also decompose more rapidly because their high N/C and P/C ratios stimulate microbial growth (Buchsbaum et al., 1991; Duarte, 1992; Enriquez et al., 1993). Therefore, the replacement of seagrass meadows by macroalgae can effect changes not only in the timing and rate of detritus production but also in decomposition rate, which in turn leads to a decline in water quality. For example, fast macroalgal growth followed by decomposition of the accumulated biomass is considered to be one of the most important factors in the occurrence of dystrophic crises in several Italian lagoons (Izzo & Hull, 1991; Sfriso et al., 1992; Viaroli et al., 1995). In addition to the above-mentioned effects, changes also occur in the composition of the primary producer community as a consequence of eutrophication. Rooted phanerogams provide gas transport throughout the surficial sediment contributing to its oxygenation, while floating macroalgae can promote physical stratification of the water (Crawford, 1992). For example, the large thalli of *Ulva* act as physical barriers separating the water column into two distinct layers: the uppermost well-oxygenated or even supersaturated layer, and the lower one, which is anoxic and highly reduced.

Hence, growth patterns (intensity, duration and rate) and distribution of macroalgae and seagrasses may explain the changes which occur in other processes within the coastal ecosystems (Sand-Jensen & Borum, 1991). For example, benthic processes are particularly sensitive and can undergo catastrophic changes as result of sudden non-linear responses to the external stresses. The structure or form of plant communities may mitigate against the drastic changes of external forcing, since aquatic ecosystems with rooted phanerogams are more resilient than those with green macroalgae. Such mitigation can contribute to the resistance and resilience of ecosystems, acting as buffers against external perturbations (Sherman, 1994).

## Aims and objectives

In the three ecosystems investigated the principal primary producers are seagrasses and macroalgae which are also responsible for the bulk of the vegetative detritus deposited on the sediment surface (Auby et al., 1994; Castel et al., 1996). It would therefore be expected that benthic fluxes are dependent primarily on growth patterns and distribution of the primary producers and decomposition rates of the plant detritus. We postulated that the intensity of water-sediment exchanges in lagoons dominated by slow growing rooted angiosperms, such as *Ruppia cirrhosa* (Pet.) Grande and *Zostera noltii* Hornem, would show less variation than those occurring in lagoons dominated by opportunistic faster- growing macroalgae, such as *Ulva rigida*. In addition to studying benthic fluxes in relation to growth patterns, we also investigated the effects of plant morphology on water stratification. It was anticipated that the development of pelagic macroalgae would lead to laminar stratification which limits water mixing and gas exchange, while the growth of rooted phanerogams would enhance vertical exchange and gas transport throughout the water column and surficial sediment.

In 1993 and 1994, fluxes of oxygen, sulphide and nutrients were measured in early and late summer. These sampling periods were selected to represent production and decomposition phases of the dominant macrophytes (Christian et al., 1996). In this paper we present data from selected experiments in which the relationships between benthic fluxes and macrophyte community structure have been studied by estimating

1. standing crop of dominant macroalgae and rooted phanerogams and the elemental and macromolecular composition of plant biomass;
2. benthic fluxes of oxygen, sulphide, nitrogen and phosphorus by means of incubation of dark and light benthic chambers;
3. water-sediment profiles of free-sulphide in sediments dominated by either rooted phanerogams (*Ruppia*) or with floating macroalgae (*Ulva*).

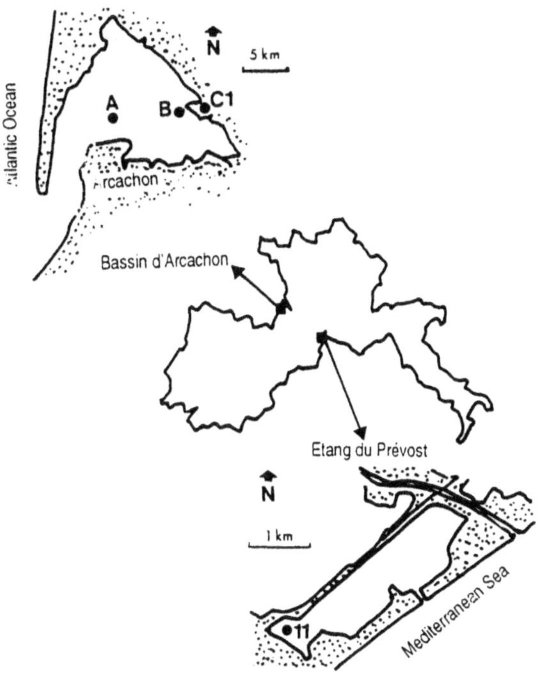


*Figure 1.* General layout of the Arcachon Bay and the Etang du Prévost, and positions of the experimental sites.

## Materials and methods

### *Study area*

Our work addressed the effects of macrophyte communities on benthic fluxes in three shallow coastal environments: Etang du Prévost (mediterranean coast of France) characterized by the large floating macroalga *Ulva rigida*; fishponds of Certes (Bassin d'Arcachon), covered by *Ruppia cirrhosa*; and the inner intertidal flat in the Bassin d'Arcachon (west Atlantic coast of France), which is dominated by *Zostera noltii* (Figure 1). Detailed descriptions of experimental sites have been reported by Auby et al. (1994). Here we refer to the Etang du Prévost as station 11, fishponds of Certes as station C1 and the inner site in the Bassin d'Arcachon as station B.

### *Biomass, and elemental and macromolecular composition of dominant macrophytes*

The biomass of macroalgae and macrophytes was determined by direct harvesting inside circular cores (30 cm i.d.) The harvested plants were sorted, washed with both lagoon and tap water, weighed and oven dried at 70 °C until they reached a constant dry weight (DW). During the sampling of August 1994 the epiphytes covering *Ruppia* (station C1) were carefully removed from the host plants by means of forceps and oven dried at 70 °C. At station C1 sestonic chlorophyll-*a* was also measured by conventional techniques (APHA, 1975).

Dried subsamples were used to determine the elemental and macromolecular composition of the harvested biomasses. Subsamples in the range 100–200 mg were analyzed with a LECO 600 CHN elemental analyzer to determine the total C and N content. About one gram of dried biomass was ignited at 550 °C in a muffle furnace to determine ash and ash free dry weight (AFDW). Total phosphorus was determined by spectrophotometry after acid extraction of the ashes (Aspila et al., 1976). Approximately 2–5 g were also analyzed for hemicellulose, cellulose and lignin (Goering & Van Soest, 1970).

### *Estimation of benthic fluxes by incubation of multiple dark and light benthic chambers*

Fluxes of nutrients, oxygen and free-sulphide were determined using dark and light cylindrical benthic chambers and bell jars with volume of about 14 litres and cross-sectional area of approximately 700 $cm^2$.

On 20 May and 31 August 1994, at Station B three dark and three light benthic chambers were incubated over the *Zostera* beds. Each experiment was carried out around midday and because of the tidal cycle incubated for only 3–4 hours.

At stations C1 and 11, each experiment began in the early afternoon and lasted approximately 50 hours. From 23 until 25 May 1994, at station C1 we compared three different conditions: bare sediment, low (70 g DW $m^{-2}$) and high (110 g DW $m^{-2}$) *Ruppia* biomass. From 23 to 25 August 1994, we studied the impact of the growth of the epiphytic community by analyzing the following conditions: bare sediment, *Ruppia* (86 $\pm$ 11 g DW $m^{-2}$) in excess of epiphytes (52 $\pm$ 14 g

*Table 1.* Average biomass (g DW $m^{-2}$) of macrophytes and epiphytes at the experimental sites. n.d.: not detectable.

| | | Growth phase | Decay phase |
|---|---|---|---|
| Station 11 | *Ulva rigida* | $520 \pm 97$ ($n$=4) | $459 \pm 55$ ($n$=4) |
| Station B | *Zostera noltii* | $101 \pm 29$ ($n$=7) | $81 \pm 38$ ($n$=13) |
| Station C1 | *Ruppia cirrhosa* | $73 \pm 27$ ($n$=8) | $67 \pm 34$ ($n$=13) |
| | epiphytes | n.d. | $101 \pm 61$ ($n$=13) |

*Table 2.* Major characteristics of the opportunistic macroalga *Ulva rigida* and the rooted phanerogams *Zostera noltii* and *Ruppia cirrhosa* and epiphytes growing on *Ruppia* at the experimental sites. Units: % dry weight; ud.: not detectable; n.d.: not determined.

| | *Ulva* | *Ruppia* | *Zostera* | Epiphytes |
|---|---|---|---|---|
| Total Carbon | 20.5–38.9 | 34.9–37.7 | 40.1–45.5 | n.d. |
| Total Nitrogen | 2.2–5.1 | 2.2–3.4 | 2.4–3.1 | 2.9–3.3 |
| Total Phosphorus | 0.11–0.68 | 0.14–0.38 | 0.24–0.29 | 0.16–0.21 |
| Ash Free Dry Weight | 66.2–82.9 | 76.3–83.7 | 77.0–82.0 | 62.2–67.1 |
| Hemicellulose | 22.3–29.3 | 19.6–26.5 | 25.3–29.2 | 100 |
| Cellulose | 7.4–14.0 | 12.0–19.1 | 15.1–20.0 | ud. |
| Lignin | ud. | 2.0–3.4 | 3.1–7.2 | ud. |

DW $m^{-2}$) (R>E) and *Ruppia* ($58 \pm 24$ g DW $m^{-2}$) less than epiphytes ($167 \pm 47$ g DW $m^{-2}$) (R<E).

From 23 to 25 August 1993, at station 11 in the Prévost lagoon we studied two conditions: benthic chambers with bare sediment and sediment covered by *Ulva*. From 16 until 18 May 1994, we studied three different conditions: benthic chambers with bare sediment and sediment covered by *Ulva*, and bell jars with *Ulva*, the plankton community but without sediment. In 1993 and 1994, the initial biomass of the macroalga inside the incubation chambers was about 500 g DW $m^{-2}$.

Water samples were collected at different times from the benthic chambers with a syringe, avoiding contact with the air. In general, the sampling frequency was higher during the first 24 hours of incubation. At the same time additional water samples were taken from a fixed station outside the chambers. Approximately 30 ml of water were fixed immediately for oxygen determination, while 30 ml were used for pH, Eh and sulphide analyses. One hundred ml were filtered on pre-rinsed filters (GF/C Whatman) for nutrient determination. The total amount of water sampled from benthic chambers during the experiments never exceeded 10% of the chamber volume. Water analyses were conducted on duplicate subsamples.

*Water-sediment profiles of free-sulphide throughout* Ruppia *and* Ulva *stands*

In August 1994, water-sediment profiles of free-sulphide in the sediment cores were determined. Four conditions were compared: sediment from station C1 either with or without *Ruppia* and sediment from station 11 either with or without *Ulva*. After sampling, *Ulva* thalli entrapped inside a mesh bag were positioned 5 cm above the sediment surface. All the cores were then kept in the dark at room temperature for approximately one day. After 24 hours, sulphide profiles were recorded during the stepwise lowering of a needle electrode with a hand-driven micromanipulator (approx. precision 0.1 mm).

We used a ion-specific needle electrode (courtesy of R. De Wit) with a reference K701 double junction electrode (Radiometer) connected to a milli-voltmeter (Micro pH 2002, Crison). In parallel, pH was also measured at the water-sediment interface, being the equilibrium among $H_2S$, $HS^-$ and $S^{2-}$ pH-dependent. Calibration of the sulphide electrode was carried out according to Van Gemerden et al. (1989). Phosphate buffer (200 mM, pH = 8.0) was sparged with nitrogen for two hours; afterwards, increasing amounts of sulphide were added with a syringe. In parallel, total sulphide concentrations were determined by the methylene-blue method (Cline, 1969). Using logarithmic regression we obtained a straight line in the range

*Table 3*. Classification of shallow environments by the metabolism based benthic trophic state index (BTSI) according to Rizzo et al. (in press). $NCP_{max}$= maximum net oxygen productivity in light incubation chambers and DR = oxygen uptake in dark incubation chambers.

| BTSI | Condition | Description |
|---|---|---|
| 0 | $NCP_{max} \leq DR$ | Totally heterotrophic |
| 1 | $DR < NCP_{max} \leq 0$ | Net heterotrophic |
| 2 | $0 < NCP_{max} \leq \|DR\|$ | Net autotrophic |
| 3 | $\|DR\| < NCP_{max}$ | Totally autotrophic |

10–1000 $\mu$M. Because ion-specific electrodes are only sensitive to the fully dissociated sulphide, total sulphide concentrations were estimated from the equilibrium at the given pH (Kuhl & Jørgensen, 1992).

*Analytical methods adopted for water analyses*
Oxygen was determined using the idometric titration of Winkler (APHA, 1975). Sulphides were analyzed by inflection-point titration with silver nitrate (TIM 90, F1212S and K711 electrodes, Radiometer). In parallel, pH (GK 2401 C electrode, Radiometer) and redox (platinum P101 and reference K711 electrodes, Radiometer) were measured with the same apparatus. Salinity was determined by means of conductometry (CDM 83, CDC 304 cell, and T 801 temperature compensation, Radiometer). Nitrite and nitrate (reduction by cadmium columns) were determined by spectrophotometry after diazotation (APHA, 1975); ammonium was measured by the indophenol-blue method (Koroleff, 1970) and soluble reactive phosphorus (SRP) by the ascorbic acid method (Valderrama, 1977).

Total dissolved phosphorus and nitrogen as well as dissolved reactive silica were also measured and were discussed in the preliminary report (Viaroli et al., 1994).

## Results and discussion

Macrophyte biomass provides fundamental information on the primary producer communities at each of the sampling sites studied (Table 1). The *Z. noltii* community was a stable community with an almost constant biomass. In the sheltered station C1, *R. cirrhosa* experienced a summer decay due to the extended epiphyte growth. At station I1, *U. rigida* reached biomass level about five times higher than those of rooted phanerogams, even though the distribution of the macroalgae was patchy due to the prevailing hydrodynamic conditions.

Differences among the macrophytes were also confirmed by differences in the elemental and macromolecular composition of the dry biomass. The slow growing seagrasses consisted of about 20% fibre (by dry weight) with significant amounts of lignin, whereas the fast growing *Ulva* consisted of less cellulose and no lignin. In contrast, *Ulva* showed luxury storage of nutrients reaching maximum tissue concentrations of N and P double those of *Zostera* and *Ruppia* (Table 2).

Oxygen, sulphide and nutrient fluxes were analyzed at the three sites by incubating different combinations of primary producers in dark and light benthic chambers and bell jars. Experiments were conducted in early and late summer in presence or absence of community components to determine their role in benthic processes. The experiments, which lasted about two days, were designed to determine the resistance of the different plant communities to sudden stress induced by the prolonged incubation periods. Prolonged experiments were not feasible at station B, where incubations lasted approximately 4 hours coinciding with high tide.

*Bassin d'Arcachon, station B*
At station B, in the inner intertidal flat of Bassin d'Arcachon, *Zostera noltii* maintained an almost constant cover from May to August 1994. On 20 May and on 31 August, three dark and three light BC were incubated at high tide over the *Zostera* beds around midday for approximately 4 hours. The net oxygen flux in the light benthic chambers showed no significant differences when sampled ($6.32 \pm 0.73$ mmol $O_2$ $m^{-2}$ $h^{-1}$ on 20 May and $6.93 \pm 1.21$ mmol $O_2$ $m^{-2}$ $h^{-1}$ on 31 August), while fluxes inside the dark ones increased from $-2.46 \pm 0.80$ mmol $O_2$ $m^{-2}$ $h^{-1}$ (20 May) to $-7.90 \pm 1.92$ mmol $O_2$ $m^{-2}$ $h^{-1}$ (31 August) (Figure 2).

*Fishponds of Certes, station C1*
The sheltered and dammed fishpond of Certes was covered by large stands of *Ruppia cirrhosa*. In May and August 1994, the *Ruppia* biomass followed a sharp horizontal gradient with the highest values along the eastern shoreline (111 g DW $m^{-2}$ on 20 May and 132 g DW $m^{-2}$ on 26 August) and bare sediment along the opposite shore. In August, the emerging fronds were covered by a thick layer of epiphytes, which reached a maximum biomass of 215 g DW $m^{-2}$ in the central part of the lagoon. The ratio of epiphytes to *Ruppia*

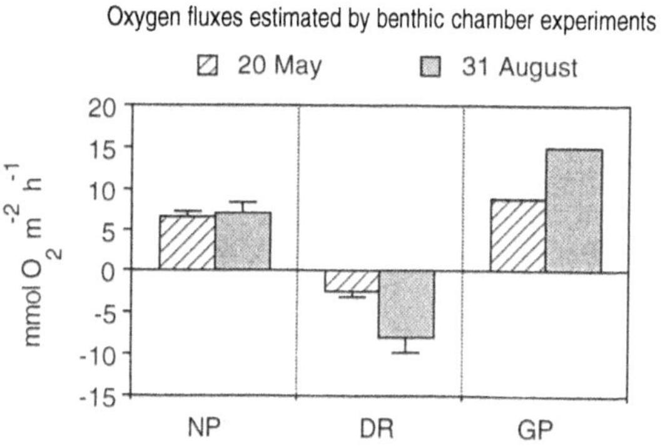


*Figure 2.* Oxygen fluxes at the community level determined by benthic chamber experiments at station B in 1994. NP = net oxygen production, DR = dark respiration and GP = gross oxygen production.

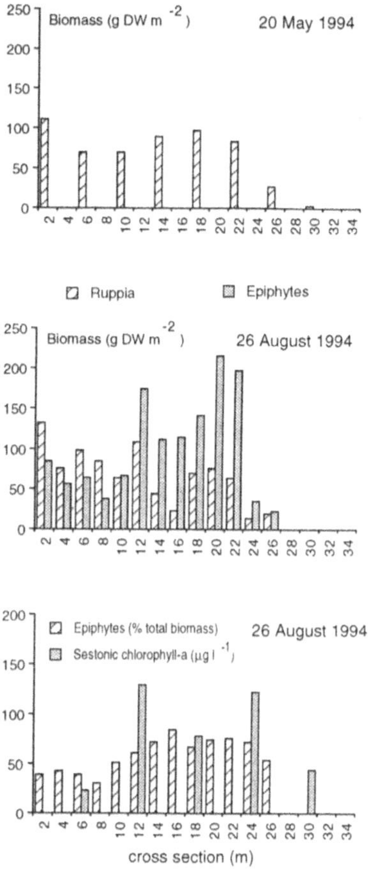


*Figure 3.* Variations in biomass of *Ruppia cirrhosa* and its epiphytes (top and intermediate panels) and relationship between epiphytes and sestonic chlorophyll-*a* (bottom panel) across a section at station C1 in 1994.

ranged from 0.30 to 0.84 on a dry weight basis. Such a thick epiphytic layer provides large quantities of suspended particulate matter, as indicated by chlorophyll-*a* concentrations reaching values of 129.61 $\mu$g $l^{-1}$ in the water column of the central area of the fishpond (Figure 3).

At station C1, from 23 to 25 May 1994 we compared three different levels of macrophyte cover using the benthic chambers: bare sediment, low (70 g DW $m^{-2}$) and high (110 g DW $m^{-2}$) *Ruppia* biomass. As expected, changes in oxygen production and consumption were closely related to the standing stock of *Ruppia*, while the role played by the plankton and microphytobenthic communities was less significant (Figure 4). Analysis of the results shows that the oxygen decrease in the sediment plus plankton chambers was initially almost as high as the other chambers, while afterwards consumption was minimal due to the lack of production. Variations in oxygen concentration in the lagoon fully conformed to those observed inside the benthic chambers in the light which contained the highest biomass and followed a clear daily trend consistent with that reported by Thimel & Labourg (1992). Under all the experimental conditions nitrate, nitrite, and soluble reactive phosphorus (SRP) remained below the level of dectection. The ammonium concentration remained constant in the light benthic chambers (around 5 $\mu$mol $l^{-1}$), while it showed significant increases principally in the dark chambers without macrophyte cover and containing a high biomass (Figure 4). However, most of the dissolved nitrogen was recycled and stored in the organic pool, as shown by dissolved organic nitrogen concentrations ranging between 50 and 240 $\mu$mol $l^{-1}$ inside all benthic chambers (data not shown).

From 23 to 25 August 1994, we studied the impact of the growth of the epiphytic community by analyzing bare sediment, and two different assemblages of macrophytes and epiphytes: *Ruppia* in excess of epiphytes (R>E) and vice-versa (R<E). In the light benthic chambers the oxygen concentration followed a clear daily trend with night anoxia, which was coupled with free-sulphide production (Figure 5). Inside the R>E chamber sulphide increased to 12.3 mmol $m^{-2}$ over about 8 hours and in R<E chamber it reached 25.9 mmol $m^{-2}$ after about 8 hours. Inside the dark benthic chambers, oxygen was depleted within a few hours. Afterwards an almost linear increase of free-sulphide was observed. After 49 hours, 58.3 (bare sediment), 86.1 (R>E) and 118.1 (R<E) mmol $S^{2-}$ $m^{-2}$ were released to the water column. Significant sul-

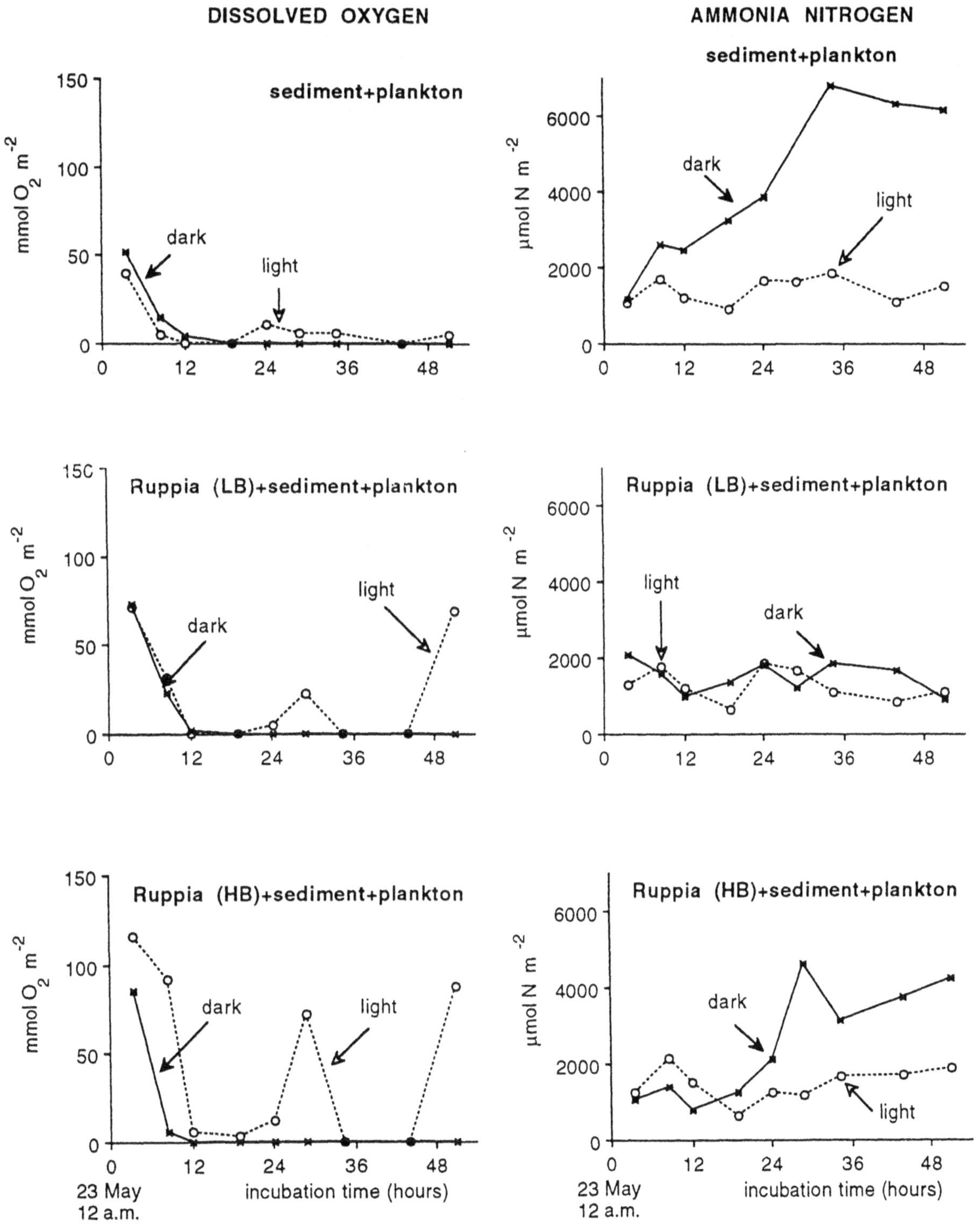


*Figure 4*. Main results of the experiment with benthic chambers carried out at station C1 from 23 until 25 May 1994. *Ruppia* biomass inside benthic chambers: 70 g DW $m^{-2}$ (LB) and 110 g DW $m^{-2}$ (HB).

phide fluxes occurred in spite of the expected buffering capacity due to the high iron content (Bartoli et al., 1996). If we consider that sulphide production was related to the *Ruppia*/epiphyte ratio we can postulate that sulphide production occurred in close proximity to epiphyte layer. Under these conditions phosphate release might have been expected but SRP release from the sediment was negligible. All the inorganic forms of both nitrogen and phosphorus in the water were below the level of detection, while dissolved organic nitrogen concentrations were similar to those observed in May (data not shown).

*Etang du Prévost, station 11*

In May 1994, the lagoon of Prévost was covered by floating mats of *Ulva rigida* with a patchy distribution of biomass (from 76 to 631 g DW $m^{-2}$) and net growth rates of $0.078 \pm 0.022$ $d^{-1}$ (Viaroli et al., 1994). Subsequently, the *Ulva* standing crop decreased dramatically due to harvesting, which started in early summer. In August, at station 11 the sediment was almost denuded, while along the southern and western shoreline there were large accumulations of filamentous green algae.

From 16 to 18 May 1994, we studied three different experimental conditions: benthic chambers with bare sediment, benthic chambers with sediment covered by *Ulva* and bell jars with *Ulva* plus the plankton community but without sediment. Macroalgal biomass inside the incubation chambers was approximately 500 g DW $m^{-2}$. Changes in oxygen fluxes were to a large extent influenced by macroalgal activity as indicated by the maximum oxygen increase in the light benthic chamber (9.69 mmol $O_2$ $m^{-2}$ $h^{-1}$) and bell jar (30.75 mmol $O_2$ $m^{-2}$ $h^{-1}$) containing *Ulva* (Figure 6). In the light chamber with bare sediment, the oxygen concentration decreased showing that consumption prevailed over production processes. A similar trend was observed in the corresponding dark chamber, where oxygen uptake was almost constant at 4.68 mmol $O_2$ $m^{-2}$ $h^{-1}$ (Figure 6). These data indicate that *Ulva* was responsible for the bulk of the oxygen production, while benthic metabolism accounted for most of the oxygen consumption. The association of metabolic activities of both macroalgae and sediment resulted in major changes in the physico-chemical properties of the water in the light and dark benthic chambers and bell jars. For example the pH increased from 7.9 to 9.7. Significant SRP fluxes were observed inside the light ($8.4 \pm 1.5$ $\mu$mol P $m^{-2}$ $h^{-1}$, $r = 0.988$, $p < 0.0001$) and dark chambers ($15.6 \pm 3.1$ $\mu$mol P $m^{-2}$ $h^{-1}$, $r = 0.984$, $p < 0.0001$) with bare sediment, and in the dark benthic chamber with *Ulva*, where SRP followed a nearly linear increase ($27.1 \pm 11.7$ $\mu$mol P $m^{-2}$ $h^{-1}$, $r = 0.933$, $P < 0.0001$). Inside the light and dark bell jars and in the light benthic chamber with *Ulva*, most of the SRP was present in the macroalgal biomass and from these data we can infer that most of SRP was obtained from the sediment. Over the same period, a significant linear increase in ammonium was observed inside the dark benthic chamber containing *Ulva* ($34.7 \pm 8.2$ $\mu$mol N $m^{-2}$ $h^{-1}$, $r = 0.978$, $p < 0.0001$). After about 12 hours ammonium produced in the dark benthic chamber with *Ulva* was almost three times higher than in the dark benthic chamber with bare sediment and the dark bell jar containing the macroalga (Figure 6). These results highlight the potential activity of *Ulva* and its epiphytes of immobilising available phosphorus and nitrogen and slowing their recycling, even under the highly reducing conditions of the dark bell jar. In contrast, such conditions appear to enhance SRP release from anoxic sediment leading to an imbalance between phosphorus and nitrogen recycling.

Benthic fluxes during the *Ulva* growth phase were compared to those measured during the decomposition phase in 1993. From 24 to 26 August 1993, the incubation of about 500 g DW $m^{-2}$ of decaying macroalgae resulted in anoxia accompanied by the production of large quantities of free-sulphide in both the light and dark benthic chambers (Figure 7). In the dark chamber without macroalgae, the sulphide flux through the bare sediment was several times lower and was delayed by almost 24 hours. The presence of free sulphide in the water of the Prévost lagoon is not unusual probably due to the large amount of organic matter undergoing rapid degradation and the low buffering capacity of the system (Caumette, 1986). Although the low buffering capacity of the sediments may explain the high sulphide fluxes recorded, the great difference in fluxes between the dark chamber containing bare sediment and those containing *Ulva* infers that sulphide production was initiated within the decomposing macroalgal mats.

The dramatic changes in redox conditions which occurred in the dark benthic chambers were accompanied by the substantial release of both ammonium and SRP. Dark recycling of P and N was clearly related to *Ulva* decay, even though significant ammonium fluxes were also observed in the bare sediment chamber. However, in the latter case the ammonium to phosphorus ratios were close to the conventional Redfield ratio (N:P = 16:1), whereas in the chamber with *Ulva* SRP

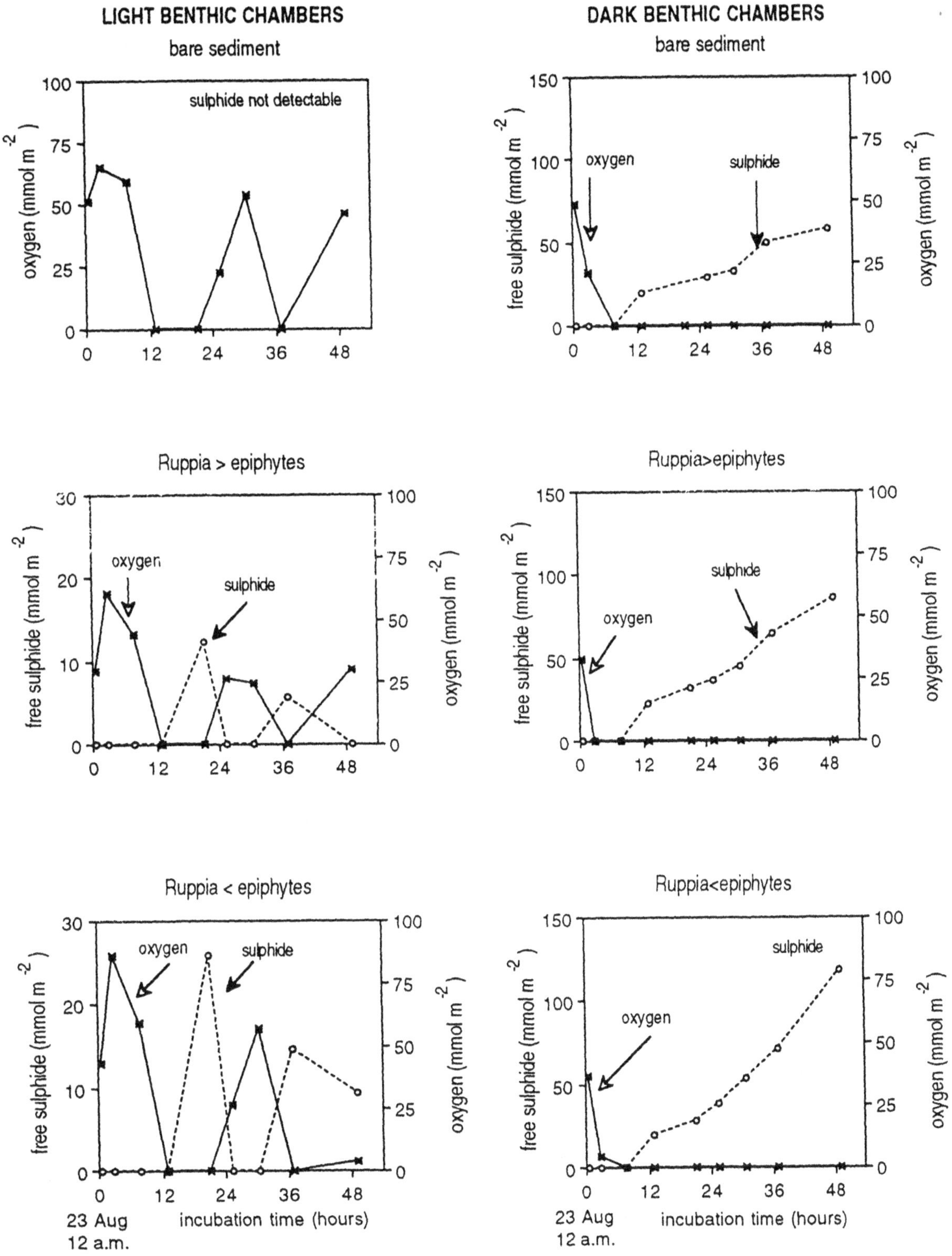


*Figure 5.* Main results of the experiment with benthic chambers carried out at station C1 from 23 until 25 August 1994 (explanation given in the text).

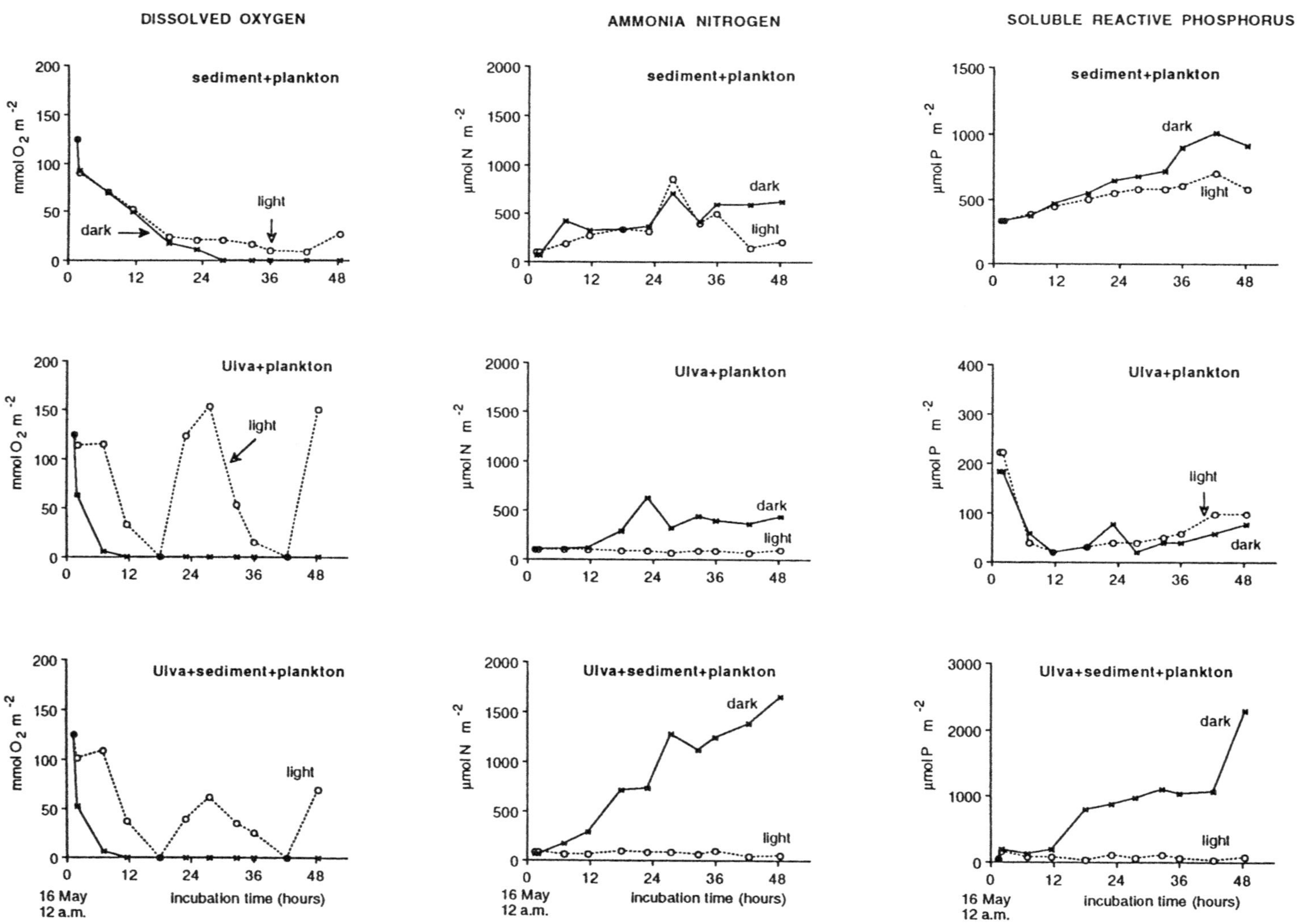


*Figure 6.* Main results of the experiment with benthic chambers carried out at station 11 from 16 until 18 May 1994.

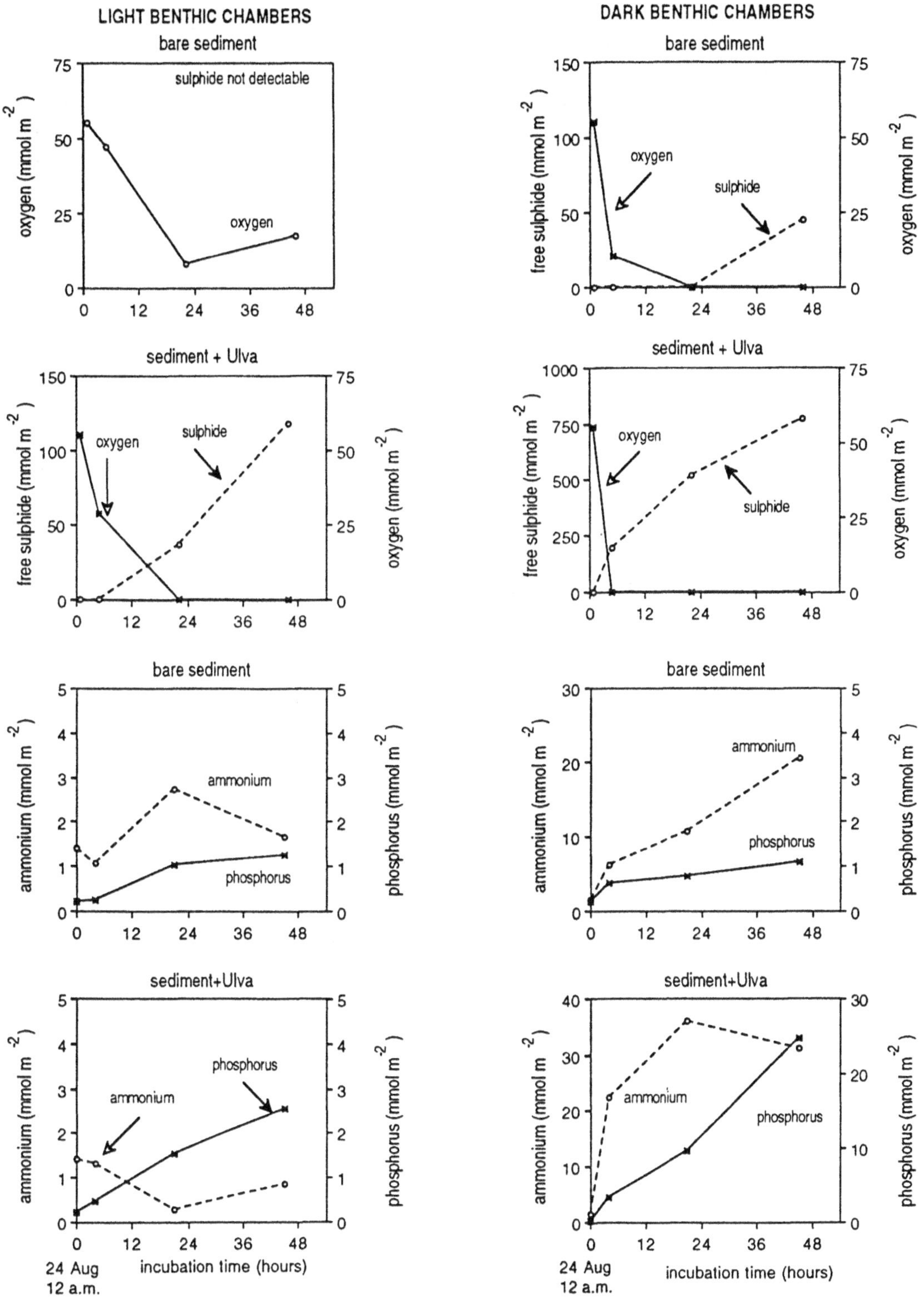


*Figure 7.* Main results of the experiment with benthic chambers carried out at station 11 from 24 until 26 August 1993.

flux was at least three times greater. This trend allow us to conclude that *Ulva* decomposition may result in an imbalance between ammonium and SRP flux, probably because ammonium was primarily recycled from the decomposing biomass, while SRP was also released from anoxic sediment (Figure 7). In contrast, net ammonium release was not recorded in the light, indicating that macroalgae may function as a nitrogen sink, even under the extreme conditions which developed after several hours of incubation.

*Intersystem comparison*

The macroscopic differences observed at the selected sites are to a large extent dependent on hydrodynamic factors: station B is well flushed by tidal currents, station 11 is slowly flushed due to the bathymetry of the Etang du Prévost, while station C1 is in a man-made lagoon which undergoes limited water exchange (Escaravage, 1990). However, we can reasonably postulate that internal factors linked to macrophyte community structure also contribute to the maintenance of such diversity (Odum, 1990; Brinson, 1993).

Standing crop data provide basic information on the structure of the primary producer communities. At intertidal station B, the *Z. noltii* meadows are characterized by slow growth and an almost constant biomass. In the more sheltered station C1, *R. Cirrhosa*, which has a morphology similar to that of *Z. noltii*, underwent a period of decomposition in the summer due to extensive growth of epiphytes. At station 11, in the Etang du Prévost, *U. rigida* reached very high biomass levels. However, ephemeral macroalgae such as *Ulva* have a patchy distribution because they are affected by prevailing hydrodynamic conditions, such as those induced by wind and currents.

The differences in structure and metabolism of the primary producers can be related to the wide range of fluxes of oxygen, nutrients, and free-sulphide which were recorded at the different sampling sites. Initially, we used a first ordination of such differences by using the metabolism based trophic index (BTSI) modified from that proposed by Rizzo et al. (in press). BTSI is based on the relative magnitude of hourly rates of maximum net oxygen fluxes measured under both light and dark conditions (Table 3). In this scheme maximum oxygen production in the light (X-axis) is plotted against oxygen uptake in the dark (Y-axis). Relative positions are indicative of degrees of autotrophy and heterotrophy. We used this approach to assign to each benthic system a categorical value which broadly reflects the extent to which each site can support processes associated with oxygen availability, i.e. benthic autotrophy. This index becomes very useful when comparing sites on a seasonal basis (Figure 8). For example, station 11 which was autotrophic during the spring growth phase clearly became heterotrophic in the summer decomposition phase. Station C1 underwent slight changes probably linked to shifts in community structure, while station B was highly autotrophic at all times. In the latter case, the relationship between oxygen production and consumption was presumably overestimated due to the short-term incubation. Moreover, high oxygen availability would be likely maintained by tidal flushing. Results from the potential sulphide fluxes induced by incubating macrophytes inside the benthic chambers show that they were proportional to oxygen consumption (Figure 9). This pattern highlights the relationship between macrophyte community and ecosystem stability. As previously discussed, data from station B should be considered carefully being sulphide fluxes affected by tidal flushing.

If we consider oxygen and sulphide fluxes as indicators of the stability and resilience of these lagoon systems, then a clear progression in macrophyte succession is evident viz.:

*Zostera* ≥ *Ruppia* > *Ruppia* + epiphytes >> *Ulva*

with the *Zostera* system more buffered than the *Ulva* one. This explains why the Etang du Prévost is subject to annual dystrophic crises (Caumette, 1986), while the fishponds of Certes experience a seasonal decay as well as summer anoxia mainly due to the massive development of epiphytes on the *Ruppia* (Escaravage, 1990; Thimel & Labourg, 1992).

A quite similar pattern can be also recognized when the fibre content of the detritus is considered, which is an index of organic matter biodegradability or recalcitrance (Godshalk & Wetzel, 1978; Mann, 1988):

*Zostera*≥healthy *Ruppia* (refractory)>

*Ruppia* epiphytes (intermediate)>*Ulva* (labile)

The large differences between the rooted and floating systems have also been demonstrated by means of vertical profiles taken along the water-sediment horizon (Figure 10). In the system with *Ulva* high sulphide concentrations were detected in the water layer beneath the floating thalli. These values were almost twice as high as those recorded either in the cores without macroalgae or with *Ruppia*. In the latter appreciable amounts of sulphide were detected only a few

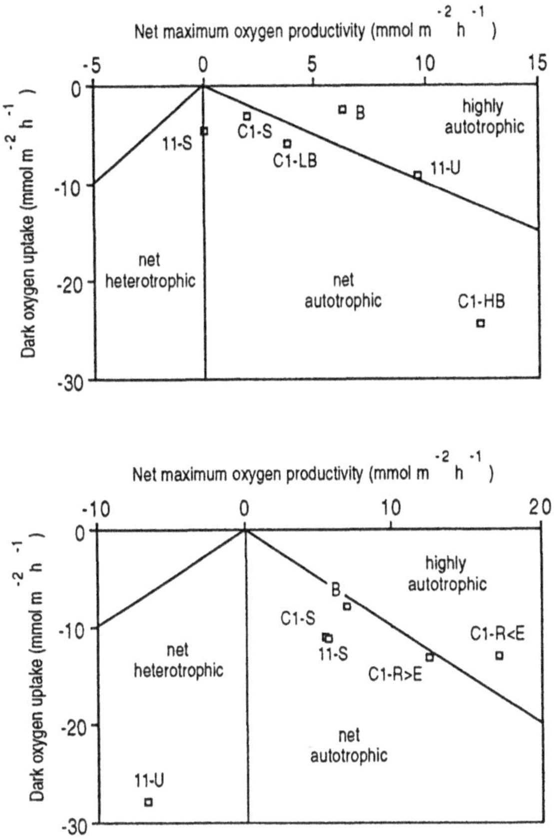


*Figure 8.* Ordination of the sites considered according to the BTSI (metabolism based trophic state index). Top panel: growth phase, bottom panel: decay phase. Symbols — station B: with *Zostera* (B); station 11: with *Ulva* (11-U) and with bare sediment (11-S); station C1: with low (C1-LB) and high (C1-HB) biomass of *Ruppia*, *Ruppia* in excess of epiphytes (R>E), epiphytes in excess of *Ruppia* (R<E), and bare sediment (C1-S). Further details are given in the text.

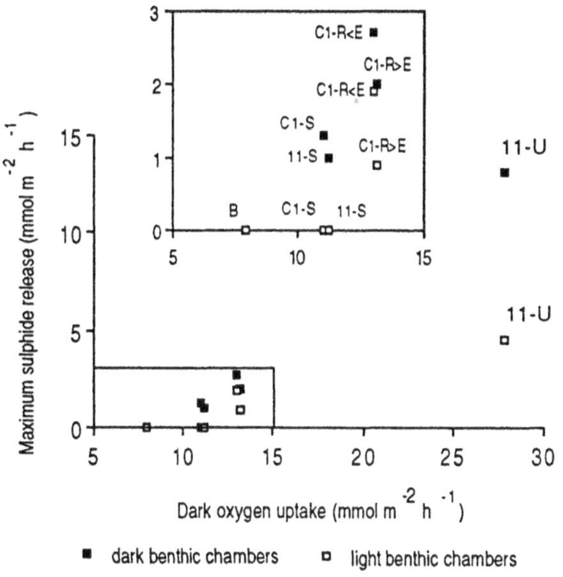


*Figure 9.* Relationships between dark oxygen uptake and maximum potential sulphide fluxes in dark and light benthic chambers. Symbols as in Figure 8.

centimeters below the sediment surface and can be explained by the deposition of a thick bed of epiphytes with an organic matter content in the range 20–30% w/w. Such organic matter consists mainly of mucilaginous aggregates which are readily degradeable and can sustain high levels of microbial activity which exceed the buffering capacity of the available iron (Bartoli et al., 1996). Likewise, the significant fluxes of sulphide observed in the dark and light benthic chambers can be explained by the production of free-sulphide by the epiphytes growing on the leaves of the *Ruppia* or by the epiphytes covering the sediment surface. These results demonstrate the importance of epiphytes in establishing very active localised systems at the sediment - water interface and around the *Ruppia* leaves. In the short-term, the epiphytic pathway may operate almost independently from that in the sediment and may thus determine internal fluxes which overlap the benthic ones. These results are consistent with those of Thimel & Labourg (1992), who recognised the key role of detritus in determining internal carbon flows.

To some extent, the epiphytic subsystem at station C1 resembles the subsystem which developed at station 11 in the Etang du Prévost due to accumulation of *Ulva*. Both appear to result in significant direct effects on water quality, even though station 11 is less protected against dystrophic crisis. These differences between the two sites, which are undergoing similar slow water reneval, can be explained primarily by the different macrophyte communities.

## Concluding remarks

*Ulva rigida*, *Ruppia cirrhosa* and *Zostera noltii* show very different growth patterns as well as production rates (Figure 11). In a number of dystrophic lagoons, *Ulva rigida* reaches huge biomasses (up to $10^3$ g DW $m^{-2}$) in few weeks (Sfriso et al., 1992). Unlike, rooted phanerogams they have annual cycles and their biomasses never exceed a few hundred g DW $m^{-2}$ (Verhoeven, 1980, Philippart, 1995).

The fast growth of *Ulva* and the very high biomass produced does not enable the top down mechanisms

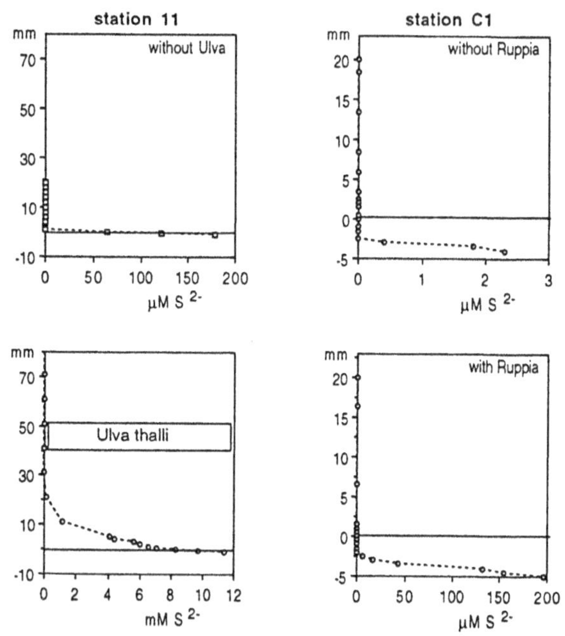


*Figure 10.* Sulphide profiles determined along a water-sediment horizon in sediment cores collected at station C1 (right) and 11 (left) in August 1994.

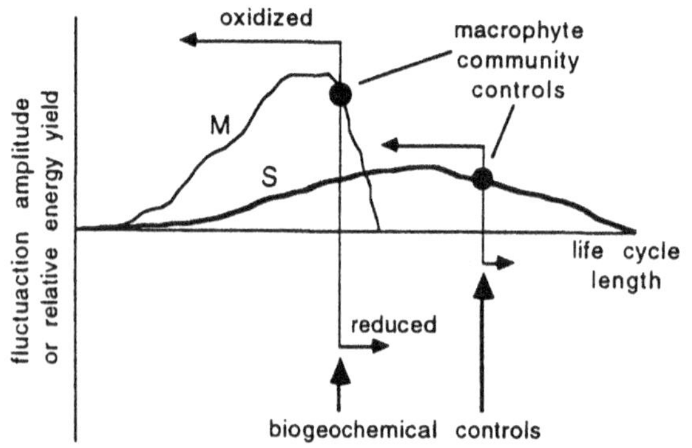


*Figure 11.* Schematic representation of intrasystem variablity in macroalgal (M) and seagrass (S) communities. Macrophyte controls: density, self shading, nutrient limitation, timing of detritus production, molecular recalcitrance, etc.

to control primary production and can lead the system organization to a shortened food chain (Smith & Horne, 1985, Pugnetti et al., 1992). Thus, energy flow is characterized by wide seasonal pulses with great organic matter accumulation and then great instability. In this manner, a change in the community structure can affect the organic matter conversion and determine a disturbance in the ecosystem functioning (e.g. dystrophy). Moreover, biogeochemical reactions (e.g. precipitation of metal sulphide) can exert a bottom-up control lowering the amplitude of such disturbance. When this kind of biogeochemical control is ineffective, the effects of dystrophic events would presumably lower the resistance as well as the resilience of the lagoon community leading to a positive feedback.

## Acknowledgments

This research was supported by the joint EU project 'CLEAN': Coastal Lagoon Eutrophication and ANaerobic processes (Contract No EV5V-CT92-0080) We gratefully acknowledge the support and facilities provided in Arcachon by Professor Pierre Caumette, Dr Jacques Castel, Dr P. J. Labourg, Dr R. de Wit and their colleagues during the course of this project. We are indebted with Dr M. Cattadori (Istituto di Ecologia, Università di Parma) who contributed to the field and laboratory work and useful information on the refractory compounds.

## References

A.P.H.A., A.W.W.A., W.P.C.F., 1975. Standard methods for the examination of water and wastewaters. 14th ed. A.P.H.A. Washington, 1193 pp.

Aspila, K. I., H. Ageminian & A. S. Y. Chau, 1976. A semiautomated method for the determination of inorganic, organic and total phosphate in sediments. Analyst 101: 187– 197.

Auby, I., F. Manaud, D. Maurrer & G. Trut, 1994. Etude de la proliferation des algues vertes dans le Bassin d'Arcachon. Rapport IFREMER-CEMAGREF-SSA-SABARC. IFREMER, Arcachon, 163 pp.

Bartoli, M., M. Cattadori, G. Giordani & P. Viaroli, 1996. Benthic oxygen respiration, ammonium and phosphorus regeneration in surficial sediments of the Sacca di Goro (Northern Italy) and two french coastal lagoons: a comparative study Hydrobiologia 329 (Dev. Hydrobiol. 117): 143–159.

Brinson, M. M., 1993. Changes in the functioning of wetlands along environmental gradients. Wetlands 13: 65–74.

Buchsbaum, R., I. Valiela, T. Swain, M. Dzierzesky & S. Allen, 1991. Available and refractory nitrogen in detritus of coastal vascular plants and macroalgae. Mar. Ecol. Prog. Ser. 72: 131–143.

Castel, J., P. Caumette & R. Herbert, 1996. Eutrophication gradients as exemplified by the Bassin d'Arcachon and the Etang du Prévost. Hydrobiologia 329 (Dev. Hydrobiol. 117): xi–xxx.

Caumette, P., 1986. Phototrophic sulphur bacteria and sulphate reducing bacteria causing red waters in a shallow brackish coastal lagoon (Prevost Lagoon, France). FEMS Microbiol.-Ecol. 38: 113–124.

Christian, R. R., E. Fores, F. A. Comin, P. Viaroli, M. Naldi & I. Ferrari, 1996. Nitrogen cycling networks of coastal ecosystems: influence of trophic status and primary producer form. Ecol. Model. 87 (in press).

Cline, J. D., 1969. Spectrophotometric determination of hydrogen sulphide in natural waters. Limnol. Oceanogr. 14: 454–459.

Crawford, R. M. M., 1992. Oxygen availability as an ecological limit to plant distribution. Adv. ecol. Res. 23: 93–185.

Duarte, C. M., 1992. Nutrient concentration of aquatic plants: patterns across species. Limnol. Oceangr. 37: 882– 889.

Enriquez, S., C. M. Duarte & K. Sand-Jensen, 1993. Patterns in decomposition rates among photosynthetic organisms: the importance of detritus C:N:P content. Oecologia 94: 457–471.

Escaravage, V., 1990. Daily cycles of dissolved oxygen and nutrient content in a shallow fishpond: the impact of water renewal. Hydrobiologia 207: 131–136.

Fong, P., T. C. Foin & J. B. Zelder, 1994. A simulation model of lagoon algae based on nitrogen competition and internal storage. Ecol. Monogr. 64: 225–247.

Giesen, W. B. J. T., M. M. Van Katwijk & C. Den Hartog, 1990. Temperature, salinity, insolation and wasting disease of eelgrass (*Zostera marina* L.) in the Dutch Wadden Sea in the 1930's. Neth. J. Sea Res. 25: 395–404.

Godshalk, G. L. & R. G. Wetzel, 1978. Decomposition of aquatic angiosperms. III. *Zostera marina* L. and a conceptual model of decomposition. Aquat. Bot. 5: 329–354.

Goering, H. K. & P. J. Van Soest, 1970. Forage fibre analysis (apparatus, reagents, procedure and some applications). U.S. Dept. Agric., Agric. Res. Ser., Agric. Hanab., 379 pp.

Izzo, G. & V. Hull, 1991. The anoxic crises in dystrophic processes of coastal lagoons: an energetic explanation. In C. Rossi & E. Tiezzi (Eds), Ecological Physical Chemistry. Elsevier Sci. Pbls. Amsterdam: 559–572.

Koroleff, F., 1970. Direct determination of ammonia in natural waters as indophenol blue. Information on techniques and methods for seawater analysis. I.C.E.S. Interlaboratory Rep. No. 3: 19–22.

Kuhl, M. & B. B. Jørgensen, 1992. Microsensor measurements of sulfate reduction and sulfide oxidation in compact microbial communities of aerobic biofilms. Appl. envir. Microbiol. 58: 1164–1174.

Mann, K. H., 1988. Production and use of detritus in various freshwater, estuarine, and coastal marine ecosystems. Limnol. Oceanogr. 33: 910–930.

Nienhuis, P. H., 1992. Eutrophication, water management, and the functioning of Dutch estuaries and coastal lagoons. Estuaries 15: 538–548.

Odum, W., 1990. Internal processes influencing the maintenance of ecotones: do they exist? In R. J. Naiman & H. Décamps (eds), The Ecology and Management of Aquatic- terrestrial Ecotones. MAB 4. The Parthenon Publishing Group, Paris: 91–102.

Philippart, C. J. M., 1995. Seasonal variation in growth and biomass of an intertidal *Zostera noltii* stand in the Dutch Wadden Sea. Neth. J. Sea Res. 33: 205–218.

Pugnetti, A., P. Viaroli & I. Ferrari, 1992. Processes leading to dystrophy in a Po River Delta lagoon (Sacca di Goro): phytoplankton-macroalgae interactions. Sci. tot. Envir. suppl. 1992: 445–456.

Rizzo, W. M., S. K. Dailey, G. J. Lackey, R. R. Christian, B. E. Berry & R. L. Wetzel, 1995. A metabolism-based trophic state index for comparing the ecological values of shallow water sediment habitats. Estuaries (in press).

Sand-Jensen, K. & J. Borum, 1991. Interactions among phytoplankton, periphyton, and macrophytes in temperate freshwater and estuaries. Aquat. Bot. 41: 137–175.

Sfriso, A., B. Pavoni, A. Marcomini & A. A. Orio, 1992. Macroalgae, nutrient cycles, and pollutants in the lagoon of Venice. Estuaries 15: 517–528.

Sherman, K., 1994. Sustainability, biomass yields, and health of coastal ecosystems: an ecological perspective. Mar. Ecol. Progr. Ser. 112: 277–301.

Smith, D. W. & A. J. Horne, 1985. Experimental measurement of resource competition between planktonic microalgae and macroalgae (seaweeds) in mesocosms simulating the San Francisco Bay-Estuary, California. Hydrobiologia 159: 259–268.

Stevenson, J. C., 1988. Comparative ecology of submersed grass beds in freshwater, estuarine, and marine environments. Limnol. Oceanogr. 33: 867–893.

Thimel, A. & P. J. Labourg, 1992. *In situ* metabolism of a benthic community in a shallow brackish lagoon (Fish-impoundments of Arcachon Bay, Atlantic coast, France). Vie Milieu 42: 185–192.

Twilley, R. R., W. M. Kemp, K. W. Staver, J. C. Stevenson & W. R. Boynton, 1985. Nutrient enrichment of estuarine submersed vascular plant communities. Algal growth and effects on production of plants and associated communities. Mar. Ecol. Prog. Ser. 23: 179–191.

Valderrama, J. C., 1977. Methods used by the Hydrographic Department of National Board of Fisheries, Sweden. In K. Grasshoff (ed.), Report of the Baltic Intercalibration Workshop. Interim Commission for the Protection of the Environment of the Baltic Sea: 14–34.

Van Gemerden, H., C. S. Tughan, R. de Wit & R. A. Herbert, 1989. Laminated microbial ecosystems on sheltered beaches in Scapa Flow, Orkney Islands. FEMS Microbiol. Ecol. 62: 87–102.

Verhoeven, J. T. A., 1980. The ecology of *Ruppia*- dominated communities in Western Europe. II. Structure and dynamics of the macroflora and macrofauna communities. Aquat. Bot. 8: 1–85.

Viaroli, P., M. Bartoli, C. Bondavalli, M. Cattadori, G. Giordani & M. Naldi, 1994. Oxygen, sulphide and nutrient fluxes in shallow eutrophic lagoons with different primary producer communities. I. Macrophyte communities and their impact on benthic fluxes. In P. Caumette (Coord.), C.L.E.A.N. Progress Report, European Commission: 327–352. European Community Environment Programme, DG XII, Brussels.

Viaroli, P., M. Bartoli, C. Bondavalli & M. Naldi, 1995. Oxygen fluxes and dystrophy in a coastal lagoon colonized by *Ulva rigida* (Sacca di Goro, Po River Delta, Northern Italy). Fresenius Envir. Bull. 4: 381–386.

Williams, S. L. & M. H. Ruckelshaus, 1993. Effects of nitrogen availability and herbivory on eelgrass (*Zostera marina*) and epiphytes. Ecology 74: 904–918.

*Hydrobiologia* **329**: 121–131, 1996.
*P. Caumette, J. Castel & R. Herbert (eds), Coastal Lagoon Eutrophication and ANaerobic Processes (C.L.E.AN.).*



# Differential anaerobic decomposition of seagrass (*Zostera noltii*) and macroalgal (*Monostroma obscurum*) biomass from Arcachon Bay (France)

Sophie Bourguès*, Isabelle Auby, Rutger de Wit & Pierre-Jean Labourg
*Laboratoire d'Océanographie Biologique, Centre d'Océanographie et de Biologie Marine, Université Bordeaux I and C.N.R.S.-U.R.A.-197, 2 rue du Pr. Jolyet, 33120 Arcachon, France*
**Corresponding author: Fax: 33-56-83 51 04*

*Key words: Zostera noltii, Monostroma obscurum,* anaerobic decomposition, nutrient recycling

## Abstract

Arcachon Bay is characterized by extensive meadows of the seagrass *Zostera noltii*. Moreover, as a consequence of eutrophication, massive proliferations of the macroalga (*Monostroma obscurum*) have occurred since the beginning of 1990s.

This paper describes the anaerobic decomposition of biomass of both species under experimental conditions by two methods. Firstly, the dynamics of decomposition were studied *in situ* using litter bags. The remaining biomass and the elemental composition of the decomposing macrophytes were monitored. Secondly, degradation was studied in experimental containers under anoxic conditions in which the release of inorganic nutrients and the development of fermentative and sulfate-reducing bacterial populations were followed.

The decomposition rate of total biomass was faster for macroalgae than for the vascular plants, thus corroborating previous observations. However, both *in situ* and laboratory experiments showed that the anaerobic decomposition of the seagrass *Z. noltii* resulted in rapid release of inorganic N and P, and increasing C/N and C/P ratios of the residual biomass. As a result, the recycling of inorganic nitrogen and phosphorus compounds was slightly more efficient for *Z. noltii* than for *M. obscurum*. Recycling of inorganic nutrients appears to be of a great importance to the whole ecosystem, because of the extensive spreading of *Z. noltii* in the bay.

## Introduction

Arcachon Bay is a triangular shaped mesotidal lagoon on the French Atlantic coast which covers an area of 156 km$^2$. Presently, primary production in this lagoon is mainly due to two macrophytes: the perennial phanerogam *Zostera noltii* Hornem. which forms extensive meadows with above and below-ground biomasses of 70–100 and 70–160 g dry weight m$^{-2}$, respectively (Auby, 1991; Auby & Labourg, in press), and the macroalga *Monostroma obscurum* Kutzing. (Monostromaceae), which proliferates during spring and early summer. The *Z. noltii* meadows cover 70 km$^2$ of the intertidal area of the lagoon (Figure 1) representing the largest seagrass meadows in Western Europe (Auby, 1991). The abundance of this small seagrass species has been described since the last century (Durieu de Maisonneuve, 1855). In contrast, the occurrence of macroalgal blooms is a recent phenomenon, which is most likely due to increased nutrient loading. Massive blooms of *Enteromorpha clathrata* (Roth) Grev. have been reported since 1982 (Ribes, 1988) and continued throughout the 1980's. However, the importance of this species declined during the early 1990's, and concomitantly, *M. obscurum*, which was first observed in 1990, started to proliferate in the Eastern, and particularly in the Southeastern part of the lagoon (Figure 1). A detailed study of this recent phenomenon has been published by Auby et al. (1994), who reported that the maximum biomass of *M. obscurum* and *Z. noltii* (including leaves, roots and rhizomes) were 2.5 × 10$^6$ kg and 14 × 10$^6$ kg, respectively. Thus, *M. obscurum* biomass is equal to approximately 20% of that of *Z. noltii* (by dry weight).

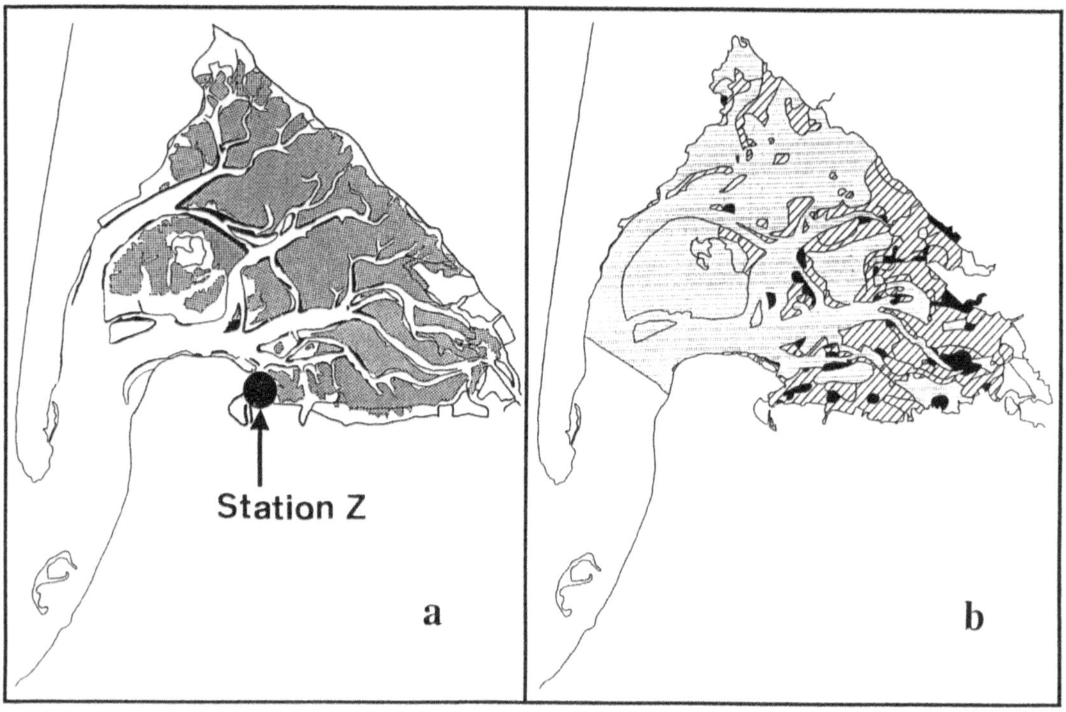


*Figure 1.* a. Distribution of *Zostera noltii* (pointed) and *Zostera marina* (black) meadows in Arcachon Bay and location of the study site Z. b. Repartition of *Monostroma obscurum* biomass in Arcachon Bay (July 1993). pointed: 0–10 g fresh weight $m^{-2}$ (intertidal) or $m^{-3}$ (subtidal) hatched: 10–500 g fresh weight $m^{-2}$ (intertidal) or $m^{-3}$ (subtidal) black: >500 g fresh weight $m^{-2}$ (intertidal) or $m^{-3}$ (subtidal).

Grazing by herbivores on *Z. noltii* is quantitatively of little importance, as comparable to other vascular plants. The major herbivores are populations of the dark-bellied Brent geese (*Branta bernicla bernicla*) which overwinter in the lagoon. Although numerous (up to 40 000 birds), their consumption represents at most 2% of the above-ground annual production of *Z. noltii* (Auby, 1991). A small part of the biomass is exported out of the bay by the currents. However, the importance of this biomass export remains unquantified. Hence, a large part of the *Z. noltii* detritus decomposes at the sediment surface or in the sediment itself, supplying large quantities of organic matter to the system. The slow decomposition rates characteristic of different seagrass species have been described extensively in the literature (Harrisson & Mann, 1975; Thayer et al., 1977; Godschalk & Wetzel, 1978; Pellikaan, 1984). This feature is often attributed to its relatively high content of refractory constituents, such as structural carbohydrates, lignin and cellulose, as well as to the C:N:P ratios (Enriquez et al., 1993).

The macroalga *M. obscurum* constitutes a food source for gastropods and amphipods. Nevertheless, it appears that the grazing rate is extremely low (Auby et al., 1994). In addition, the export of this macroalga by currents is presumably of minor importance, because its production takes place predominantly in the confined parts of the bay. As a result, a large fraction enters the detritus food chain and is mineralized in the sediments. Macroalgae are therefore important in nutrient cycles in coastal lagoons, because they act as both nutrient sinks and sources, during growth and decay phases, respectively (Lavery & McComb, 1991). The relatively fast decomposition rate of macroalgae provides also a rapid recycling of nutrients which may be an important factor to maintain high productivity (Hanisak, 1993).

This paper reports the results of a comparative study of the decomposition of *Z. noltii* and *M. obscurum* detritus under anoxic conditions. The research was conducted by both *in situ* and laboratory incubation experiments. The decay rates were compared and attention was paid to the importance of inorganic nutrient recycling.

## Materials and methods

### *Sites of sampling and in situ experiments*

In 1992–1993, *Z. noltii* leaves and *M. obscurum* thalli were sampled at station Z which was located at mean tide level (M.T.L.) on an intertidal mudflat, in Arcachon Bay (Figure 1). The *in situ* decomposition experiments were performed adjacent to the sampling station.

The freshly collected macrophytes were transported to the laboratory in coolboxes and immediately sorted and rinsed with sea water. During spring 1992, decomposition experiments were conducted both with leaves of *Z. noltii* and thalli of *M. obscurum*; additional experiments were performed in other seasons with *M. obscurum* thalli only. At the start of the experiments, replicate litter bags of 1 mm nylon mesh (21 × 21 cm) were filled with the same amount of freshly harvested and rinsed leaves or thalli (30 to 50 g wet weight) and buried 1 cm deep in the sediment. Minielectrode measurements showed that permanently anoxic conditions occurred at these sediment depths. The entire *in situ* experiments covered the periods listed in Table 1. At selected time intervals, one or two bags were harvested for subsequent analyses of remaining dry weight and elemental composition (i.e., C, N and P contents). Prior to these analyses, the harvested detritus was gently rinsed with sea water to remove sediment particles and subsequently with fresh water to remove salt.

### *Laboratory experiments*

Laboratory experiments were performed to investigate inorganic nutrient release. To simulate environmental anoxic dark conditions, aquaria (21 × 21 × 22 cm) were wrapped in aluminium foil and sealed with silicone to prevent oxygen ingress. A volume of 30 $cm^3$ of sediment from station Z was added, the aquaria completely filled with seawater (9.7 l) and amended as follows: (1) control treatment containing sediment alone. (2) Sediment plus 100 g wet weight of *M. obscurum* (equivalent to 10.78 g dry weight), and (3) Sediment plus 100 g wet weight of *Z. noltii* (equivalent to 14.23 g dry weight). Treatments 2 and 3 were established in duplicate. Incubation was performed at ambient laboratory temperature (20–25 °C) for about 2 months. During incubation, water samples for chemical analyses and bacterial enumerations were withdrawn through a sample port with syringes (50 ml) to prevent atmospheric contamination.

### *Analytical procedures*

Dry weight of macrophyte detritus was determined after drying at 60 °C for 48 hours. Analysis of Carbon, Nitrogen and Phosphorus content were performed at the Institut Européen de l'Environnement de Bordeaux (IEEB). Organic carbon content was measured according to Le Corre (1983), nitrogen content with the Kjeldahl method (Allen, 1970) and phosphorus content by a colorimetric method after acidic digestion with $HNO_3$-$H_2SO_4$ (APHA, AWWA, WPCF, 1980).

Water samples were filtered (Whatman membrane, 0.45 $\mu$m pore size) and stored frozen in polyethylene vials until analysed for $NH_4^+$, $NO_2^-$+$NO_3^-$ and $PO_4^{3-}$ using a Technicon autoanalyser at the Laboratory of CREMA (l'Houmeau), according to Tréguer & Le Corre (1975). Sulfide concentrations were determined colorimetrically according to Cline (1969).

### *Bacterial enumerations*

Sulfate-reducing bacteria (SRB) were enumerated using the Most Probable Number technique in liquid medium with 10 mM sodium lactate and 10 mM sodium acetate (Pfennig et al., 1981). The medium was dispensed aseptically in Hungate's tubes and the growth of SRB was determined by the formation of a brown precipitate after addition of $CuSO_4$/HCl solution (5 mM/50 mM) (Caumette, 1986).

Fermentative bacteria were enumerated by counting Colony Forming Units (C.F.U.) on agar plates. The growth medium consisted of filtered (0.22 $\mu$m) sea water, 5 g peptone, 5 g yeast extract, 15 g agar-agar. The pH was adjusted to 7.2–7.4. The medium was autoclaved and dispensed into Petri dishes, which were spread using a portion from aseptically prepared decimal dilution series and incubated in anaerobic jars.

### *Curve-fitting*

The time courses of *in situ* biomass decomposition were modelled according to first order kinetics using the exponential function:

$$W_t = W_0 \times e^{-kt} \qquad \text{(Eq.1)}$$

or according to the logistic function:

$$W_t = \frac{W_0}{1 + b \times e^{kt}}, \qquad \text{(Eq.2)}$$

in which $W_t$ is the remaining biomass at any time $t$, $W_o$ is the biomass at the start of the experiment, $k$

*Table 1.* Main parameters characterizing the in situ anaerobic decomposition of *Monostroma obscurum* and *Z. noltii*.

| Season (Julian day) Year | Loss of dry weight at the end of the experiment | Model fitted | Calculated half time (days) | Average temperature of sediment | Initial C/N ratio |
|---|---|---|---|---|---|
| SPRING (73–143) 1992 *Z. noltii* | 77% in 66 days | $W_0 = 102.7 \times e^{-0.019t}$ $r^2 = 0.973$ $p < 0.001$ | 37 | 12.2 °C | 12.6 |
| SPRING (73–103) 1992 *M. obscurum* | 76% in 25 days | $W_0 = 93.3/(1 + 0.0002 \times e^{0.433t})$ $r^2 = 0.891$ $p < 0.001$ | 20 | 12.2 °C | 8.6 |
| SUMMER (191–209) 1992 *M. obscurum* | 73.5% in 18 days | $W_0 = 97.0 \times e^{-0.082t}$ $r^2 = 0.847$ $p < 0.01$ | 8.5 | 24.3 °C | – |
| AUTUMN (309–345) 1992 *M. obscurum* | 81% in 36 days | $W_0 = 102.9/(1 + 0.0020 \times e^{0.223t})$ $r^2 = 0.970$ $p < 0.001$ | 28 | 12.7 °C | 4.43 |
| WINTER (18–42) 1993 *M. obscurum* | 44% in 30 days | not significant | – | 9.3 °C | 5.17 |

*Table 2.* Enumeration of sulfate-reducing and fermentative bacteria (cells $ml^{-1}$) in the water of aquaria during the laboratory decomposition experiment.

| | | Control container | *M. obscurum* containers | *Z. noltii* containers |
|---|---|---|---|---|
| Sulfate-reducing bacteria | $t$=0 | $2.5 \times 10^1$ | $6 \times 10^1$ | $25 \times 10^3$ |
| | $t$ = 43 days | $2.5 \times 10^2$ | $5.5 \times 10^7$ | $1.9 \times 10^6$ |
| Fermentative bacteria | $t$ = 0 | $1.1 \times 10^2$ | $8.4 \times 10^1$ | $8.5 \times 10^5$ |
| | $t$ = 49 days | $1 \times 10^2$ | $6 \times 10^6$ | – |
| | $t$ = 76 days | $1.23 \times 10^2$ | – | $6.4 \times 10^6$ |

is the specific degradation rate (per day) and $b$ is a coefficient specific to the logistic function only, which is related to a time delay. Curve fitting was achieved by an iterative procedure to minimize the sum of squares of the residuals (Sokal & Rohlf, 1981).

## Results

### In situ *experiments*

*In situ* time course experiments of decomposition of *Z. noltii* and *M. obscurum* biomass differed considerably (Figure 2). Initially the decomposition rate of *M. obscurum* biomass was low, but increased substantially after a lag period of 15 days. The initial dry weight of 5.40 g decreased by 76% to 1.30 g over the 25 days incubation period. For *Z. noltii*, the initial dry weight of 7.11 g decreased to 1.64 g after 66 days, corresponding to a loss of 77%. Thus, the phanerogam biomass decomposed to a similar extent in a period which was double that found for *M. obscurum*.

The decomposition processes of macrophytes can generally be modelled by first order kinetics (exponential function) or if a significant time lag is observed, by a logistic function according to equation 1 and 2, respectively (see Method section). We found that first order kinetics provided a good description for

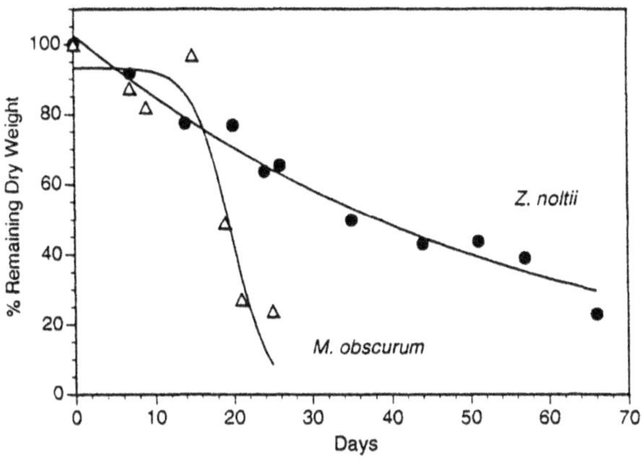


*Figure 2.* Loss of dry weight of Monostroma obscurum (open triangles) and *Zostera noltii* (solid circles) in litter bags, as percentage of remaining dry weight. Spring experiment (March 1992).

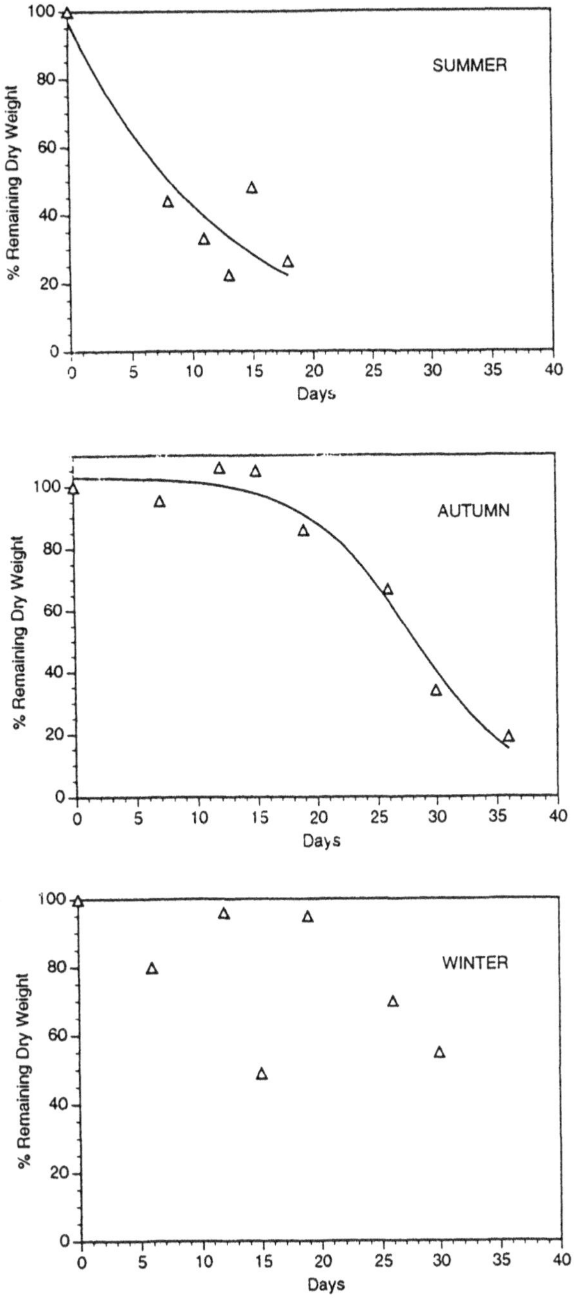


*Figure 3.* Seasonal loss of dry weight of *Monostroma obscurum* during decomposition in litter bags, as percentage of remaining dry weight. Summer (July 1992), autumn (November 1992) and winter (January 1993) experiments.

the decomposition of *Z. noltii* but not for *M. obscurum* which was best modelled by the logistic equation. Accordingly, the calculated times required for 50% loss of dry weight were 20 and 37 days for *M. obscurum* and *Z. noltii*, respectively.

The seasonal variations of the *in situ* decomposition was studied in more detail for *M. obscurum* (Figure 3). The decay process appeared very efficient in summer: exponential loss started without lag phase and 73.5% of the initial dry weight was lost in 18 days. However, in autumn, a lag phase characterized the first two weeks of decomposition, similarly as was observed during the spring experiment. After 36 days, only 19% of the initial algal detritus dry weight remained in the litter bags. Finally, the process of decay appeared to be very slow in winter, albeit it was difficult to quantify because neither exponential nor logistic curve fitting gave significant regression ($p>0.05$). Using the selected mathematical functions, the time required to lose 50% of the initial dry weight equalled 8.5 and 28 days for summer and autumn, respectively (Table 1).

During the spring experiment, we followed the time courses of carbon, nitrogen and phosphorus contents in the decomposing detritus (Figure 4). The initial carbon contents were 32.9% and 42.0% of detritus dry weight for *M. obscurum* and *Z. noltii*, respectively. Loss of carbon was more pronounced for the macroalga than for the seaweed. Carbon content of decaying *M. obscurum* detritus was 22.9% of dry weight towards the end of the experiment.

The initial nitrogen content of *M. obscurum* and *Z. noltii* were 4.45 and 3.90% of dry weight, respectively. This difference was small considering the reported general trends among for phanerogam and seaweed species (Duarte, 1992). However, this parameter showed markedly different time courses for both macrophytes during the decay process. The N-content of *M. obscurum* decreased strongly from 4.45 to 2.40%

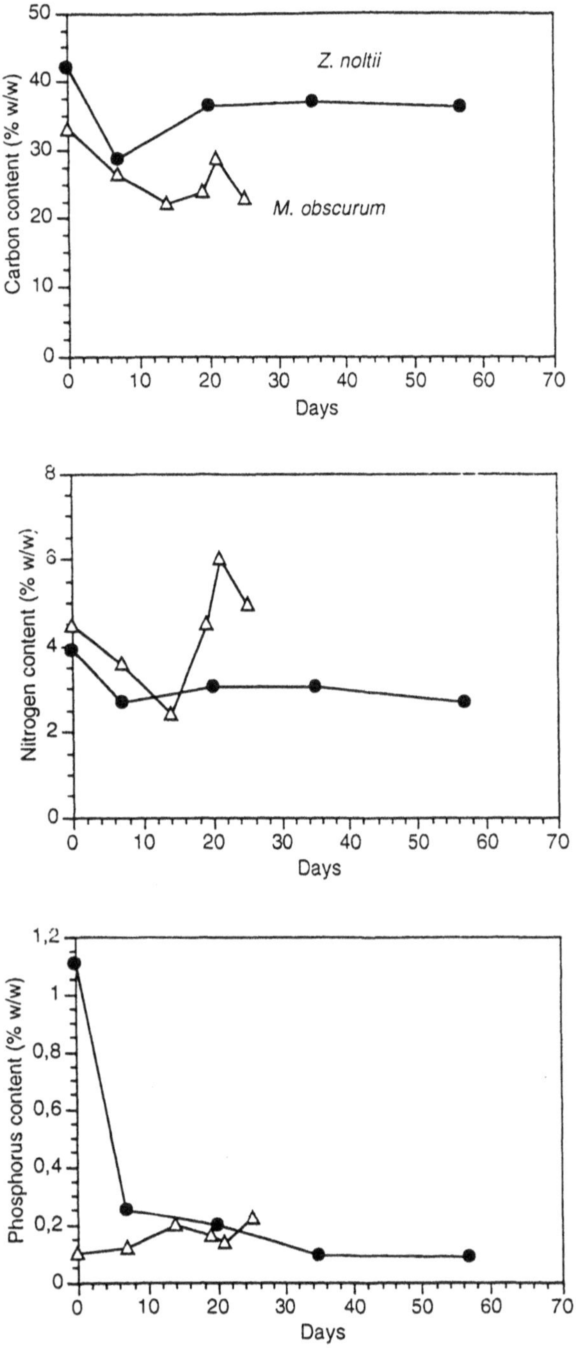


*Figure 4*. Carbon, nitrogen and phosphorus content as percentage (weight/weight) of dry weight during the spring decomposition of *Monostroma obscurum* (open triangles) and *Zostera noltii* (solid circles) in litter bags.

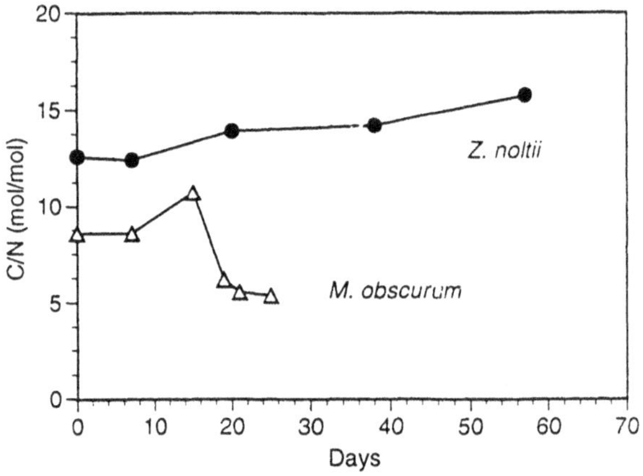


*Figure 5*. C/N atomic ratios during the spring decomposition of *Monostroma obscurum* (open triangles) and *Zostera noltii* (solid circles) in litter bags.

during the first two weeks while total biomass remained virtually constant (lag phase). Afterwards, whilst a rapid loss of remaining biomass occurred, N-content increased again to 6% (after 21 days) (Figure 4 and Table 1). In contrast, a minor but continuing decrease of nitrogen content was found for *Z. noltii*.

Initial phosphorus contents were 0.10% and 1.10% for *M. obscurum* and *Z. noltii*, respectively. Nevertheless, P-content of the latter appeared to be exceptionally high in March 1992. Comparably, Viaroli (pers. comm.) found values ranging from 0.2 to 0.4% in May and August 1993 at station B in Arcachon Bay, which is a more sheltered than station Z. After a week of incubation, P-content was 0.25% and then decreased further to 0.087%. In contrast, phosphorus content of *M. obscurum* detritic material increased during the decay process.

At the beginning of the spring experiment, the C/N ratios were 8.6 and 12.6 for *M. obscurum* and *Z. noltii*, respectively (Figure 5). Algal detritus C/N ratio increased during the first two weeks of the experimental period and then decreased (C/N = 5.39 after 25 days). On the other hand, the C/N ratio of seagrass leaves detritus gradually increased during the incubation up to 15.75.

The similarity of the decomposition process of *M. obscurum* in spring and autumn was also confirmed by the time course of the C/N ratio, which followed the same trend. In fact, C/N, in autumn, equaled to 4.43 at the beginning of the experiment. This ratio increased up to 5.18 during the two first weeks and then decreased (C/N = 3.82 after 30 days).

*Laboratory experiments*

Figure 6 represents the time course of dissolved inorganic nutrients ($NH_4^+$, $NO_2^-$ + $NO_3^-$, $PO_4^{3-}$) in the control container (no macrophyte addition). At the beginning of the experiment, $NH_4^+$ was not detectable in the water phase. A peak of 17 $\mu$M was observed after 13 days but this concentration declined rapidly and fell to 0.23 $\mu$M by the end of the incubation period. Concomitantly with the decrease of the $NH_4^+$ concentration, the concentration of ($NO_2^+$_$NO_3^-$) increased from 2.34 to 10.42 $\mu$M, indicating that ammonium was being nitrified ($NH_4^+$ – >$NO_2^-$ + $NO_3^-$). The absence of sulfide and the occurrence of nitrification indicated that oxygen remained present during the incubation period. Phosphate concentration (1.23 $\mu$M at $t$=0) increased concomitantly with $NH_4^+$ and dropped to 0.22 $\mu$M after 55 days.

Ammonium was the major inorganic nitrogen compound released into the water column during the degradation of both macrophytes. Furthermore, $NH_4^+$ showed the same evolution for both macrophytes (Figure 7). In containers amended with *M. obscurum*, $NH_4^+$ concentration increased with time and reached a plateau at approximately 2 mM. The same trend was observed in the containers amended with *Z. noltii*, but the plateau concentrations were higher (about 3.5 mM). The sum of $NO_2^-$ + $NO_3^-$ water column concentrations, which were initially 1.51 and 1.47 $\mu$M for *M. obscurum* and *Z. noltii* containers, respectively, increased with time. Although the final values were 3 and 2-fold higher than in the control experiment (i.e., up to 29.2 $\mu$M for *M. obscurum* containers), these amounts were negligible compared to the quantities of $NH_4^+$ released.

Laboratory incubations rather than field experiments in litter bags enabled the measurement of dissolved inorganic nitrogen recycling from the macrophyte detritus during the decay process. The original weight of *M. obscurum* was 10.78 g dry weight per container, with a nitrogen content of 4.45% thus corresponding to an N-amendment of 3.53 mmol $l^{-1}$. At the end of the experiment, detritus dry weight was 5.30 g with 3.6% of nitrogen, still representing 1.40 mmol $l^{-1}$, whilst the maximum concentration of ammonium in the water column was 2.21 mM. Thus, at the end of the experiment, the sum of N in both compartments corresponded to the totality (102%) of the original N added with biomass and dissolved ammonium was equal to 62.5% of the original amendment. In comparison, for the experiment with *Z. noltii* the recovery was 114%

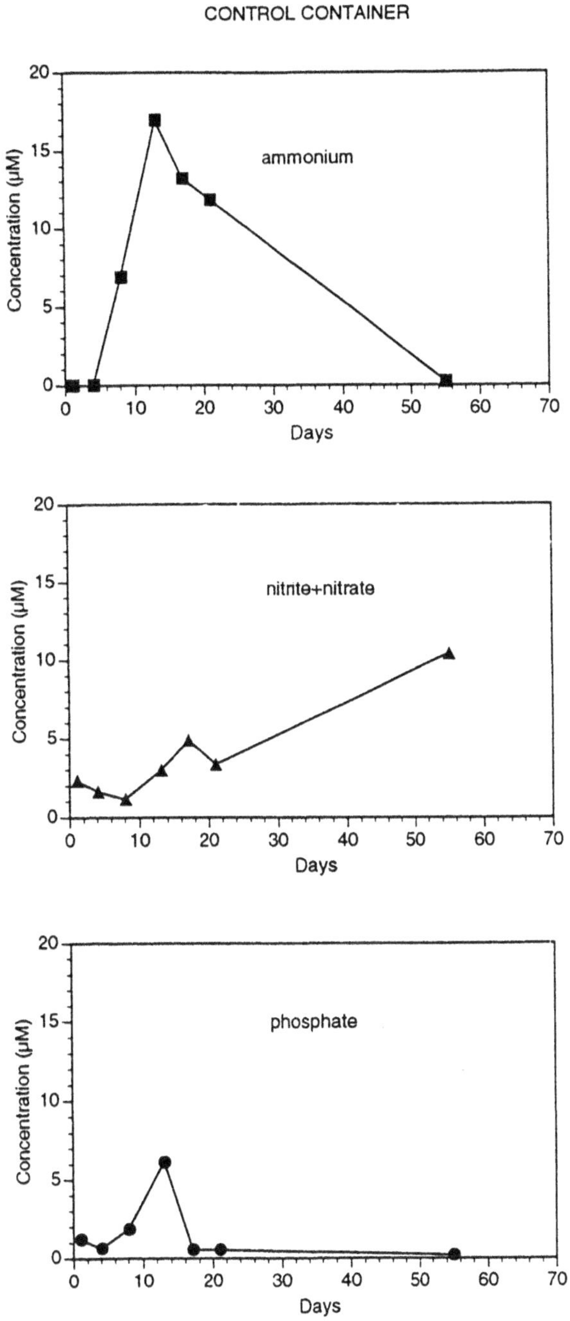


*Figure 6.* Variation of ammonium (solid squares), nitrite + nitrate (solid triangles) and phosphate (solid circles) concentrations in the water of control container during the laboratory degradation experiment.

and dissolved ammonium corresponded to 86% of the original N added with biomass.

The high phosphate concentrations observed in the *Z. noltii* containers (183 $\mu$M at $t$=8 days) can

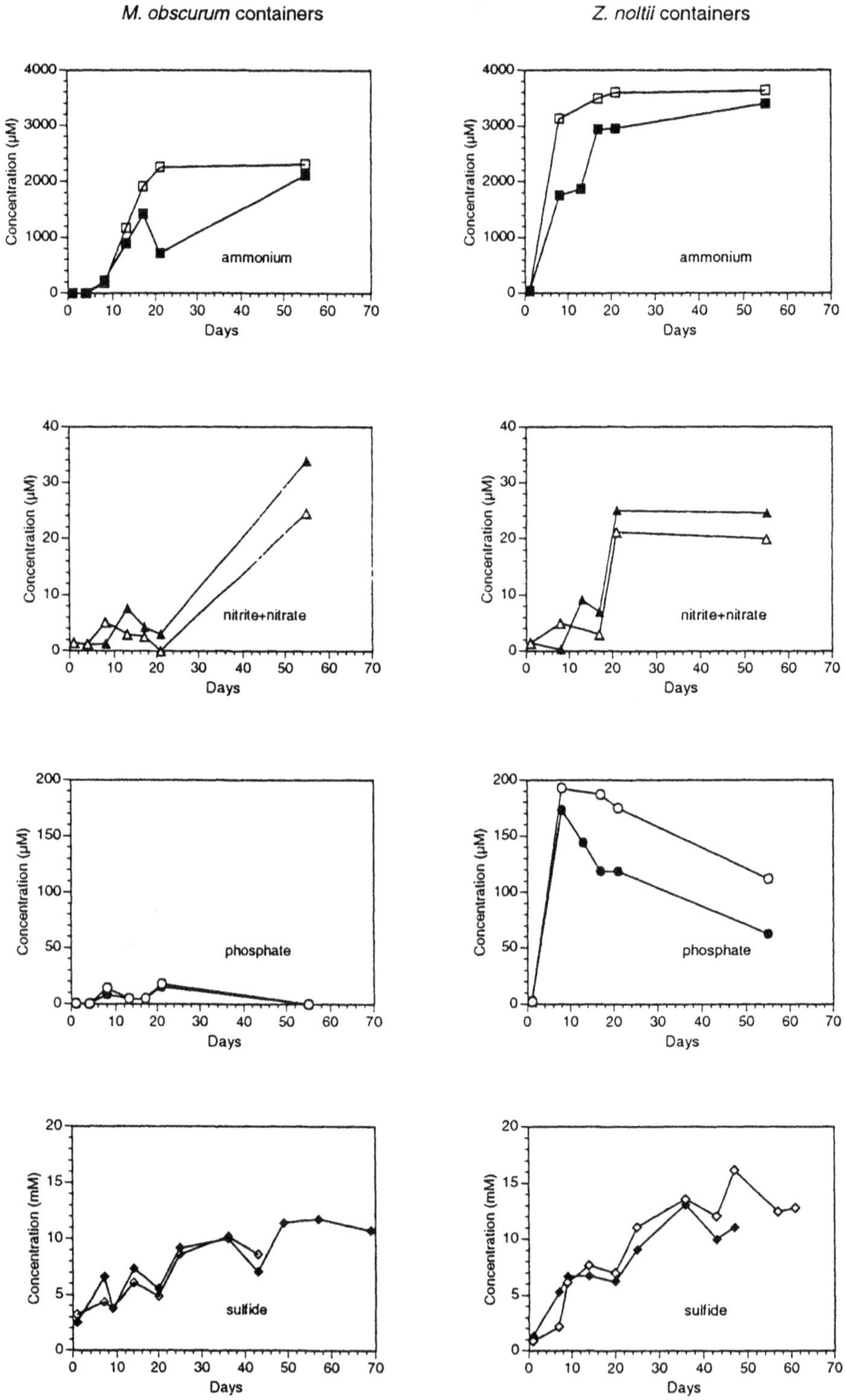


*Figure 7.* Variation of ammonium (solid and open squares), nitrite + nitrate (solid and open triangles), phosphate (solid and open circles) and sulfide (solid and open lozenges) concentrations in the water of aquaria containing *M. obscurum* and *Z. noltii*.

be explained by efficient P-release and the higher P-content. Maximal phosphate concentration reached 16.6 $\mu$M in the *M. obscurum* containers after the two first weeks of incubation but this value decreased with time to 0 after 55 days. A concentration of 87.4 $\mu$M was observed in the *Z. noltii* aquarium water at the end of incubation. Conceivably, the 'mass-balance' calculations for phosphorus are more difficult than for nitrogen compounds, because of the relative importance of P release or binding to sediment particles.

In both macrophyte incubations, $H_2S$ was detected from the first day of incubation. $H_2S$ concentrations increased gradually with time up to 10.4 mM in the *M. obscurum* container and to 16.3 mM in the case of *Z. noltii*.

The number of sulfate-reducing bacteria (SRB) increased with time in all treatments (Table 2). However, the increase was higher when macrophyte material was present. After 43 days of incubation SRB numbers reached $5.5 \times 10^7$ and $1.9 \times 10^6$ bacteria/ml for *M. obscurum* and *Z. noltii* containers, respectively. In contrast, trends in the density of fermentative bacteria were different. Fermentative bacteria were present in high numbers at the start but did not increase significantly in the *Z. noltii* combination, whereas a significant increase was detected in the *M. obscurum* container.

## Discussion

*In situ*, the decomposition rate of *M. obscurum* was about twice as fast as for *Z. noltii*. Even slower degradation rates have been found for *Z. noltii* (Hemminga & Nieuwenhuize, 1991). *M. obscurum* decomposition rates were however similar to those reported for the green alga *Caulerpa cupressoides* (Williams, 1984). Our findings thus corroborate previous studies that decomposition of marine vascular plants is a relatively slow process (Fenchel, 1977; Harrisson & Mann, 1975; Thayer et al., 1977; Godschalk & Wetzel, 1978; Pellikaan, 1984).

According to Rice & Tenore (1981), the rate of decomposition is largely dependent on the chemical composition of the macrophyte. Macroalgal tissues are commonly richer in nitrogen than vascular plant material (Duarte, 1992). The nitrogen content measured in spring for both macrophytes were however rather similar (4.45% and 3.90% for *M. obscurum* and *Z. noltii*, respectively). The difference between *M. obscurum* and *Z. noltii* decomposition rates can thus not be explained by this factor. Phosphorus appears to be present in labile form in *Z. noltii* and its leaching is very efficient, confirming the results of Auby (1991). Another factor influencing *in situ* decomposition rates may be detritus fragmentation. In fact, *Chironomus* larvae were detected from the first week in the litter bags containing decaying *M. obscurum* but were never found in the *Z. noltii* bags. The mechanical activity, facilitating the detritus breakdown in particles of small size, is likely an important factor for the colonization by microorganisms (Pellikaan, 1982). *Z. noltii* leaves are composed of more than 40% of cellulose, hemicellulose and lignin (relative to total dry weight: 14.6%, 19.1% and 10.2%, respectively; Auby, 1991). Macroalgae are characterized by lower contents of these components which are between 10% and 34% for *Ulva lactuca* Lin. (Brouard, 1983; Faucille, 1984). This difference in refractory structural constituents can explain the slow decay of phanerogam relative to macroalgal decomposition. In addition, vascular plants are known to produce toxic substances, which are liberated early during the decomposition process, thus causing a delay of microbial colonization resulting in a slower degradation rate. It has been reported that *Z. marina* contained bioactive compounds (phenolic acid and flavonoids) shown to inhibit microbial hydrolytic enzymes (Horner et al., 1988; Swain, 1979).

Three phases are generally recognized in seagrass decomposition processes based on changes in decay rate with time (Rice & Tenore, 1981; Pellikaan, 1982). During the first phase a rapid loss of dry weight does occur, presumably corresponding to the leaching of soluble compounds. The second phase is characterized by microbial colonization, reflected by an increase of nitrogen content. Finally, the third one corresponds to the loss of refractory material. These 3 phases should imply corresponding changes in C/N ratio. Nevertheless we were not able to detect these stages during *in situ Z. noltii* decay, as it followed first order kinetics and C/N ratios increased throughout the experiment. Unambiguous interpretations of the time course of C/N ratios are, however, difficult, because the leaching of soluble compounds, element enrichment of the detritus by microbial colonization and protein binding reactions may counteract whilst they occur simultaneously (Pellikaan, 1984).

Surprisingly, degradation of *M. obscurum* followed first order kinetics only in summer and a lag period of two weeks has been found both in spring and autumn. During the lag phase a decrease in N-content took place, whilst this parameter increased again after degradation had started at a high rate.

In contrast to the *in situ* observations, the laboratory decomposition experiments showed an almost equal loss of dry weight for both macrophytes (50.8% and 46.8% for *M. obscurum* and *Z. noltii*, respectively) at the end of incubation period (55 days). Thus, for *M. obscurum*, the decay process studied in the laboratory was slower than observed *in situ*. Firstly, the high sulfide concentrations measured in the overlying water in the experimental containers partially explain this difference. $H_2S$ is toxic for many bacterial species, but in the field this compound diffuses away and can be trapped by sedimentary iron. The iron content in Arcachon Bay is particularly high (Stal et al., 1996; Viaroli et al., 1996) and minielectrode measurements show that free $H_2S$ is rarely present or if present at a concentration below 0.5 mM (data not shown). Secondly, *Chironomus* larvae were never found in laboratory containers, whereas they were present in *M. obscurum* litter bags (see above), thus suggesting that the degradation of this macrophyte is indeed enhanced by detritus fragmentation caused by these animals.

In the laboratory experiments, populations of sulfate-reducing bacteria were well developed and actively participated in the decomposition. Assuming that two moles of sulfide are produced for each mole of carbon mineralized (Jørgensen, 1977), we have estimated that sulfate reduction accounted for around 75% of the mineralization of macrophyte carbon.

The laboratory experiments demonstrated that anaerobic decomposition of both *M. obscurum* and *Z. noltii* results in a high inorganic nutrient release. Water column concentration of dissolved N and P species reflect the net result of sediment flux and macrophyte nutrient release minus the uptake due to bacterial processes. We found that the concentrations of inorganic N-compounds in the water column in containers amended with *M. obscurum* or *Z. noltii* provide a good measure of the N-recycling during the decomposition process. A more complicated pattern was observed for dissolved phosphate, because this compound reached a maximum and subsequently decreased.

Ammonium was the dominant form of dissolved nitrogen released from decaying macrophyte detritus in marine environments. Both Williams (1984) and Hanisak (1993) reported the same conclusions. Due to the anoxic conditions in the containers, bacterial nitrification was arrested and ammonium was therefore not converted into nitrite and nitrate. Phosphate concentrations released from both decaying macrophytes were also very high. The release of inorganic phosphorus from the sediment, favoured by the reducing conditions prevailing in the containers, could partially explain the quantities measured during our experiment. The regeneration, due to the *in situ* decomposition of *Ulva rigida* under anaerobic conditions, accounted for more than 50% of the increase of phosphorus concentrations observed in the Sacca di Goro lagoon during dystrophic crises, when particulate + dissolved phosphorus reached values of 14.5 $\mu$M in the water column. However, nitrogen regeneration was less efficient in the latter system, thus causing an imbalance between N and P release (Viaroli et al., 1992).

In Arcachon Bay, the release of inorganic nutrients from decaying detritus is of a great importance for both macrophytes. According to Iizumi et al. (1982), the mechanism of ammonium regeneration from organic nitrogen in sediments with eelgrass beds constitutes the main pathway for supplying nitrogen to the whole ecosystem. Our study demonstrated the potential importance of the decomposition of the dominant macrophytes of Arcachon Bay, mainly *Z. noltii* phanerogam, as an important nitrogen and phosphorus source in the ecosystem. Nitrogen recycling may become less efficient due to losses caused by coupled nitrification-denitrification processes. However, coupled nitrification denitrification were inhibited in our experiments, but were also very low at comparable intertidal stations in the bay (Rysgaard et al., 1996). Thus, this coupled process is likely not important in affecting N-regeneration in the field. Seagrasses appear to be the most important factor in determining nutrient availability because of their huge biomass in this ecosystem. We demonstrated that the release of nutrients from *Zostera noltii* was a relatively fast process. *Monostroma obscurum*, characterized by a faster decay enhances this nutrient recycling even though it accounts only for the 20% of the total plant biomass.

**Acknowledgments**

The authors highly acknowledge Miss M. Durin for her contribution to the *Zostera noltii* degradation experiment and Dr A. Herbland, director of the laboratory of CREMA (L'Houmeau, La Rochelle, France), for the use of the Technicon autoanalyser. We are also grateful to the staff of IFREMER Arcachon, especially to G. Trut for his kindness and assistance during nutrients measurements and his valuable help for drawing

Figure 1. Finally, we thank E. Buffan, Y. Dubau for helping us in the data presentation and D. Welsh for his help with the English presentation. This work was supported by a E. U. Environment programm (CLEAN contract EV5V-CT92-0080).

## References


Allen, S. E., 1970. Chemical analysis of ecological materials. Blackwell Scientific Publications, Oxford: 185–186.

APHA, AWWA, WPCF, 1980. Standard methods for the examination of water and wastewater. 15th edn. 1134 pp.

Auby, I., 1991. Contribution à l'étude des herbiers de *Zostera noltii* dans le Bassin d'Arcachon. Thèse Doctorat, Université Bordeaux I, 234 pp.

Auby, I. & P. J. Labourg, 1996. Seasonal dynamics of *Zostera noltii* Hornem. in the Bay of Arcachon (France). J. Sea Res. (in press).

Auby, I., F. Manaud, D. Maurer & G. Trut, 1994. Etude de la prolifération des algues vertes dans le Bassin d'Arcachon. Rapport SIBA: 163 pp.

Brouard, F., 1983. Digestion aérobie de la biomasse végétale aquatique. Thèse de Docteur Ingénieur, INSA Toulouse, 175 pp.

Caumette, P., 1986. Phototrophic sulfur bacteria and sulfate-reducing bacteria causing red waters in a shallow brackish coastal lagoon. FEMS Microbiol. Ecol. 38: 113–124.

Cline, J. D., 1969. Spectrophotometric determination of hydrogen sulfide in natural waters. Limnol. Oceanogr. 14: 454–458.

Duarte, C. M., 1992. Nutrient concentration of aquatic plants: patterns across species. Limnol. Oceanogr. 37: 882–889.

Durieu de Maisonneuve, M., 1855. Notes détachées sur quelques plantes de la flore de la Gironde, et description d'une nouvelle espèce d'*Avena*. Actes Soc. linn. Bordeaux. 20: 1–83.

Enriquez, S., C. M. Duarte & K. Sand-Jensen, 1993. Patterns in decomposition rates among photosynthetic organisms: the importance of detritus C:N:P content. Oecologia 94: 457–471.

Faucille, S., 1984. Digestion anaérobie de végétaux aquatiques. Végétaux marins et d'eau douce. Thèse Institut national Polytechnique de Lorraine, 100 pp.

Fenchel, T., 1977. Aspects of the decomposition of seagrasses. In C. P. McRoy & C. Helfferich (eds), Seagrasses ecosystems: a scientific perspective. Marcel Dekker, New York: 123–145.

Godschalk, G. L. & R. G. Wetzel, 1978. Decomposition of aquatic angiosperms. c. *Zostera marina* L. and a conceptual model of decomposition. Aquat. Bot. 5: 329–354.

Hanisak, M. D., 1993. Nitrogen release from decomposing seaweeds: species and temperature effects. J. appl. Phycol. 5: 175–181.

Harrisson, P. G. & K. H. Mann, 1975. Detritus formation from eelgrass (*Zostera marina* L.): the relative effects of fragmentation, leaching and decay. Limnol. Oceanogr. 20: 924–934.

Hemminga, M. A. & J. Nieuwenhuize, 1991. Transport, deposition and *in situ* decay of seagrasses in a tropical mudflat area (Banc d'Arguin, Mauritania). Neth. J. Sea Res. 27: 183–190.

Horner, J. D., J. R. Gosz & R. G. Cates, 1988. The role of carbon-based secondary metabolites in decomposition. Am. Nat. 132: 869–883.

Iizumi, H., A. Hattori & C. P. McRoy, 1982. Ammonium regeneration and assimilation in eelgrass (*Zostera marina*) beds. Mar. Biol. 66: 59–65.

Jørgensen, B. B., 1977. The sulfur cycle of a coastal marine sediment (Limfjorden, Denmark). Limnol. Oceanogr. 22: 814–832.

Lavery, P. S. & A. J. McComb, 1991. Macroalgal-sediment nutrient interactions and their importance to macroalgal nutrition in a eutrophic estuary. Estuar. coast. shelf Sci. 32: 281–295.

Le Corre, P., 1983. Dosage du carbone organique particulaire. In CNEXO (ed.), Manuel des analyses chimiques en milieu marin. BNDO/Documentation, Brest (France): 203–208.

Pellikaan, G. C., 1982. Decomposition processes of eelgrass, *Zostera marina* L. Hydrobiol. Bull. 16: 83–92.

Pellikaan, G. C., 1984. Laboratory experiments on eelgrass (*Zostera marina* L.) decomposition. Neth. J. Sea Res. 18: 360–383.

Pfennig, N., F. Widdel & H. G. Trüper, 1981. The dissimilatory sulfate-reducing bacteria. In M. P. Starr, H. Stolp, H. G. Trüper, A. Balows & H. G. Schlegel (eds), The prokaryotes, a handbook on habitats, isolation and identification of bacteria. Springer-Verlag, Berlin, vol. 1: 926–940.

Ribes, E., 1988. Contribution à l'étude de la prolifération des algues vertes dans le Bassin d'Arcachon. Contrat I.F.R.E. MER 875527053. 31 pp.

Rice, D. L. & K. R. Tenore, 1981. Dynamics of carbon and nitrogen during the decomposition of detritus derived from estuarine macrophytes. Estuar. coast. shelf Sci. 13: 681–690.

Rysgaard, S., N. Risgaard-Petersen & N. P. Sloth, 1996. Nitrification, denitrification and nitrate ammonification in sediments of two coastal lagoons in Southern France. Hydrobiologia **000** (Dev. Hydrobiol. **000**): **000–000**.

Sokal, R. R. & F. J. Rohlf, 1981. Biometry. W. H. Freeman & Company, San Fransisco, 859 pp.

Stal, L. J., S. B. Behrens, M. Villbrandt, S. van Bergeijk & F. Kruyning, 1996. The biogeochemistry of two eutrophic marine lagoons and its effect on microphytobenthic communities. Hydrobiologia (this volume).

Swain, T., 1979. Tannins and lignins. In G. E. Rosenthal & D. H. Janzen (eds), Herbivores: their interactions with secondary plant metabolites. Academic Press, New York: 657–682.

Thayer, G. W., D. W. Engel & M. W. Lacroix, 1977. Seasonal distribution and changes in the nutritive quality of living, dead and detrital fractions of *Zostera marina* L. J. exp. mar. Biol. Ecol. 30: 109–127.

Tréguer, P. & P. Le Corre, 1975. Manuel d'analyse des sels nutritifs dans l'eau de mer (utilisation de l'autoanalyseur II Technicon R). 2e ed. Laboratoire d'Océanologie Chimique, Université de Bretagne occidentale. 110 pp.

Viaroli, P., M. Bartoli, C. Bondavalli, R. R. Christian, G. Giordani & M. Naldi, 1996. Macrophyte communities and their impact on benthic fluxes of oxygen, sulphide and nutrients in shallow eutrophic environments. Hydrobiologia 329 (Dev. Hydrobiol. 117): 105–119.

Viaroli, P., A. Pugnetti & I. Ferrari, 1992. *Ulva rigida* growth and decomposition processes and related effects on nitrogen and phosphorus cycles in a coastal lagoon (Sacca di Goro, Po River Delta). In G. Colombo, I. Ferrari, V. U. Ceccherelli & R. Rossi (eds), Marine eutrophication and population dynamics. Olsen & Olsen, Fredensborg (Denmark): 77–84.

Williams, S. L., 1984. Decomposition of the macroalga *Caulerpa cupressoides* (West) C. Agardh: field and laboratory studies. J. exp. mar. Biol. Ecol. 80: 109–124.

*Hydrobiologia* **329**: 133–141, 1996.
*P. Caumette, J. Castel & R. Herbert (eds), Coastal Lagoon Eutrophication and ANaerobic Processes (C.L.E.AN.).*




# Nitrification, denitrification, and nitrate ammonification in sediments of two coastal lagoons in Southern France

Søren Rysgaard*[1], Nils Risgaard-Petersen & Niels Peter Sloth
*Department of Microbial Ecology, Institute of Biological Sciences, University of Aarhus, Bd. 540, DK-8000 Aarhus C, Denmark; [1]National Environmental Research Institute, Vejlsøvej 25, 8600 Silkeborg, Denmark*
**Corresponding author*

*Key words:* Nitrification, denitrification, nitrate, ammonification, sediments, isotope, $^{15}N$, ammonium, flux

**Summary**

Seasonal and diurnal variations in sediment-water fluxes of $O_2$, $NO_3^-$, and $NH_4^+$ as well as rates of nitrification, denitrification, and nitrate ammonification were determined in two different coastal lagoons of southern France: The seagrass (*Zostera noltii*) dominated tidal Bassin d'Arcachon and the dystrophic Etang du Prévost. Overall, denitrification rates in both Bassin d'Arcachon (<0.4 mmol $m^{-2}$ $d^{-1}$) and Etang du Prévost (<1 mmol $m^{-2}$ $d^{-1}$) were low. This was mainly caused by a combination of low $NO_3^-$ concentrations in the water column and a low nitrification activity within the sediment. In both Bassin d'Arcachon and Etang du Prévost, rates of nitrate ammonification were quantitatively as important as denitrification.

Denitrification played a minor role as a nitrogen sink in both systems. In the tidal influenced Bassin d'Arcachon, *Z. noltii* was quantitatively more important than denitrification as a nitrogen sink due to the high assimilation rates of the plants. Throughout the year, *Z. noltii* stabilized the mudflats of the bay by its well- developed root matrix and controlled the nitrogen cycle due to its high uptake rates. In contrast, the lack of rooted macrophytes, and dominance of floating macroalgae, made nitrogen cycling in Etang du Prévost more unstable and unpredictable. Inhibition of nitrification and denitrification during the dystrophic crisis in the summer time increased the inorganic nitrogen flux from the sediment to the water column and thus increased the degree of benthic-pelagic coupling within this bay. During winter, however, benthic microalgae colonizing the sediment surface changed the sediment in the lagoon from being a nitrogen source to the overlying water to being a sink due to their high assimilation rates. It is likely, however, that this assimilated nitrogen is liberated to the water column at the onset of summer thereby fueling the extensive growth of the floating macroalgae, *Ulva* sp. The combination of a high nitrogen coupling between sediment and water column, little water exchange and low denitrification rates resulted in an unstable system with fast growing algal species such as phytoplankton and floating algae.

**Introduction**

Coastal regions often receive large anthropogenic inputs of nitrogen and phosphorus that cause eutrophication. The impact is especially strong in estuaries and coastal lagoons characterized by limited water exchange with larger bodies of water. Such increased nutrient loading often promotes growth of phytoplankton and fast growing floating macroalgae, while benthic rooted plants (seagrasses) and benthic microalgae are suppressed due to reduced light availability (Duarte, 1995). This shift from benthic to pelagic primary production introduces large diurnal variations in oxygen conditions from high rates of photosynthesis during day followed by high respiration rates at night. In addition, oxygen consumption within the sediment increases following the deposition of easy degradable algal material on the sediment. In semi-enclosed basins with minor water exchange such high rates of consumption may regularly cause anoxic bottom water and sulfide emissions from the sediment to the water column resulting in mass mortality of infauna and fish.

In temperal coastal areas, nitrogen typically limits primary productivity, which makes processes reducing nitrogen availability of primary concern in reversing the effects of eutrophication. This can be accomplished by either lowering the input of combined nitrogen, increasing water exchange, or managing the area in a way where nitrate removal by denitrification is enhanced. However, it is important to gain detailed information about nitrogen cycling in different ecosystems, suffering from different eutrophication stress, if a successful management is to be obtained.

In this paper we describe the benthic nitrogen transformations in two coastal lagoons of southern France: Bassin d'Arcachon on the Atlantic coast and Etang du Prévost on the Mediterranean coast. The biological structure of the two lagoons was very different due to different nutrient loading and hydrodynamics. Bassin d'Arcachon was dominated by benthic primary producers, thus representing less eutrophic conditions whereas Etang du Prévost was dominated by pelagic primary producers and represented a dystrophic ecosystem.

## Materials and methods

### *Experimental sites and sediment sampling*

Bassin d'Arcachon has an area of 155 $km^2$, with a 3-km wide connection to the sea. The tidal range in the area is 2–3 m, and the bay is dominated by large tidal mudflats occupying about 70% of the total area. A thorough description of the hydrology of the basin was published by Robert et al. (1987) who reported salinities varying from 28–35‰ in the outermost zone to 20–35‰ in the inner zone. The water temperature varies between 5 °C during winter and 25 °C during summer. Most of the tidal flats are covered by the widgeon grass *Zostera noltii* Hornem. The sampling site (Station A) was located on a tidal flat in the central part of the basin about 1 km north of the city of Arcachon.

The lagoon Etang du Prévost has an area of 4 $km^2$ and is linked to the Mediterranean by a single 12 m wide channel. The tidal range is about 20 cm, and most of the lagoon is permanently water-covered with an average depth of 150 cm. The salinity is uniform throughout the lagoon, but varies between 25‰ during winter and 42‰ during dry periods in the summer. The temperature ranges from 5 °C during winter to 27 °C during summer. The lagoon is characterized by large spring blooms of green macroalgae (*Ulva* sp.) and by periods of dystrophic crisis during summer, when most of the water column goes anoxic. The sampling site (Station 11) was located at 60 cm water depth in the western and very sheltered part of the lagoon.

Undisturbed sediment cores were collected at each locality in June and September 1993 and in January 1994. In the Etang du Prévost, 30 cm long Plexiglas core tubes with an i.d. of 5.2 cm were used for sediment sampling and incubation, whereas 40 cm long core tubes with an i.d. of 10 cm were used for sampling and incubation of sediment cores in the Bassin d'Arcachon to contain intact eelgrass plants.

### *Flux and denitrification measurements*

Within a few hours of sampling, the sediment cores were transported to the laboratory and placed in an open tank filled with air-saturated *in situ* water and maintained at *in situ* temperatures. The sediment in all cores was adjusted to an equal height of 8 cm, and a 3-cm long Teflon-coated magnet was suspended 5 cm above the sediment surface. When seagrass was present, the magnet was placed 5 cm above the leaves. Momentum for rotation of the small magnets was provided by a large external magnet rotating at 60 r.p.m. to ensure a homogenous mixing of the water column. Five cores were illuminated at a light irradiance of 400 $\mu$mol photons $m^{-2}$ $s^{-1}$ in the 400–700 nm range, and 5 cores were incubated in darkness.

Measurements of oxygen and combined nitrogen fluxes across the sediment-water interface were initiated by closing the sediment cores with floating glass lids. The lids could be removed during the incubation period and water samples collected with a syringe. Samples were collected 3 times at regular time intervals during the incubation period. The total incubation time ranged from 1 to 8 h, resulting in a 10–20% change from the initial $O_2$ concentration. Samples for $O_2$ determinations were collected in glass vials (Exetainer, Labco, High Wycombe, UK) and Winkler reagents were added immediately (Strickland & Parsons, 1972). Samples collected for ammonia and nitrate determinations were filtered and frozen in 20 ml plastic vials.

The $^{15}N$ experiment for determining anaerobic nitrate reduction rates was initiated by adding $K^{15}NO_3^-$ (99.9 atom%) to the water in the open incubation tank to a final concentration of 30 $\mu$M $^{15}NO_3^-$. A syringe was used to exchange the water in each sediment core with $^{15}NO_3^-$-rich water from the reservoir in order to obtain a uniform mixing of the added isotope in all cores. Initial samples from the water column were taken after the water had equilibrated with the sediment for 15 min. The core-tubes were then closed and incubated as described above. Samples for analysis of concentrations and $^{15}N$ enrichment of $NO_3^-$ were collected from the water column in the core tubes with a syringe and frozen in 20 ml plastic vials. Samples for $^{15}N$ analysis of $N_2$ and $NH_4^+$ were collected from both the water column and the porewater. The latter was accomplished by carefully mixing the sediment and water column with a metal stick after the addition of 250 $\mu$l 7M $ZnCl_2$ to the sediment surface.. Water samples and sediment-water suspension samples for $N_2$ isotope analysis were preserved in gas-tight vials (Exetainer, Labco, High Wycombe, UK.) with 2% v/v $ZnCl_2$ solution, and samples for $NH_4^+$ concentrations and isotope analysis were immediately frozen in 50 ml plastic vials.

*Analysis of concentrations and isotopic composition of nitrogen species*

Nitrate was determined using a standard method (Grasshoff et al., 1983) on a flow injection analyzer (Tecator, Höganäs, Sweden). Ammonium was analyzed colorimetrically as described by Bower & Holm-Hansen (1980). The concentration and labeling of the $^{15}N$ labeled $N_2$ isotopes ($^{14}N^{15}N$, $^{15}N^{15}N$) were analyzed on a combined gas chromatrograph-mass spectrometer (RopoPrep-G+, in line with Tracermass, Europa Scientific, Crewe, UK) as described by Risgaard-Petersen and Rysgaard (1995). The $^{15}N$ atom% of $NO_3^-$ was measured by mass spectrometer after biological reduction to $N_2$ (Risgaard- Petersen et al., 1993). The concentrations of $^{15}NH_4^+$ were obtained by use of the micro- diffusion technique (Blackburn, 1993) on KCL extracts of the sediment-water suspensions as described by Risgaard-Petersen & Rysgaard (1995).

*Calculations*

Fluxes ($F$) of $O_2$, $NH_4^+$, and $NO_3^-$ across the sediment-water interface were calculated as

$$F = \frac{\alpha V_1}{A}, \tag{1}$$

where $\alpha$ is the slope of the regression line obtained by plotting the concentration of the relevant species as a function of incubation time, $V_1$ is the volume of the water column, and $A$ is the surface area. Production rates of $^{15}N$ isotopes were calculated as

$$p^{15}N_x = \frac{(C_{\text{sed}} - C_{\text{ini}})(\phi V_{\text{sed}} + V_2) + (C_{\text{water}} - C_{\text{ini}})(V_1 - V_2)}{tA}, \tag{2}$$

where $p^{15}N_x$ is the production rate of the relevant isotope ($^{15}NH_4^+$, $^{14}N^{15}N$ or $^{15}N^{15}N$), $C_{sed}$ and $C_{water}$ are the concentrations of the isotope in the sediment-water suspension and the water column, respectively, $C_{ini}$ is the initial concentration of the isotope, $V_{sed}$ is the volume of the sediment core, $\Phi$ is the sediment porosity, $V_1$ and $V_2$ are the volumes of the water column before and after water-samples are collected, respectively, $t$ is the incubation time, and $A$ is the surface area. Rates of denitrification per $m^2$ were estimated from the equations derived by Nielsen (1992):

$$D_{15} = p(^{14}N^{15}N) + 2p(^{15}N^{15}N),$$

$$D_{14} = \frac{p(^{14}N^{15}N)}{2p(^{15}N^{15}N)} * D_{15}, \tag{3}$$

where $D_{15}$ and $D_{14}$ are the rates of denitrification of $^{15}NO_3^-$ and $^{14}NO_3^-$, respectively, $p(^{14}N^{15}N)$ and $p(^{15}N^{15}N)$ are the production rate of the two $^{15}N$ labeled $N_2$ species $^{15}N^{14}N$ and $^{15}N^{15}N$, respectively.

The *in situ* rate of denitrification of nitrate supplied from the water column ($D_w$) was calculated from $D_{15}$ and the $^{15}N$ atom% of $NO_3^-$ ($^{15}N\ atom\%\ NO_3^-$):

$$Dw = (\frac{100}{^{15}Natom\%NO_3^-} D_{15}) - D_{15}. \tag{4}$$

Coupled nitrification-denitrification ($D_n$) was calculated as the difference:

$$D_n = D_{14} - D_w. \tag{5}$$

*In situ* rates of $NO_3^-$ ammonification based on nitrate originating from the water column ($DNRA_w$) was estimated from the difference:

$$DNRA_w = (\frac{100}{^{15}Natom\%NO_3^-} p^{15}NH_4^+) - p^{15}NH_4^+. \tag{6}$$

## Results and discussion

*N-cycling at the Etang du Prévost*

The shallow Etang du Prévost lagoon was a highly unstable system, and underwent a complex series of changes during the three sampling periods. In June

1993, a high abundance of *Ulva* sp. was observed in the water column and the surface sediments were well-oxidized and inhabited by dense populations of benthic infauna. In late June–early July 1993, however, the lagoon underwent dystrophic crisis with anoxia and occurrence of sulfide in the water column (Viaroli et al., 1996), resulting in a substantial mortality of benthic infauna. In the months following the September 1993 campaign, the system stabilized and by January 1994 the lagoon was colonized by a dense community of benthic microphytes. These changes in the overall ecosystem structure were reflected in both the exchange rates of combined nitrogen across the sediment-water interface and in the processes of nitrification and nitrate reduction within the sediment (Tables 2 and 3).

The highly abundant infauna present in June 1993 ventilated the uppermost few centimetres of the sediment and may therefore have been responsible for the high sediment $O_2$ uptake rates recorded in the dark (Table 2). Although low concentrations of $NO_3^-$ in the overlying water during the June campaign (Table 1) could only support low rates of denitrification ($D_w$, Table 3), high rates of nitrification within the sediment created suitable conditions for significant coupled nitrification-denitrification activity ($D_n$, Table 3). A minimum estimate of the nitrification rate can be calculated as the sum of the rates of $NO_3^-$ release and coupled nitrification-denitrification (Rysgaard et al., 1993). The estimated nitrification rate during darkness was 3.3 mmol $m^{-2}$ $d^{-1}$, a high rate compared to nitrification activities measured at other marine shallow-water locations (e.g., Henriksen et al., 1981; Nishio et al., 1983; Henriksen & Kemp, 1988; Kemp et al., 1990; Sloth et al., 1992; Rysgaard et al., 1993). This high nitrification activity was probably caused by a combination of the presence of a dense benthic infaunal population and high rates of $NH_4^+$ regeneration (Table 2) originating from decomposition of organic matter. Stimulation of nitrification and coupled nitrification-denitrification by the presence of ventilating animals in the sediment has also been described by other investigators (Aller, 1988; Kristensen, 1988; Henriksen & Kemp, 1988; Binnerup et al., 1992; Peligri et al., 1994), and the mechanism responsible for this stimulation has been attributed to an increase in sediment surface area caused by animal irrigation (Aller, 1988; Kristensen, 1988).

Fluxes of DIN ($NO_3^-$ + $NH_4^+$) were directed from the sediment towards the water column, indicating that the sediment was a nitrogen source for pelagic primary production. Most of this primary production will be deposited within the lagoon because only a relatively small amount will be lost by tidal exchange. Thus, degradation of this organic material increases the internal nitrogen cycling between water and sediment. The nitrification capacity in the sediment made it possible, however, that a minor fraction of nitrogen was lost from the system via coupled nitrification-denitrification during each mineralization/assimilation cycle. About 22% of the $NH_4^+$ produced by mineralization ($NH_4^+$ flux + nitrification) was nitrified and 27% of the $NO_3^-$ formed was subsequently denitrified. Thus, 8% of the $NH_4^+$ liberated by mineralization underwent coupled nitrification-denitrification. If nitrification-denitrification activity occurs continuously, approximately 50 kg N would be removed from the lagoon each day, which represents an internal capacity for recovering eutrophication stress if external loading is reduced.

The dystrophic crisis in the Etang du Prévost in the early summer that resulted in sulfide in the water column and mass mortality of benthic fauna, completely inhibited sediment nitrification and thus the basis for coupled nitrification- denitrification. Nitrate from the overlying water was therefore the only source for denitrification during the September field campaign (Table 3). Since the rate of denitrification based on water column $NO_3^-$ ($D_w$) is proportional to the $NO_3^-$ concentration in the water column (Nielsen et al., 1990), the ~ 4 x higher $NO_3^-$ concentration in the overlying water in September compared with June caused the rate of $D_w$ to increase also by ~ 4 x (Table 3). However, the total rate of denitrification in September was only about half of that measured in June because nitrification did not occur.

Nitrate ammonification was a significant process in September and accounted for approximately 33% of the anaerobic nitrate reduction activity in the sediment (Table 3). Although there was a net removal of $NO_3^-$ from the water column to the sediment, a substantial part of this $NO_3^-$ was then returned to the water column as $NH_4^+$ through the dissimilatory reduction pathway. We did not measure nitrate ammonification rates in June (i.e. before the dystrophic crisis) and thus it is difficult to evaluate whether the occurrence of nitrate ammonification was a consequence of the dystrophic crisis. However, determinations of the most probable numbers of nitrate ammonifiers and denitrifiers (Welsh & Herbert, pers. comm.) in the sediments showed that the former group was dominant in the uppermost sediment layers in both June and September.

*Table 1.* Concentrations of $NO_3^-$ and $NH_4^+$, and temperature of the water column at the sampling stations.

| Locality | $[NO_3^-]$ ($\mu$M) | | | $[NH_4^+]$ ($\mu$M) | | | Temperature (°C) | | |
|---|---|---|---|---|---|---|---|---|---|
| | June | September | January | June | September | January | June | September | January |
| Etang du Prévost | 1.2 | 4.9 | 17.1 | 5.7 | 10.5 | 39.6 | 21 | 21 | 6 |
| Bassin d'Arcachon | 0.32 | 2.0 | 25.5 | 2.4 | 2.0 | 5.0 | 21 | 21 | 10 |

*Table 2.* Net sediment uptake and release rates of $O_2$ $NO_3^-$, and $NH_4^+$ in light and dark incubated cores. Positive and negative values represent net release and net uptake rates, respectively. Standard errors are shown in parenthesis (n = 5). Nitrate uptake rates were not determined at Bassin d'Arcachon in January due to analytical problems.

| Locality | $O_2$ (mmol $m^{-2}$ $d^{-1}$) | | | $NO_3^-$ (mmol $m^{-2}$ $d^{-1}$) | | | $NH_4^+$ (mmol $m^{-2}$ $d^{-1}$) | | |
|---|---|---|---|---|---|---|---|---|---|
| | June | September | January | June | September | January | June | September | January |
| Etang du Prévost, Light | −83 (24) | −33 (23) | 172 (12) | 2.3 (0.4) | −1.1 (0.2) | −6.3 (0.2) | 7.6 (1.1) | 2.2 (2.5) | −8.1 (0.5) |
| Etang du Prévost, Dark | −222 (46) | −111 (5) | −32 (0.4) | 2.3 (0.3) | −1.2 (0.3) | −6.5 (0.3) | 15.6 (4.2) | 10.3 (2.8) | −2.3 (0.6) |
| Bassin d'Arcachon Light | 52 (9) | 99 (25) | 28 (4) | 0.1 (0.1) | −2.0 (0.5) | nd | −2.6 (1.4) | −2.0 (0.5) | −1.0 (0.1) |
| Bassin d'Arcachon Dark | −44 (6) | −84 (6) | −35 (5) | 0 (0) | −1.4 (0.2) | nd | −1.6 (0.7) | −1.4 (0.2) | 0.8 (0.1) |

Since the release of $NH_4^+$ from the sediment of Etang du Prévost in September 1993 was much higher than the $NO_3^-$ uptake from the water column, a net efflux of DIN from the sediment to the water column supplied the pelagic primary production with nitrogen. As in June, a significant internal N-cycling between the sediment and the water column at the Etang du Prévost was possible. The lack of nitrification within the sediment, however, disconnected the link between N-mineralisation and N-removal via coupled nitrification-denitrification that was present in June. Consequently, nitrogen could only be lost from the lagoon through denitrification based on water column $NO_3^-$, through export of organic and inorganic nitrogen by tidal exchange to the sea, or by permanent burial of refractory N- compounds within the sediment.

The dominance of benthic microalgae in January had a strong influence on both $O_2$ exchange rates across the sediment- water interface and the inorganic N-metabolism within the system, as has also been shown in other shallow marine systems, e.g. Andersen et al., 1984; Nielsen et al., 1990; Sundbäck et al., 1991. In contrast to the situation in June and September, the sediment was net oxygen producing and net $NO_3^-$ and $NH_4^+$ consuming (Table 2), indicating that benthic microalgae assimilated nitrogen from the water column. Denitrification was based on both $NO_3^-$ diffusing from the water column and on $NO_3^-$ produced by nitrification within the sediment (Table 3). If no benthic microalgal assimilation of $NO_3^-$ occurred, it would be expected that the rate of denitrification based on water column $NO_3^-$ ($D_w$) would be higher than the rate obtained during the September campaign because the water column $NO_3^-$ concentration was considerably higher during winter (Table 1 and 3). However, the measured rate of $D_w$ in January was 25% of the rate during September and this accounted for <2% of the net $NO_3^-$ uptake by the benthic microalgae. The rate of coupled nitrification- denitrification ($D_n$) in January was only 5% of the activity measured in June, despite a much lower sediment uptake of $O_2$ and higher $NH_4^+$ concentrations in the water column in January (Table 1). This suggests that benthic microalgae suppressed both $D_w$ and $D_n$ activity. Other studies have shown that benthic microphytes may significantly reduce both the nitrification activity and the rate of coupled nitrification-denitrification in marine sediments (Henriksen & Kemp, 1988; N. Risgaard-Petersen, in prep; Dalsgaard et al., in prep). The mechanisms responsible for this reduction are not satisfactorily explained, but it has been suggested that the release of bacteriostatic compounds by the benthic microalgae,

*Table 3.* Rates of denitrification of $NO_3^-$ supplied from the water column ($D_w$), coupled nitrification-denitrification ($D_n$), and nitrate ammonification of $NO_3^-$ supplied from the water column $DNRA_w$, in light and dark incubated cores. Standard errors are shown in parenthesis (n = 5).

| | $D_w$ (mmol $m^{-2}$ $d^{-1}$) | | | $D_n$ (mmol $m^{-2}$ $d^{-1}$) | | | $DNRA_w$ (mmol $m^{-2}$ $d^{-1}$) |
|---|---|---|---|---|---|---|---|
| Locality | June | September | January | June | September | January | September |
| Etang du Prévost, Light | 0.09 (0.02) | 0.44 (0.16) | 0.05 (0.03) | 1.01 (0.18) | 0 (0) | 0.01 (0.01) | 0.13 |
| Etang du Prévost, Dark | 0.08 (0.01) | 0.44 (0.13) | 0.17 (0.01) | 0.79 (0.2) | 0 (0) | 0.09 (0.06) | 0.31 |
| Bassin d'Arcachon Light | 0.001 (0.001) | 0.001 (0.001) | 0.082 (0.015) | 0.03 (0.02) | 0 (0) | 0.14 (0.03) | 0.07 |
| Bassin d'Arcachon Dark | 0.001 (0.001) | 0.009 (0.002) | 0.277 (0.051) | 0.02 (0.02) | 0.01 (0.01) | 0.19 (0.01) | 0.13 |

fluctuating pH and $O_2$ conditions, and competition for $NH_4^+$ may be important factors (Henriksen & Kemp, 1988).

Nitrate ammonification was not quantified during January, but an estimate of the maximum rate of anaerobic $NO_3^-$ reduction based on the actual $NO_3^-$ and $O_2$ concentrations in the water column, and the sediment $O_2$ uptake in the dark (Christensen et al., 1990), suggested that nitrate ammonification accounted for less than 10% of the net uptake of $NO_3^-$.

Fluxes of DIN ($NO_3^-$ and $NH_4^+$) were directed from the water column towards the sediment surface in January, leading to a net removal of inorganic nitrogen from the water above the sediments, and most of this DIN uptake was attributed to algal assimilation. In contrast to the June and September field campaigns, a close and efficient coupling between $NH_4^+$ regeneration and algal N-assimilation within the sediment was established during the January field campaign. The close coupling between $NH_4^+$ regeneration and algal N-assimilation in the sediment, that prevented DIN release to the overlying water and suppressed the coupled nitrification- denitrification activity may contribute to a prolonged retention of nitrogen in the sediment during the winter months. This may then be liberated to the water column during the spring months when floating macroalgae recolonize the lagoon and shade out the benthic microalagae. Consequently, the loss of benthic microalgae will increase the efflux of nutrients from the sediment to the water column and thus further stimulate the growth of *Ulva* sp..

*N-cycling in sediments at the Bassin d'Arcachon*

The intertidal sediments at the bay of Arcachon were much more stable than the sediments at the Etang du Prévost throughout all three sampling periods. The sediment was inhabited by productive eelgras (*Zostera noltii*) populations throughout all three field campaigns, as shown by the net oxygen production during light incubations of intact plants (Table 2). Concentrations of $NO_3^-$ and $NH_4^+$ were low in the water column during June and September (2–5 $\mu$M), during January the $NO_3^-$ concentration was elevated (25 $\mu$M) due to increased riverine input.

Rates of denitrification of $NO_3^-$ supplied from the water column ($D_w$) and rates of coupled nitrification-denitrification ($D_n$) were highest in January (Table 3). The lowest rate of $D_w$ occurred in June, while the lowest rate of $D_n$ occurred in September. The observed seasonal pattern of $D_w$ may be explained by differences in the $NO_3^-$ load to the system, as $D_w$ is proportional to the $NO_3^-$ concentration in the overlying water (Nielsen et al.,1990). The observed pattern of $D_n$ may be linked directly to the seasonal variation in the nitrification activity. The highest rate of potential nitrification in sediments is usually observed during winter when oxygen penetrates deep into the sediment and when competition for $NH_4^+$ from benthic primary producers is weak (Hansen et al., 1981). Oxygen consumption in Arcachon Bay, was lowest in January, indicating a deeper $O_2$ penetration depth and thus better conditions for nitrifying bacteria (Jensen et al., 1993). $NH_4^+$ release from the sediment was only observed during January suggesting that the $NH_4^+$ availability for nitrifiers was higher because uptake by seagrass was lower during the winter period.

In the results presented thus far we have focused solely on nitrification-denitrification processes associated with the sediment surface. Rooted benthic macrophytes may also influence these processes in deeper sediment strata surrounding the rhizosphere. The release of $O_2$ via the roots (Sand-Jensen et al., 1982) to the sediment may stimulate nitrification and thus provide an additional $NO_3^-$ source for denitrification (Iizumi et al., 1980; Christensen & Sørensen, 1986; Reddy et al., 1989; Caffrey & Kemp, 1990). However, our measurements of coupled nitrification- denitrification activity associated with the rhizosphere of *Z. noltii* showed no evidence for $^{15}N$ enrichment of the sediment $N_2$ pool even when sediment cores containing intact plants were incubated in the light (400 $\mu$mol photones $m^{-2}$ $s^{-1}$) for 7–24 h and enriched with $^{15}NH_4^+$ (final sediment concentration of $^{15}NH_4^+$ was 1 mM). Coupled nitrification- denitrification activity was therefore exclusively associated with the sediment surface in the Bassin d'Arcachon.

Overall, the sediment in the bay of Arcachon had a low N- removal capacity via denitrification. Rates of both denitrification based on water column $NO_3^-$ and coupled nitrification-denitrification were extremely low during summer and winter (Table 3) compared with rates measured at other localities (e.g. Seitzinger, 1988; Kemp et al., 1990; Rysgaard et al., 1993; Rysgaard et al., 1995; Nielsen et al., 1995). One factor contributing to the low denitrification activity was the dominance of bacteria with a nitrate ammonification pathway (Welsh & Herbert, pers. comm.) compared to denitrifiers. In September we investigated the importance of nitrate ammonification, and in accordance with the quantitative dominance of ammonifiers over denitrifiers observed by Welsh & Herbert (pers. comm.), nitrate ammonification accounted for 95% of the total dissimilatory reduction of $NO_3^-$ supplied from the water column (Table 3). It has been suggested that nitrate ammonification is a secondary metabolic pathway for various groups of bacteria including *Clostridia* sp. and sulphate reducing bacteria (Tiedje, 1988). In Bassin d'Arcachon, sulphate reduction rates were very high in the sediment (Isaksen & Finster, Submitted) and sulphate reducing bacteria with a capacity for nitrate ammonification were highly abundant (Finster, pers. comm). Recently, it has been shown that sulphate reducing bacteria may have a higher affinity for $NO_3^-$ at lower $NO_3^-$ concentrations than denitrifiers (Dalsgaard & Bak, 1994). Thus, low $NO_3^-$ input to the anoxic sediment strata due to relatively low $NO_3^-$ concentrations in the water column (Table 1), competing $NO_3^-$ assimilation by *Z. noltii*, and extremely low abundance of nitrifiers (Welsh & Herbert, pers. comm.) may contribute to the low denitrification activity relative to nitrate ammonification activity.

During June and September, there was a net uptake of inorganic nitrogen from the water column during light and dark conditions, although there was a small net release of $NO_3^-$ in June. In January, $NH_4^+$ was only released to the overlying water in the dark (Table 2). In September, when the net fluxes of $NO_3^-$ and $NH_4^+$ were directed towards the sediment, less than 1% of the $NO_3^-$ uptake from the water column was due to denitrification (Table 2 and 3). The low denitrification capacity of the bay of Arcachon suggests that only a minor fraction of the $NO_3^-$ input to the basin is removed from the system via denitrification as $N_2$ gas, and that most $NO_3^-$ is assimilated by *Z. noltii* directly either from the water column or from the sediment via anaerobic nitrate-ammonification. The small nitrification capacity within the sediment prevented a significant N removal from decomposing plant or algal material via coupled nitrification-denitrification. Since inorganic combined nitrogen was not released from the sediment during summer, an efficient coupling between sediment $NH_4^+$ regeneration and assimilation by *Z. noltii* retained nitrogen within the sediment.

## Conclusions

In the present study the nitrogen cycle was described for two very different lagoons, the seagrass-dominated Bassin d'Arcachon and the algal-dominated dystrophic Etang du Prévost. Denitrification rates in both lagoons were low and played a minor role as a nitrogen sink. Throughout the year, the seagrass in Bassin d'Arcachon controlled the nitrogen cycle due to its high nitrogen uptake rates. Consequently, nitrogen fluxes were directed into the sediment from the water column, and the sediment in this lagoon acted as a significant nitrogen sink. In contrast, the sediment of the hypereutrophic lagoon Etang du Prévost acted as a significant N source during the summer months, due to extensive mineralization and lack of benthic primary production. During winter, benthic microalgae colonized the sediment surface and changed the sediment in the lagoon from being a nitrogen source to the overlying water to being a sink due to their high assimilation rates. It is likely, however, that this assimilated nitrogen is liberated to the water column at the onset of summer thereby fueling the extensive growth of the floating macroalgae, *Ulva* sp.. Thus, nitrogen cycling in Etang du Prévost is much more unstable and unpredictable as compared with Bassin d'Arcachon.

## Acknowledgments

We thank N. P. Revsbeck and Silvia Pelegri for helpful discussions concerning this manuscript. This work was financially supported by the EU Programme EV5V-CT93-0245, the Danish Research Academy and the Danish National Science Research Council.

## References


Aller, R. C., 1988. Benthic fauna and biogeochemical processes in marine sediments: The role of burrow structures. In T. H. Blackburn & J. Sørensen [eds], Nitrogen cycling in coastal marine environments. Wiley: 301–338.

Andersen, T. K., M. H. Jensen & J. Sørensen, 1984. Diurnal variation of nitrogen cycling in coastal, marine sediments. Mar. Biol. 83: 171–176.

Binnerup, S. J., K. Jensen, N. P. Revsbech, M. H. Jensen & J. Sørensen, 1992. Denitrification, dissimilative reduction of nitrate to ammonium, and nitrification in a bioturbated estuarine sediment as measured with $^{15}N$ and microsensor techniques. Appl. Envir. Microbiol. 58: 303–313.

Blackburn, T. H., 1993. Turnover of $^{15}NH_4^+$ tracer in sediments. In Kemp, P. F., Sherr, B. F., Sherr, E. B. & Cole, J. J. (eds) Handbook of methods in aquatic microbial ecology. Lewies Publishers, Boca Raton, USA: 643– 648.

Bower, C. E. & T. Holm-Hansen, 1980. A salicylate- hypochlorite method for determining ammonia in seawater. Can. J. Fish. Sci. 37: 794–798.

Caffrey, J. & W. M. Kemp, 1990. Nitrogen cycling in sediments with estuarine populations of *Potamogeton perfoliatus* and *Zostera marina*. Mar. Ecol. Prog. Ser. 66: 147–160.

Christensen, P. B., L. P. Nielsen, J. Sørensen & N. P. Revsbech, 1990. Denitrification in nitrate-rich streams: Diurnal and seasonal variation related to benthic ocygen metabolism. Limnol. Oceanogr. 35: 640–651.

Christensen, P. B. & J. Sørensen, 1986. Temporal variation of denitrification in plant-covered, littoral sediment from Lake Hampen, Denmark. Appl. Environ. Microbiol. 51: 1174–1179.

Dalsgaard, T. & F. Bak, 1994. Nitrate reduction in a sulfate-reducing bacterium, *Desulfovibrio desulfuricans*, isolated from rice paddy soil: Sulfide inhibition, kinetics, and regulation. Appl. Envir. Microbiol. 60: 291–297.

Duarte, C. M., 1995. Submerged aquatic vegetation in relation to different nutrient regimes. Ophelia 41: 87–112.

Grashoff, K., M. Erhardt & K. Kremling, 1983. Methods of seawater analysis, 2nd edn. Verlag Chemie, Weinheim, Germany.

Hansen, J. I., K. Henriksen & T. H. Blackburn, 1981. Seasonal distribution of nitrifying bacteria and rates of nitrification in coastal marine sediments. Microb. Ecol. 7: 297–304.

Henriksen, K. & W. M. Kemp, 1988. Nitrification in estuarine and coastal marine sediments. In T. H. Blackburn, and J. Sørensen [eds], Nitrogen cycling in coastal marine environments. Wiley: 201-249.

Henriksen, K., J. I. Hansen & T. H. Blackburn, 1981. Rates of nitrification, distribution of nitrifying bacteria and nitrate fluxes in different types of sediment from different types of sediment from danish waters. Mar. Biol. 61: 299–304.

Iizumi, H., A. Hattori & C. P. McRoy, 1980. Nitrate and nitrite in interstitial waters of eelgrass beds in relation to the rhizosphere. J. Exp. Mar. Biol. Ecol.: 191–201.

Isaksen, M. F. & K. Finster, 1995. Sulphate Reduction in the Root Zone of a Seagrass Bed (*Zostera noltii*) in a Tidal Area, The Basin of Arcachon, France. Mar. Ecol. Prog. Ser. (in press).

Jensen, K., N. P. Revsbech & L. P. Nielsen, 1993. Microscale distribution of nitrification activity in sediment determined with a shielded microsensor for nitrate. Appl. Envir. Microbiol. 59: 3287–3296.

Kemp, W. M., P. Sampou, J. Caffrey & M. Mayer, 1990. Ammonium recycling versus denitrification in Chesapeake Bay sediments. Limnol. Oceanogr. 35: 1545–1563.

Kristensen, E., 1988. Benthic fauna and biogeochemical processes in marine sediments: microbial activities and fluxes. In T. H. Blackburn & J. Sørensen [eds], Nitrogen cycling in coastal marine environments. Wiley: 275–299.

Nielsen, K., Nielsen, L. P. & P. Rasmussen, 1995. Estuarine nitrogen retention independently estimated by the denitrification rate and mass balance methods: a study of Norsminde Fjord, Denmark. Mar. Ecol. Prog. Ser. 119: 275–283.

Nielsen, L. P., 1992. Denitrification in sediment determined from isotope pairing. FEMS Microb. Ecol. 86: 357– 362.

Nielsen, L. P., P. B. Christensen, N. P. Revsbech & J. Sørensen, 1990. Denitrification and photosynthesis in stream sediments studied with microsensor and whole core techniques. Limnol. Oceanogr. 35: 1135–1140.

Nishio, T., I. Koike & A. Hattori, 1983. Estimates of denitrification and nitrification in coastal and estuarine sediments. Appl. Envir. Microbiol. 45: 444–450.

Peligri, S. P., L. P. Nielsen & T. H. Blacknurn, 1994. Effect of *Corophium volutator* (Pallas) on denitrification in artificial microcosms, measured by $^{15}N$ isotope pairing technique. Mar. Ecol. Prog. Ser.: 285–290.

Reddy, K. R., W. H. Patrick Jr. & C. W. Lindau, 1989. Nitrification-denitrification at the plant root-sediment interface in wetlands. Limnol. Oceanogr. 34: 1004–1013.

Risgaard-Petersen, N., S. Rysgaard & N. P. Revsbech, 1993. A sensitive assay for determination of $^{14}N/^{15}N$ isotope distribution in $NO_3^-$. J. Microbiol. Meth. 17: 155–164.

Risgaard-Petersen, N. & S. Rysgaard, 1995. Nitrate reduction in sediments and waterlogged soil measured by $^{15}N$ techniques. In Alef & Nannipieri (eds) Methods in Applied Soil Microbiology and Biochemistry Academic Press. : 277–288.

Robert, R., N. Guillocheau & Y. Collos, 1987. Hydrobiological parameters during an annual cycle in the Arcachon Basin. Mar. Biol. 95: 631–640.

Rysgaard, S., P. B. Christensen & L. P. Nielsen, 1995. Seasonal variation in nitrification and denitrification in estuarine sediment colonized by benthic microalgae and bioturbating infauna. Mar. Ecol. Prog. Ser. 126: 111–121.

Rysgaard, S., N. Risgaard-Petersen, L. P. Nielsen & N. P. Revsbech, 1993. Nitrification and denitrification in lake and estuarine sediment measured by $^{15}N$ dilution technique and isotope pairing. Appl. Environ. Microbiol. 59: 2093–2098.

Sand-Jensen, K., C. Prahl & H. Stockholm, 1982. Oxygen release from roots of submerged aquatic macrophytes. Oikos 38: 349–354.

Sloth, N. P., L. P. Nielsen & T. H. Blackburn, 1992. Nitrification in sediment cores measured with acetylene inhibition. Limnol. Oceanogr. 37: 1108–1112.

Seitzinger, S. P., 1988. Denitrification in freshwater and coastal marine ecosystems: Ecological and geochemical significance. Limnol. Oceanogr. 33: 702–724.

Strickland, J. D. H. & T. R. Parsons, 1972. A practical handbook of seawater analysis. Bull. Fish. Res. Bd Can. 167: 311 pp.

Sundbäck, K., V. Enoksen, W. Granéli & K. Petterson, 1991. Influence of sublittoral microphytobenthos on the oxygen and nutrient flux between sediment and water: A laboratory continuous-flow study. Mar. Ecol. Prog. Ser. 74: 263–279.

Tiedje, J. M., 1988. Ecology of denitrification and dissimilatory nitrate reduction to ammonium. In A. J. B. Zender (ed.), Biology of anaerobic microorganisms Wiley and Sons: 174–244.

Viaroli, P., M. Bartoli, C. Bondavalli, R. R. Christian, G. Giordani & M. Naldi, 1996. Macrophyte communities and their impact on benthic fluxes of oxygen, sulphide and nutrients in shallow eutrophic environments. Hydrobiologia 329 (Dev. Hydrobiol. 117): 105–119.

Welsh, D. T., S. Bourguès, R. de Wit & R. A. Herbert, 1996. Seasonal variation in rates of heterotrophic nitrogen fixation (acetylene reduction) in *Zostera noltii* meadows and uncolonized sediments of the Bassin d'Arcachon, South-West France. Hydrobiologia 329 (Dev. Hydrobiol. 117): 161–174.

*Hydrobiologia* **329**: 143–159, 1996.
*P. Caumette, J. Castel & R. Herbert (eds), Coastal Lagoon Eutrophication and ANaerobic Processes (C.L.E.AN.).*



# Benthic oxygen respiration, ammonium and phosphorus regeneration in surficial sediments of the Sacca di Goro (Northern Italy) and two French coastal lagoons: a comparative study

Marco Bartoli, Matteo Cattadori, Gianmarco Giordani & Pierluigi Viaroli
*Dipartimento di Scienze Ambientali, Università degli Studi di Parma, 43100 Parma, Italy*

*Key words:* Sediment, nitrogen, phosphorus, regeneration, iron availability

## Abstract

During 1994 net sediment-water fluxes of oxygen, ammonium and inorganic phosphorus as well as sediment profiles of organic matter, nitrogen, phosphorus and iron were determined in three shallow eutrophic environments. Investigations were conducted monthly from March to December at five stations in the Sacca di Goro (Po River Delta, Italy). In the late summer, samples were collected from a single site in the Prévost lagoon (French Mediterranean coast) and three stations in the Bassin d'Arcachon (French Atlantic coast). In the Sacca di Goro, water–sediment exchanges of $O_2$, $NH_4^+$ and $PO_4^{3-}$ were estimated by means of core incubation in the dark. Benthic fluxes for the French lagoons were in part determined experimentally using benthic chambers and in part from the literature.

In general in the Sacca di Goro the highest oxygen uptake and nutrient release rates were found at the central sites, affected by macroalgal growth. At the sampling site adjacent to the freshwater inlet, sediment–water exchanges were principally influenced by tidal activity. In terms of organic matter and nutrient levels, sediments from the Sacca di Goro and from the Prévost lagoon, both colonised by the floating macroalga *Ulva rigida* C. Agardh, were similar. Sediments from the inner sheltered site in the Bassin d'Arcachon, invaded by the rooted macrophyte *Ruppia cirrhosa* (Pet.) Grande, showed the highest total N and P content ($363 \pm 157$ $\mu$mol N cm$^{-3}$ and $15 \pm 2$ $\mu$mol P cm$^{-3}$ as average values in the top 10 cm of sediment), but were low in pore water ammonium and orthophosphate probably due to the high sequestering capacity of the system and/or efficient coupling between bacterial nutrient regeneration and assimilation by the plant roots. In addition the outer tidal stations in the Bassin d'Arcachon, invaded by rooted macrophytes, were low in pore water nutrients. A different trend was evident in the Prévost lagoon where the concentrations of exchangeable inorganic phosphorus and ammonium were appreciable ($0.28 \pm 0.07$ $\mu$mol P cm$^{-3}$ and $2.4 \pm 1.4$ $\mu$mol N cm$^{-3}$ as average values in the top 10 cm of sediment). High amounts of dissolved organic nitrogen were found in the pore water at all the sites investigated showing the key role of the organic nitrogen in the recycling of nitrogen in these systems.

The hypothesis that iron is a key factor in controlling phosphorus release is discussed since the Sacca di Goro, which is subject to dystrophic crises, is richer in iron than the Bassin d'Arcachon, which is a more buffered system.

## Introduction

In coastal areas, the most common symptom of eutrophication is the change in macrophyte community structure, mainly due to the appearance of macroalgal and cyanobacterial blooms (Lavery & Mc Comb, 1991; Sfriso et al., 1992; Nienhuis, 1992). In the successional stages of eutrophication, macrophytes are considered to play an important role as regulators of benthic fluxes (Sand-Jensen & Borum, 1991; Viaroli et al., 1996). Since rooted plants assimilate pore–water nutrients efficiently sediment–water fluxes are negligible, while attached or floating seaweeds can cause pulses in benthic activity. Concurrently, organic matter enrichment from both phanerogams and macroalgae settling on the sediment surface leads to different turnover rates depending upon whether it is labile or highly refractory (Buchsbaum et al., 1991). Thus, the quality of organic matter can be viewed as a regulator of phosphorus and nitrogen regeneration or burial into the deep sediment (Johnston, 1991). However in shallow environments such as lagoons or impoundments sedimentation rates of particulate matter are so high that the water column and the sediment surface may be considered to have a physical continuity. As a matter of fact detritus accumulation leads to the formation of a thick organic layer which links benthic macroalgae to the sediment surface. Despite this macroscopic overlapping, chemical structure of such interfaces tends to have abrupt changes and redox discontinuities through a few tens of micrometers mainly due to the enhanced microbial activities and reduced oxygen penetration into the deeper sediment. These discontinuities result in significant changes in pore water chemistry and reactivity (Blackburn & Blackburn, 1993).

A number of attempts have been made to predict benthic fluxes on the basis of pore–water chemistry combined with diffusion models, *in situ* measurements, and factors controlling regeneration rates (Hopkinson, 1987; Boynton & Kemp, 1985; Rizzo et al., 1992; Enoksson, 1993). In the last few years, studies have been undertaken to determine the buffering capacity of stressed benthic systems in the final stages of eutrophication, i.e. dystrophy. Dystrophic events can be considered as indicators of the saturation of the 'buffering' capacity of the system, i.e. the ability to resist perturbation. At present predictive models have only been successfully applied in a limited number of environments such as deep water systems where the effects of anthropogenic inputs and physical disturbance are minimal (Val Klump & Martens, 1981).

This study has investigated benthic fluxes of $O_2$, $NH_4^+$ and $PO_4^{3-}$ in relation to sediment chemistry under a wide range of environmental conditions, e.g. salinity gradients and different primary producer communities. A special attention was paid to macroalgae which in the past decade were largely prevailing on the others primary producers, namely phytoplankton (Pugnetti et al., 1992) and microphytobenthos. The main objective was to determine the contribution of sediment quality to oxygen and nutrient cycling within dystrophic lagoons in comparison with more healthy systems. Particular emphasis was paid to N/P ratios as regulators of N relative to P mobility and iron availability as regulator of phosphorus fluxes.

## Study area

The Sacca di Goro, a microtidal lagoon approximately triangular in shape, has a surface area of 26 km$^2$, with an average depth of about 1.5 m. Located in the southernmost region of the Po delta, the lagoon receives nutrient rich freshwater primarily from the Po di Goro and the Po di Volano (Figure 1). The Sacca di Goro is connected to the sea by a 1.5 km wide channel and several smaller channels recently opened along the southern sand banks; tidal amplitude is normally 50 cm. Water exchange in the lagoon has been estimated by models based on temperature and salinity gradients (O'Kane et al., 1992). The shallower, sheltered areas are low energy habitats, mainly due to slow water exchanges whereas the central region is most influenced by tidal currents whilst the westernmost part of the lagoon is influenced by freshwater inflow. Since 1993, a large channel has been established through the southern sand bank and this has led to major changes in the hydrodynamics of the Sacca di Goro.

In the recent past the Sacca di Goro has undergone summer dystrophic crises following massive growth of the macroalga *Ulva rigida* C. Agardh. For a more detailed description of the Sacca di Goro see Viaroli et al. (1992, 1993).

The Bassin d'Arcachon is a large mesotidal lagoon located 80 km Southwest of Bordeaux along the Atlantic coast of France. The tidal range is high and at low water approximately 75% of the total surface (156 km$^2$) is exposed to the air. The fishponds of Certes, located in the eastern region of the Bassin d'Arcachon, are a man-made complex of reservoirs and channels used in the past for aquaculture: the water level at the sampling site is relatively constant since it is not affected by tidal activity. A more detailed report of the Bassin d'Arcachon and the fishponds is described by Castel et al. (1996) and by Escaravage (1990).

The Étang du Prévost is one of a chain of small shallow lagoons which are located along the French Mediterranean coast, approximately 20 km Southwest of Montpellier. It is an extremely eutrophic system due to agricultural run off, limited water exchange and low water depth (on average 1.5 m); the Grau du Prévost, a small narrow channel, is the only connection between the lagoon and the sea. During the summer the lagoon is invaded by the floating macroalga *Ulva rigida* and undergoes several dystrophic events. For a more detailed description see Caumette (1986).

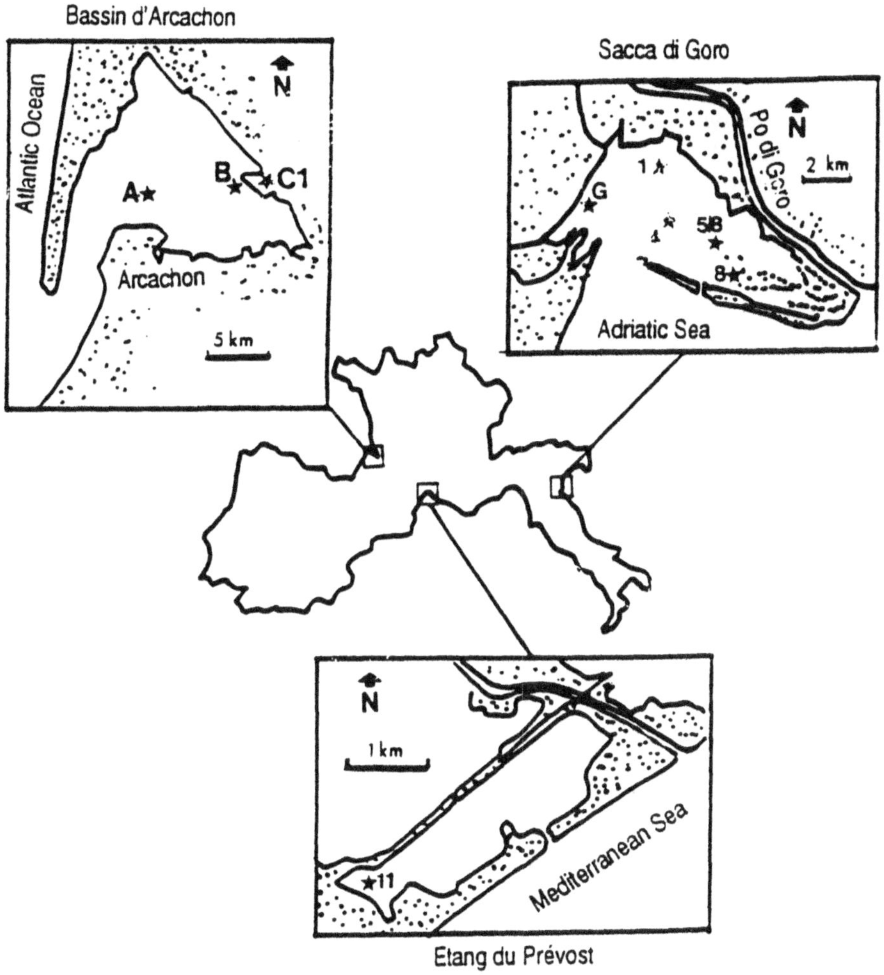


*Figure 1.* Map of the considered lagoons.

## Material and methods

Samples were collected monthly from March to December 1994 at five stations in the Sacca di Goro (Po River Delta, Italy). All the investigated sites (stations G, 1, 4, 5/8, 8) are mesotidal and the water depth ranges between 0.7 and 1.5 m. In August/September 1994, samples were also collected at station 11 in the Prévost lagoon (French Mediterranean coast) and at three stations in the Bassin d'Arcachon (French Atlantic coast).

In the Sacca di Goro, exchanges of oxygen, ammonium and soluble reactive phosphorus (SRP) across the water–sediment interface were measured by means of core incubations in the dark. Benthic fluxes for the French lagoons were in part determined experimentally in benthic chambers and in part from the literature. Along with sediment–water exchanges, density, porosity, organic matter content, total Kjeldhal nitrogen, inorganic and organic P, exchangeable $NH_4^+$ and $PO_4^{3-}$, dissolved organic N, total ferrous and ferric iron were determined in sediment cores sliced into 2 cm sections from the surface to 10 cm depth.

In the Sacca di Goro sediments were collected by means of Plexiglass cores (height 30 cm, i.d. 5 cm); bottom water samples were collected in a 1 l Ruttner bottle. Samples were collected on 16 March, 12 April, 5 May, 7 June, 15 July, 8 August, 9 September, 13 October and 14 December at the five sites shown in Figure 1. The sampling dates were chosen so that they would express a seasonal representativeness. Sediments were not collected at station 5/8 in July, at station 4 in August, at station 1 and at station 8 in December. In the French lagoons sediment cores were collected by hand at low tide: on 19 August at station 11 (Prévost lagoon), on 23 August at station C1 (Certes fishponds), on 2 September at station A and on 4 September at station B (Bassin d'Arcachon).

*Sacca di Goro: core incubation*
Three cores collected from each station were transported to the laboratory and after removal of the top bung placed in a tank filled with aerated water from the same site and equilibrated at the *in situ* temperature for two hours prior to starting the incubation. The core tubes were then sealed with a rubber bung and left in the dark for a period ranging from two to four hours without stirring the overlying water. This because of the occurrence of an easily resuspendable thick nepheloyd layer in the surficial sediment (Miller-Way et al., 1994; Rizzo et al., 1992). Oxygen, ammonium and reactive dissolved phosphorus were determined at zero time and at the end of the incubation period.

*Methods adopted for sediment analyses*
The organic matter content of the sediment was determined on 2–3 g of finely powdered sediment, previously dried at 70°C followed by ignition in a muffle furnace at 550°C until constant weight was reached.

Total nitrogen was estimated on fresh and dried sediment using the Kjeldhal method which only measures reduced forms of nitrogen. Approximately 15 ml of 96% sulphuric acid and 10 ml of 40% hydrogen peroxide were added to 1 g of sediment; the mixture was brought to ebullition until the complete mineralization of the sample. The liberated ammonium was collected by distillation, fixed as ammonium sulphate and determined by titration (APHA, 1975). In this type of sediment $NO_3^-$ and $NO_2^-$ are usually below the level of detection since oxygen only diffuses in the first few millimeters (Sloth et al., 1993).

Total phosphorus was extracted with 20 ml of 1 M HCl from 1 g of dried sediment after ignition in a muffle furnace at 550°C; total inorganic phosphorus was extracted with 20 ml of 1 M HCl from 1 g of dried sediment; total organic phosphorus was finally calculated by difference (Aspila et al., 1976). This method underestimates the organic phosphorus pool mainly due to the quality of the organic matter and its capability of being hydrolysed by HCl (Barbanti et al., 1993).

Pore water was separated from the sediment by centrifugation of about 30 $cm^3$ of wet sediment at $1800 \times g$ for 15 minutes; the supernatant was then filtered and analysed for ammonium, soluble reactive phosphorus, dissolved organic nitrogen and dissolved organic phosphorus. Extractable ammonium and extractable reactive phosphorus were kept in solution washing 2 or 3 times 1 g about of wet sediment respectively with 10 ml of 1 M KCl and with 10 ml of 1 M $MgCl_2$ (pH 7.6) for 1 hour on a multiple magnetic stirring platform. The extracted amounts were corrected for the contributions from pore water. Sediment samples of about 1 ml were immediately fixed for iron determination in 5 ml of 0.5 M HCl when slicing the cores.

*Methods adopted for water analyses*
Dissolved oxygen was determined following the Winkler method (APHA, 1975). Ammonium was determined using the indophenol-blue method (Koroleff, 1970) and reactive phosphorus was determined by the ascorbic acid method (Valderrama, 1981). Total dissolved nitrogen was determined reacting 10 ml of the sample with 1.4 ml of a potassium persulfate solution in an autoclave for 1 hour (Valderrama, 1981). The nitrate liberated was then reduced with a cadmium column (APHA, 1975) and determined as nitrites (diazotation; APHA, 1975). Total dissolved phosphorus was determined as reactive phosphorus after autoclave oxidation (Valderrama, 1981). Ferric iron was reduced to ferrous iron with hydroxylamine and then reacted with o-phenanthroline (APHA, 1975).

*Intersystem comparison*
Intersystem comparisons were performed by Factor Analysis using the Harris image procedure (Statview, Macintosh). Factor analysis was carried out on logarithm $[\ln(x+1)]$ transformed values of seven representative variables relating to the August-early September 1994 sampling period: SOD, SRP, $NH_4^+$ fluxes and the average content of organic matter, total nitrogen, total phosphorus and total iron in the top 10 cm of sediment.

## Results and discussion

The relationship between benthic fluxes of oxygen, dissolved ammonium, soluble reactive phosphorus and physico-chemical characteristics of the surficial sediment was determined at five sites in the Sacca di Goro in 1994. Fluxes measured in late summer were then compared with those determined at three sites located in the Bassin d'Arcachon and at one site in the Étang du Prévost. The key differences at the sampling stations in these different lagoon systems were salinity, dominant macrophytes and trophic condition.

*Sacca di Goro*

In 1994 seasonal trends in the hydrology of this lagoon were different from those observed in previous years. At the designated sampling sites, dissolved oxygen concentrations were relatively constant and ranged from 0.13 to 0.42 mM, reaching a minimum at station 4 (Figure 2). However, continuous monitoring conducted in the southernmost part of the lagoon from late June onwards showed that there was almost constant hypoxia with several episodes of night anoxia (Viaroli et al., 1996). This pattern was consistent with the expansion of the *Ulva rigida* beds, which reached their maximal development in the south-central part of the lagoon during this period. Nevertheless, this area did not undergo summer dystrophy, although oxygen concentrations continued to be low until early August. This situation was quite different from that observed from 1989 to 1992, where the sheltered part of the lagoon, east of station 4, experienced severe summer anoxia accompanied by substantial recycling of phosphorus (Viaroli et al., 1992, 1993). However, since 1993 the summer dystrophic episodes have been reduced in intensity and duration, due to the increased seawater inflow through a large channel cut through the southern sand bank. As a consequence, growth of *Ulva* was restricted to the south-central area of the lagoon leading to several changes in benthic processes at the other sites.

Dissolved inorganic nitrogen was dominated by nitrate mainly due to the discharge of the Po di Volano, whose water is particularly rich in this nutrient (Alvisi et al., 1993). Moreover, $NO_3^-$ in the bottom water followed a clear gradient, with decreasing concentrations from station G to stations 5/8 and 8 (Figure 2). At the latter two sites nitrate was depleted in the warmer months due in part to a decreased freshwater flow and in part to enhanced biological activity. However, nitrate depletion was less pronounced than in previous years probably due to reduced macroalgal growth (Viaroli et al., 1992, 1993). In 1994, ammonium concentrations were highest at station G with a peak of 20.4 $\mu$M, while at the other sites they were always less than 6 $\mu$M (Figure 2). Reactive phosphorus concentration was generally low with the exception of station G where it reached 0.74 $\mu$M. On average, the relative importance of dissolved inorganic N to soluble reactive P showed that the bottom water at the sampling sites was phosphate limited. This pattern was consistent with previous monitoring which indicated that the bulk of the inorganic P was carried by the Po river as adsorbed P on particulate matter (Barbanti et al., 1992). In addition there has been a sharp reduction in

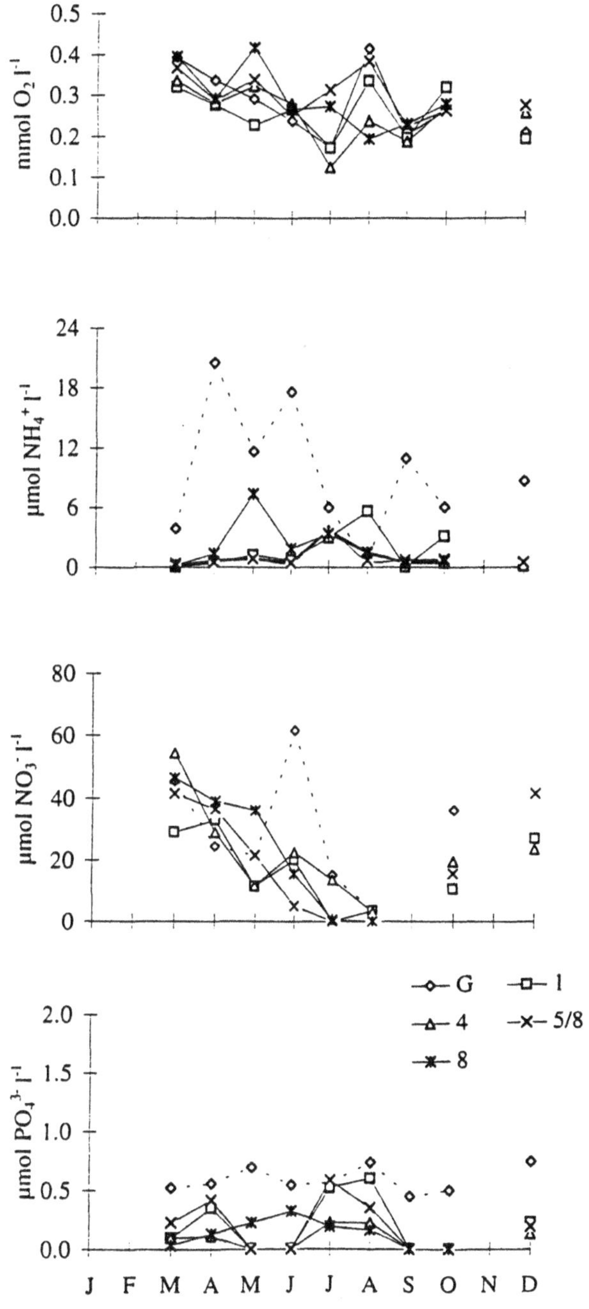


*Figure 2.* Seasonal variations in dissolved oxygen, ammonium, nitrate and soluble reactive phosphorus concentrations in the bottom water at the five sampling sites in the Sacca di Goro (March-December 1994).

recent years of the P discharged into the lagoon from the Po di Volano, which has decreased from about 200 tons $y^{-1}$ in 1989 (Viaroli et al., 1992) to 45 tons $y^{-1}$ in 1992 (Alvisi et al., 1993).

Variations in both dissolved inorganic P and N did not follow a well defined seasonal trend, although oxygen minima at stations 1, 4 and G, as well as nitrate depletion at sites 5/8 and 8 occurred in the warmer period. In the western area of the lagoon seasonal variations might also have overlapped with short-term fluctuations (e.g. daily changes) depending on increased salt water intrusions which entered the lagoon when the freshwater discharge was low.

Sediment oxygen demand (SOD) followed a clear seasonal trend with pronounced peaks in the warmer months (Figure 3). At station G, close to the Po di Volano discharge, SOD reached 80 mmol $O_2$ $m^{-2}d^{-1}$ in March and 365 mmol $O_2$ $m^{-2}d^{-1}$ at the beginning of August. SOD followed a bi-modal trend consistent with that of phytoplankton chlorophyll-$a$ (data not shown). For example, at the beginning of March in the surface waters of the Po di Volano chlorophyll-$a$ concentrations reached 109.2 $\mu$g $l^{-1}$ and particulate organic carbon 7.4 mg $l^{-1}$. It is probable that organic matter sustained microbial respiration in the sediment (Hansen & Blackburn, 1991; Enoksson, 1993). However, the highest value found in August at station G can be explained by the richness of benthic infauna rather than by microbial activity. It was noted that the selected site was colonised by several thousands of benthic Harpacticoids per square meter (Ceccherelli, pers. comm.). In August, SOD values ranging from 200 to 350 mmol $O_2$ $m^{-2}d^{-1}$ were measured at stations 1, 4 and 5/8, while in the more sheltered site 8, SOD never exceeded 150 mmol $O_2$ $m^{-2}d^{-1}$. Maximum values observed in the central part of the lagoon (station 4) were significantly higher than those measured both in 1991 (Viaroli & Naldi, 1992) and 1992 (Viaroli et al., 1993). At station 5/8, which in 1994 was affected to a slight degree by *Ulva* growth, SOD values were similar to those recorded in previous years. In contrast at station 8, the most sheltered which in the past was affected by severe summer anoxia and dystrophy, oxygen demand was relatively low. It is evident that the cutting of a new channel to the sea in 1993 has led to a major change in the water quality of this isolated area of the lagoon.

Ammonium and soluble reactive phosphorus (SRP) regeneration was clearly affected by benthic oxygen uptake and followed the same seasonal pattern as in previous years, with pronounced summer peaks except for station 8 (Figure 3). At the latter site, the amounts of $NH_4^+$ and SRP released from sediment under dark conditions were almost undetectable. The highest values of SRP (4.3 mmol P $m^{-2}d^{-1}$) and $NH_4^+$ (16.7 mmol N $m^{-2}d^{-1}$) were determined in July at station 1 and at station G respectively. At the latter site, fluxes of ammonium and SRP did not follow a well defined trend, being negative in April and June (i.e. from water to sediment) or highly positive (i.e. from sediment to overlying water). These results can be explained by considering tidal fluxes as well as mixing of fresh and marine water. At low tide ammonium and SRP concentrations were probably higher in the water overlying the sediment than in the porewater. Thus, there would be a top down flux with $NH_4^+$ and SRP entering the sediment, whereas at high tide, the opposite would occur since the overlying sea water is poorer in nutrients. However the peak observed in July can also be explained by enhanced microbial activities due to high temperature as well as organic matter inputs.

Organic matter, nitrogen and phosphorus pools profiles were determined in the surficial sediment (10 cm). The values obtained were then averaged and the standard deviations calculated. On average, organic matter concentration did not show significant changes with depth. At station G, the annual series revealed two peaks, in April and in August, which appear to be related to phytoplankton seasonality (Figure 4). Moreover, this trend was consistent with SOD fluctuations. The highest values were recorded at station 1, where in the late summer the organic matter content was 12.1% as dry weight, while the lowest average values were found at station 8, a site dominated by sand (Dal Cin & Pambianchi, 1991).

Total phosphorus content in the surficial sediment was clearly related to riverine discharge and, to a lesser degree, to macroalgal growth. Maximum concentrations were measured at station G, while low concentrations were recorded at station 1 (Figure 5). In May, at all the sites sampled significant peaks occurred, with organic P accounting for approximately 50% of the total. In the eastern part of the lagoon (stations 5/8 and 8) organic P was quite important also in August, when it became the dominant form at station 8. Similarly, at station 1, organic P content continued to be high from May to September. The rapid increase in phosphorus concentrations at the beginning of May can be possibly accounted for the increased riverine discharge following heavy rainfall that occurred in mid April (see also Viaroli et al., 1996) as well as by the deposition of phytoplankton blooms. However, the latter hypothesis

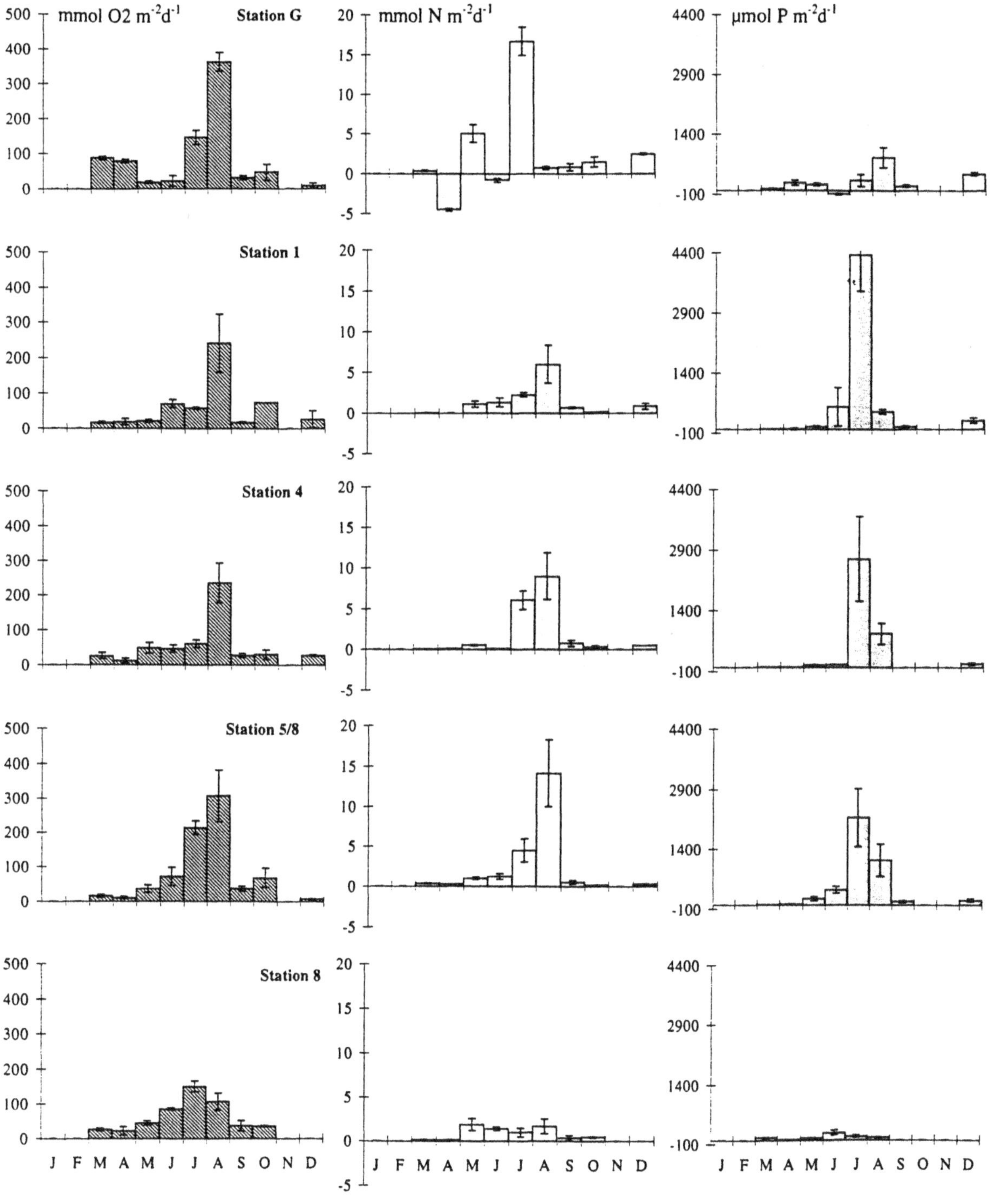


*Figure 3.* Benthic fluxes of dissolved oxygen (left), ammonium (central) and soluble reactive phosphorus at the five sites in the Sacca di Goro (March-December 1994). Values are the average of three cores; bars indicate one standard deviation.

was not confirmed by a parallel increase in nitrogen concentrations (Figure 6). Increases in organic-P content of the surficial sediment in the summer period might be explained by the burial of macroalgal detritus. It was observed that stations 1 and 5/8 were located at the edge of the zone where *Ulva* had the maximum

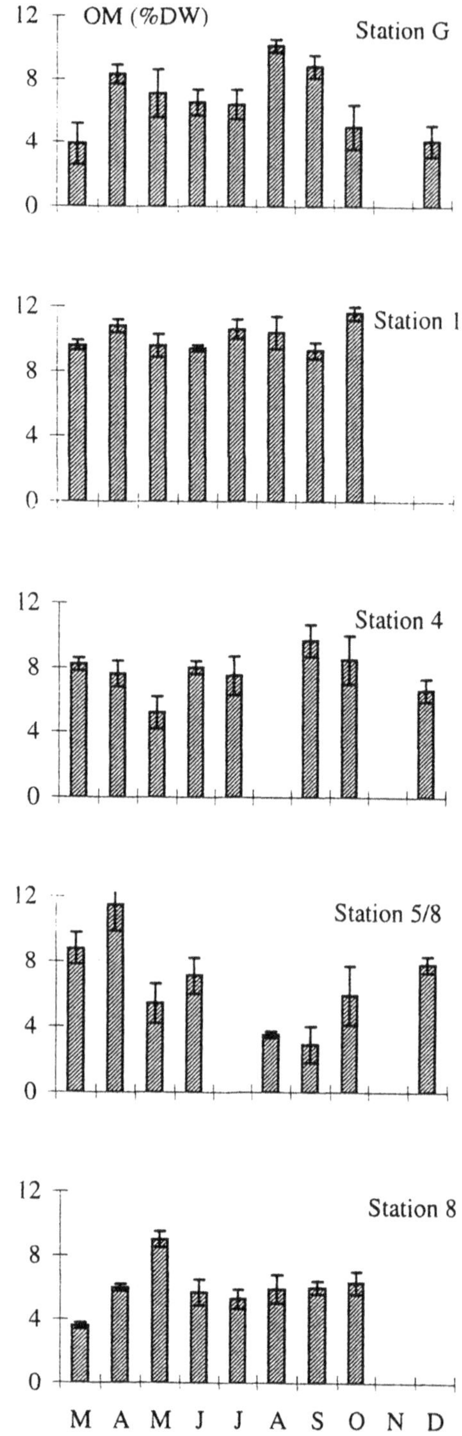


*Figure 4.* Organic matter content (% dry weight) in the top 10 cm of sediment at the five sites in the Sacca di Goro (March-December 1994). Values are the average of the first 10 cm of sediment; bars indicate one standard deviation.

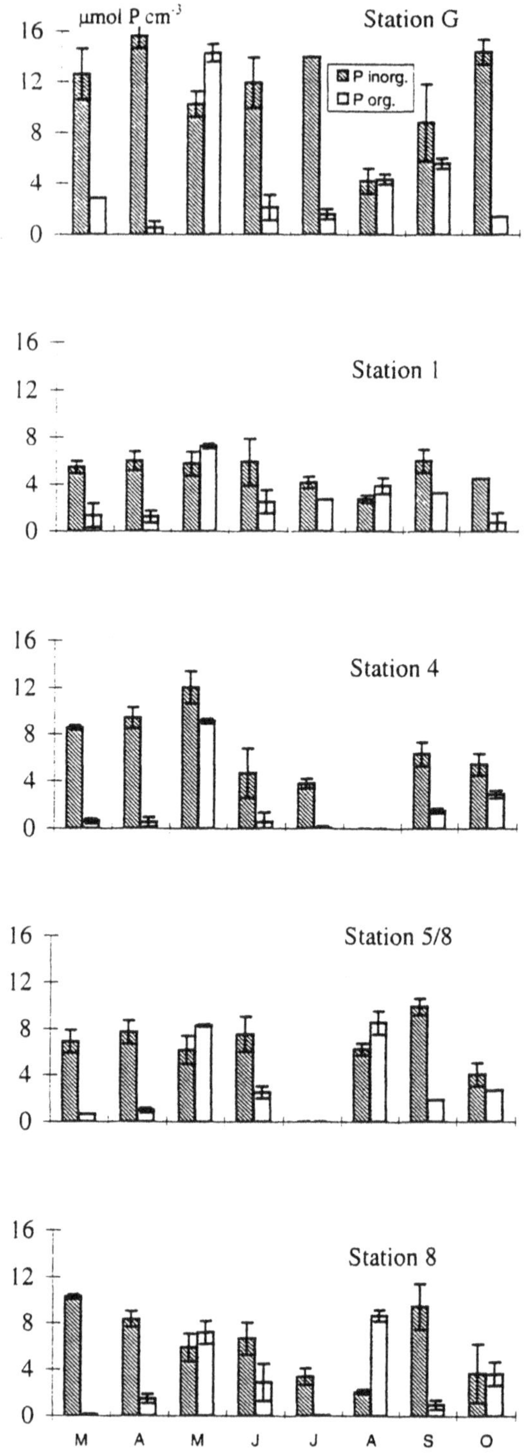


*Figure 5.* Total inorganic and organic phosphorus in the top 10 cm of sediment at the five sites in the Sacca di Goro (March-October 1994). Values are the average of the first 10 cm of sediment; bars indicate one standard deviation.

growth, while station 8 was sometime affected by drifting macroalgae transported by tidal currents. Similar patterns, characterized by organic-P deposition have been described for sediments covered by oyster beds (Mesnage & Picot, 1995).

On average, total nitrogen content in the top sediment was higher in the sites influenced by riverine discharge, even if this gradient was not so clear as it was for phosphorus. In a similar manner, seasonal patterns were not well defined, even though the spring maxima were followed by a sharp decrease in May and June (Figure 6). However, dissolved organic nitrogen (DON) showed a clear trend, principally at stations G, 1 and 4, where it progressed to a summer maximum. Although few measurements have been made of the DON flux from the sediment surface, it has been proposed that DON can play a key role especially during periods of deposition of organic detritus directly on the sediment surface (Hansen & Blackburn, 1991; Enoksson, 1993). In particular, we can assume that DON flux becomes of paramount importance in controlling benthic nitrogen recycling throughout decomposition and leaching of soluble compounds from buried macroalgae. This has become quite clear considering that extractable ammonium was on average 10 times lower than DON (Figure 7).

Potential fluxes of ammonium as well as of soluble reactive phosphorus can be considered to be controlled by complex mechanisms rather than the simple availability of dissolved inorganic P and N in the pore water (Figure 7). For example, nitrogen fluxes are affected by several microbially-mediated processes as well as plant-assimilation and immobilisation into refractory compounds (Mann, 1988; Blackburn & Blackburn, 1993). In contrast the phosphorus cycle appears to be primarily controlled by physical and chemical mechanism; in particular, inorganic phosphate can be accumulated through two main processes: adsorption onto ferric hydroxide (FeOOH) and precipitation with calcium ions (Golterman, 1995). Thus, phosphorus may be in short supply in carbonate-rich coastal waters (Lapointe et al., 1992) as well as in iron rich sediments (Golterman, 1995).

In 1994, in the Sacca di Goro, concentrations of iron in the sediment were quite high and reached values greater than 200 $\mu$mol cm$^{-3}$ (Figure 8). Seasonal trends were not well defined, probably due to the spatial heterogeneity of the system. However, the rapid decrease in ferrous-iron during the summer period leaves open the question of iron mobility in reduced

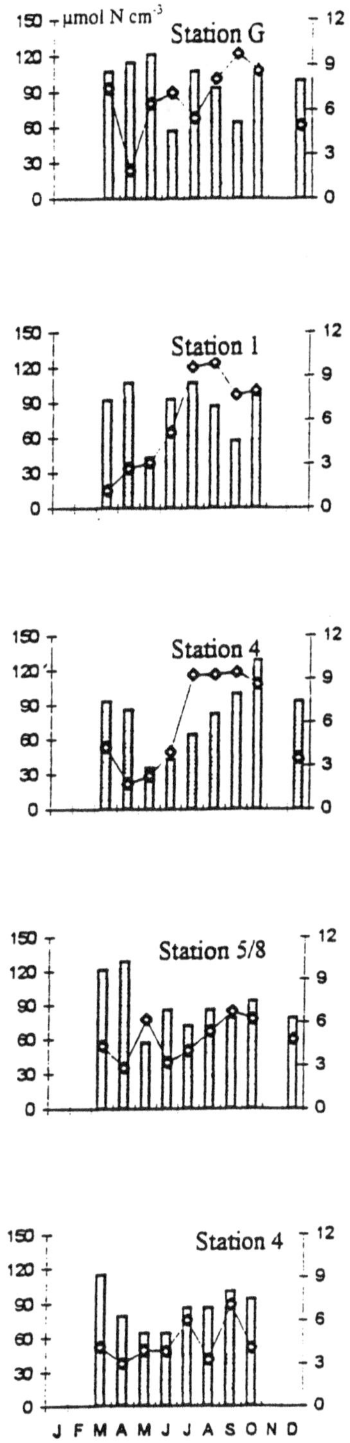


*Figure 6.* Total Kjeldhal nitrogen (bars) and total dissolved nitrogen (lines) in the top 10 cm of sediment at the five sites in the Sacca di Goro (March-December 1994). Values are the average of the first 10 cm of sediment.

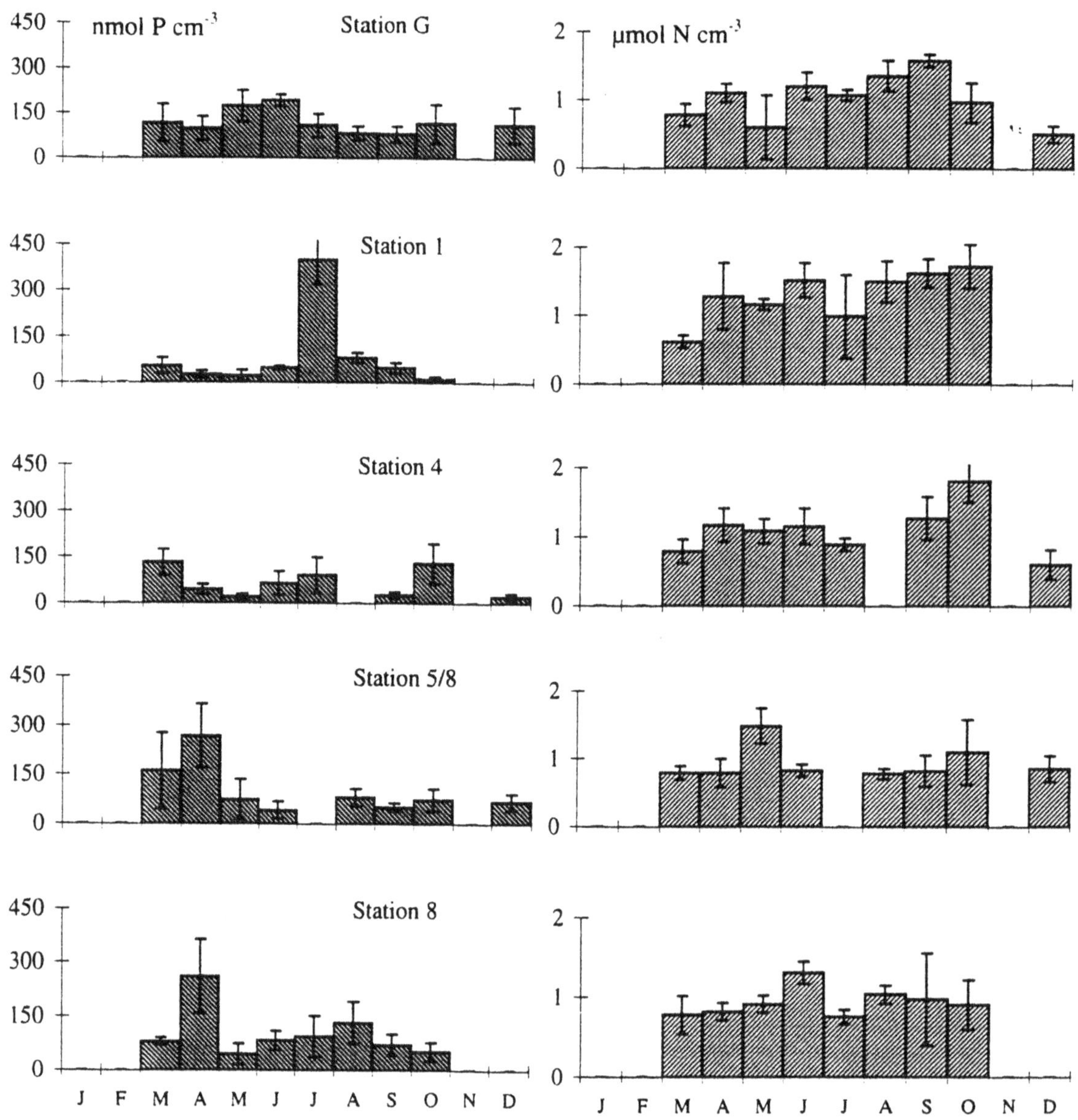


*Figure 7.* Exchangeable reactive phosphorus and ammonium and ammonium at the five sites in the Sacca di Goro (March-December 1994). Values are the average of the first 10 cm of sediment; bars indicate one standard deviation.

sediment as a possible mechanism for the observed P-remobilization.

The behaviour of nitrogen and phosphorus in the sediments has been tentatively explained by considering N/P ratios at different levels (Figure 9). At station G, total N to total P ratio (TN/TP) was always less than 10, being usually around 4. Values in the same range, although somewhat higher, were observed also at station 8. At stations 1, 4 and 5/8, TN/TP ratios were approaching the conventional Redfield ratio (N/P=16), even if they were affected by great variability. In the latter sites, slight seasonal trends were also observed, with peaks in spring and late summer. Considering these patterns, we can point out that stations G and 8 were accumulating phosphorus, while stations 1, 4 and 5/8 seemed to accumulate nitrogen mainly in spring and autumn. Moreover the slight decrease of TN/TP ratio in the warmer months was consistent with nitrogen depletion in the water column.

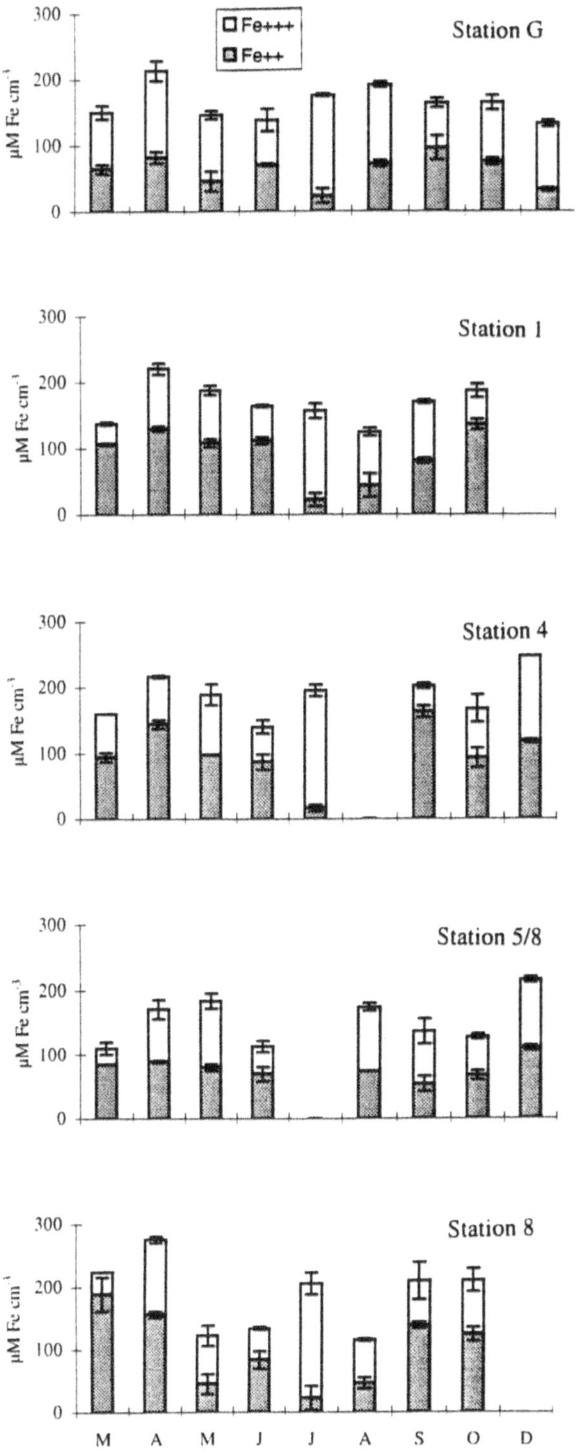


*Figure 8.* Total ferrous and ferric iron in the top 10 cm of sediment at the five sites in the Sacca di Goro (March-October 1994). Values are the average of the first 10 cm of sediment; bars indicate one standard deviation.

Exchangeable ammonium to exchangeable phosphorus ratios (N Exch/P Exch) followed a gradient which was, to some extent, similar to that observed for TN/TP (Figure 10). At station G, N Exch/P Exch never exceeded the balanced Redfield ratio, with most values less than 10. Ratios somewhat higher, but on average less than 16, were also found at stations 8 and 5/8. At station 1 and 4 N Exch/P Exch followed clear seasonal trends with pronounced peaks in spring and late autumn, and very low values in July. These patterns, show that the mobility of inorganic P relative to ammonium was at its maximum in July, while the converse occurred in spring as well as in late autumn. In the latter period the increased availability of exchangeable ammonium coincided with high concentration of dissolved inorganic nitrogen (DIN) in the water column. In contrast in July, the increased importance of exchangeable inorganic P was probably linked to DIN depletion in the water column as well as to the reducing conditions of the surficial sediment.

At station G and, to a lesser extent at stations 5/8 and 8, exchangeable phosphorus was usually in excess of exchangeable ammonium; these results infer that the sediment was accumulating phosphorus not only in the particulate form but also in the mobile one. This figure is, to some extent, consistent with that obtained by considering the ratios estimated from benthic fluxes (Nf/Pf) (Figure 11). Station G did not show predictable trends being benthic fluxes largely affected by freshwater discharge as well as by tidal currents. At stations 1, 4 and 5/8 Nf/Pf ratios, on average low, in June-July dropped in the range 1–5, putting in evidence that fluxes from sediments were mainly characterized by an imbalance with SRP exceeding $NH_4^+$. This imbalance was also evidenced in previous years during dystrophic episodes and was explained by means of the inefficiency of N mineralization due to nitrogen storage in refractory compounds (Pugnetti et al., 1992; Viaroli & Naldi, 1992) according to Mann (1988) and Buchsbaum et al. (1991).

*French lagoons*

The role played by sedimentary pools in determining benthic fluxes was also analysed at one site in the Prévost lagoon and three sites in the Bassin d'Arcachon to test environmental conditions quite different from those observed in the Sacca di Goro. As a matter of fact, Étang du Prévost is usually experiencing severe dystrophic events all through the summer (Caumette, 1986); the outer sites in the Bassin d'Arcachon seem

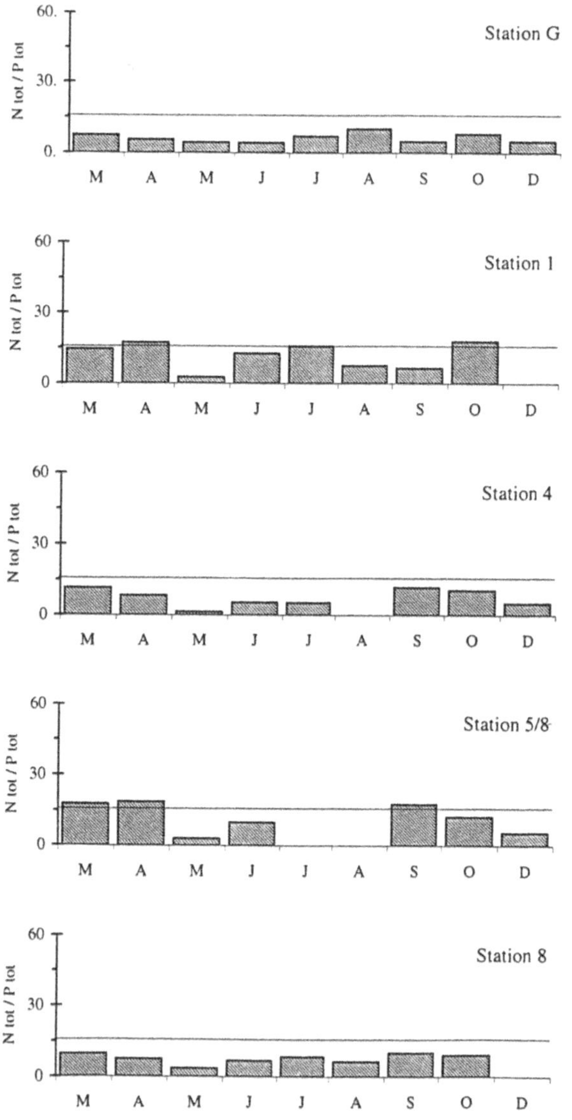


*Figure 9.* Total nitrogen (Ntot) to total phosphorus (Ptot) molar ratio in the top 10 cm of sediment in the five sites of the Sacca di Goro (March-December 1994).

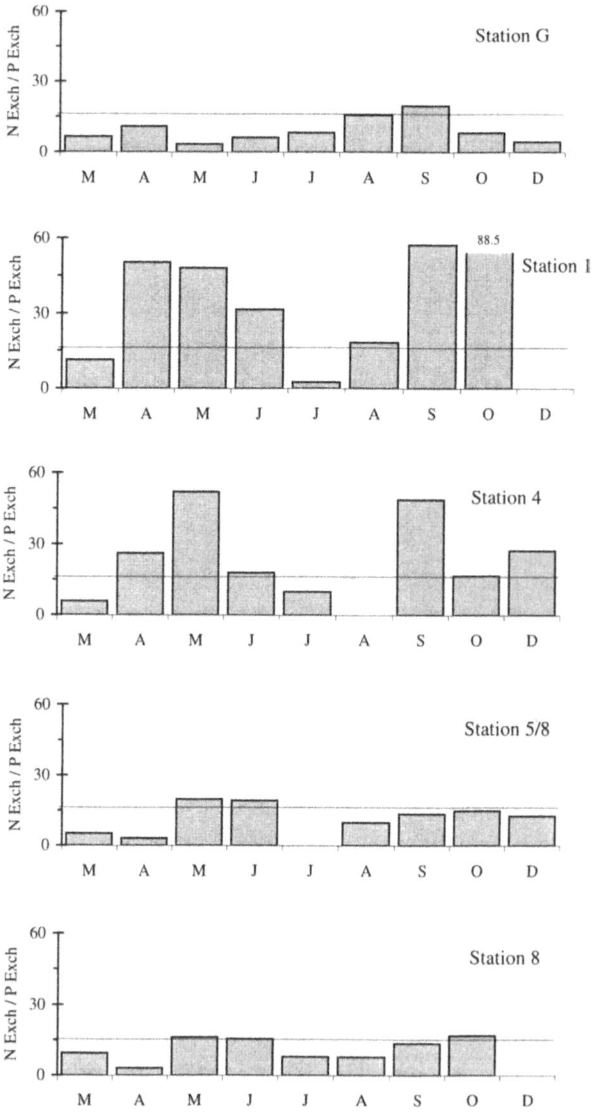


*Figure 10.* Molar ratio of exchangeable ammonium (N Exch) to exchangeable inorganic phosphorus (P Exch) at the five sites of the Sacca di Goro (March-December 1994).

to be more healthy due to the large seagrass meadows (Bachelet et al., 1994) while the impoundment at Certes shows an intermediate behaviour (Labourg, 1975; Escaravage, 1990; Castel et al., 1996).

The high content and the shape of the organic matter profiles in the sediment at the sampling stations in the Bassin d'Arcachon are probably determined by *Zostera* and *Ruppia* (Table 1). This was particularly evident at station C1, where in the first 6 cm OM was over 20%; such high values can be explained by the settling of large amounts of epiphytes growing on *Ruppia* stems. Station 11, free of rooted plants but generally invaded by floating beds of *Ulva*, did not show a significant OM profile probably because of sediment disturbance from both infaunal and human activity.

Total Kjeldhal nitrogen (TN) profiles were to a some extent overlapping those obtained for OM. Thus, it can be argued that most of the nitrogen was stored in the sediment throughout OM accumulation. As a matter of fact, in the surficial sediment at station C1, TN

*Table 1.* Organic matter (O.M.), total Kjeldhal nitrogen (TN), exchangeable ammonium (N Exch), total inorganic (TIP) and organic phosphorus (TOP) and total ferrous and ferric iron content in the top 10 cm of sediment at the station 11 in the Etang du Prévost and stations A, B and C1 in the Bassin d'Arcachon. Average values and standard deviations are reported. (1, 4, 5/8 and G), Etang du Prévost (11) and Bassin d'Arcachon (A, B, C1).

| Station | O.M. (% d.w.) | TN ($\mu$mol cm$^{-3}$) | Exch. N ($\mu$mol cm$^{-3}$) | TIP ($\mu$mol cm$^{-3}$) | TOP ($\mu$mol cm$^{-3}$) | $Fe^{2+}$ tot ($\mu$mol cm$^{-3}$) | $Fe^{3+}$ tot ($\mu$mol cm$^{-3}$) |
|---|---|---|---|---|---|---|---|
| 11 | $10.1 \pm 1.4$ | $73.6 \pm 9.5$ | $2.4 \pm 1.4$ | $6.7 \pm 1.4$ | $0.9 \pm 0.3$ | $25.4 \pm 11.0$ | $16.6 \pm 7.3$ |
| A | $9.9 \pm 5.0$ | $124.2 \pm 28.6$ | $2.4 \pm 0.4$ | $6.2 \pm 1.5$ | $4.8 \pm 1.3$ | $27.0 \pm 4.6$ | $37.2 \pm 4.9$ |
| B | $16.2 \pm 3.3$ | $112.7 \pm 14.7$ | $2.1 \pm 0.5$ | $4.0 \pm 0.5$ | $3.2 \pm 0.5$ | $44.6 \pm 22.4$ | $46.8 \pm 10.4$ |
| C1 | $19.9 \pm 6.8$ | $363.1 \pm 157.0$ | $3.0 \pm 0.2$ | $9.1 \pm 1.9$ | $5.6 \pm 1.9$ | $99.3 \pm 23.6$ | $34.6 \pm 15.3$ |

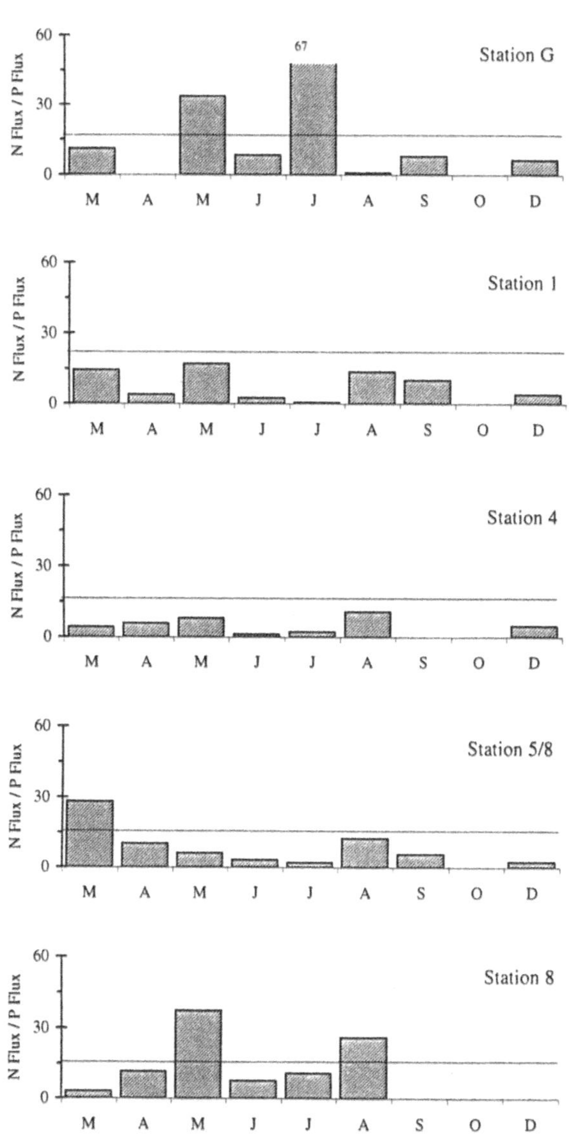


*Figure 11.* Molar ratio of ammonium (N Flux) to soluble reactive phosphorus (P Flux) estimated from benthic fluxes at the five sites of the Sacca di Goro (March-December 1994).

reached concentrations over 400 $\mu$mol N cm$^{-3}$ which is approximately five times higher than that measured at the other sites.

Most of the total phosphorus (TP) was in the inorganic form (Table 1). Sediment from station C1 showed the same pattern as for OM and TN reaching the highest TP content (about 15 $\mu$mol cm$^{-3}$). This trend can be explained in terms of either OM accumulation or sediment capacity to immobilise available P, perhaps because of the abundance of complexing agents. In this context, it should be interesting to better dinstinguish the different fractions of sediment-associated phosphorus, for example with selective extraction techniques (Golterman & Booman, 1988). At station 11, the organic P pool was generally low compared to the inorganic one, while at stations A, B and C1 it accounted for almost the 50% of the total. This may be due to different pathways in the degradation of organic P, in turn following different quality of the organic matter: easier degradable *Ulva* at station 11, more refractory *Zostera* and *Ruppia* at stations A, B and C1 (Viaroli et al., 1996).

In the sediment of the selected sites, inorganic nitrogen was always represented by ammonium, since nitrite and nitrate did not occur significantly due to both slow diffusion and high oxygen consumption. On average, exchangeable $NH_4^+$ was higher at station C1 than at the other sites (Table 1). At station 11, profiles of both exchangeable and pore water ammonium may indicate a remarkable bottom regeneration of inorganic nitrogen, even if the values were much lower than expected from preceding analyses. A possible explanation is that the conditions at station 11 in August-September '94 were quite different from those of the previous field campaigns. As a matter of fact in August 1994 the sediment surface, free of *Ulva*, was light brown and well oxidized while in May and in August '93 and in May '94 the sediment surface, covered with a thick layer of macroalgae, was black, highly reduced and releasing sulphide to the water; in this frame considerable benthic fluxes of ammonia (165 mmol N $m^{-2}d^{-1}$) were determined (Sloth et al., 1993; Viaroli et al., 1996).

The large storage of exchangeable ammonium in the sediment of station C1 can be explained either by the accumulation and availability of OM or by the sediment binding capacity. A preliminary analysis of pore water was also attempted. In the pore water, dissolved organic nitrogen (DON) clearly prevailed over the dissolved inorganic (DIN) one. On average, DON concentration, reaching 4–7 mmol N $l^{-1}$ in the top sediment at all the selected sites, was about 3 order of magnitude higher than DIN. This pattern seems to indicate that nitrogen was firstly recycled and then stored as DON (Bartoli et al., 1994). Soluble reactive (SRP) and dissolved organic (DOP) phosphorus were almost undetectable in the pore water at stations A and B, while at station C1 appreciable amounts of DOP were detected in the top sediment layer. At station 11, SRP was largely exceeding DOP and reached concentrations up to 400 $\mu$mol $l^{-1}$ in the deepest layer (8–10 cm) (Bartoli et al., 1994). At the latter site pore water profiles of SRP and $NH_4^+$ clearly overlapped and are indicative of rapid mineralisation followed by accumulation of the released inorganic N and P. At the sites in the Bassin d'Arcachon the mineralised P can be immediately immobilised by complexation with Ca and Fe oxides, or be assimilated by plant roots. This hypothesis dealing with the role played by iron as a sink for phosphorus is suggested by differences of Fe concentrations at the different sites. In particular, the three stations of the Bassin d'Arcachon are, on average, richer in iron than station 11 in the Prévost lagoon (Table 1).

*Intersystem comparison*

In the Sacca di Goro as well as the French lagoons, seagrasses and macroalgae have a recognised and important role in driving fluxes on a vast scale (see for example Viaroli et al., 1996). However, it may be expected that the magnitude of variations (seasonal and interannual) of macrophyte-mediated processes are related to the size and availability of sedimentary nitrogen, phosphorus and iron pools.

This hypothesis has been tested by means of factor analysis (Figure 12). The first three factors explain 84.4% of the total variance. The ordination of the sites from the first two factors appears to be linked to SRP fluxes (Factor 1) and total iron content in the top 10 cm of sediment (Factor 2). A weaker grouping results from the third factor which is related to the total nitrogen and phosphorus content of the surficial sediment. The high SRP and $NH_4^+$ fluxes recorded at station 11 in the Prévost lagoon can be explained in terms of water renewal and macroalgal impact. In this lagoon the water mass usually undergoes slow renewal and *Ulva* biomass can exceed 5 kg $m^{-2}$ wet weight (Bachelet et al., 1994; Viaroli et al., 1996). Thus the nutrient loads reaching this lagoon due to runoff or direct discharge of wastewater are slowly exported to the sea and cannot be totally removed either in the algal biomass or in the sediment, especially in summer, when N and P regeneration is rapid and the bottom waters become anoxic. For example in 1993 the onset

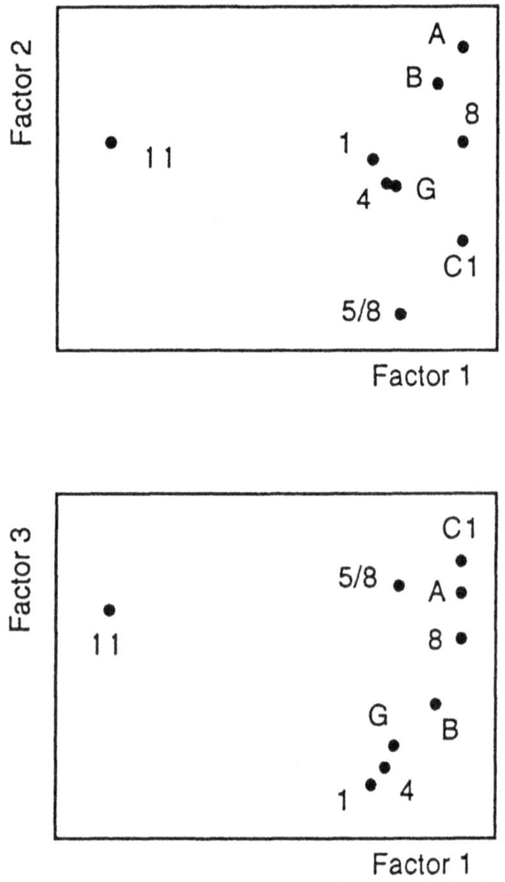


*Figure 12.* August-early September 1994: Factor analysis of sediment variables and benthic fluxes at the different sampling stations in the Sacca di Goro (1, 4, $\frac{5}{8}$ and G) Etang du Prévost (11) and Bassin d'Azcachon (A, B, C1). First factor (FACTOR 1) is inversely related to SRP fluxes, second factor (FACTOR 2) is inversely related with total iron content of sediment and third factor (FACTOR 3) is directly related to total nitrogen and phosphorus content of sediment.

of dystrophic crisis with white water was observed at the end of August, when the water layer between the sediment and the algae was anoxic with ammonium and phosphorus concentrations up to 23 $\mu$mol N $l^{-1}$ and 7 $\mu$mol P $l^{-1}$ respectively.

All the sites in the Bassin d'Arcachon, as well as station 8 in the Sacca di Goro, are characterised by a negligible net exchange of nutrients with the water column. It appears that the pools of dissolved nutrients in the water column were almost depleted since mineralisation processes are coupled to efficient uptake by plant roots or by microbial mats or even immobilisation in the sediments. The remaining sites of the Sacca di Goro show an intermediate behaviour which is in agreement with the benthic fluxes previously described. These patterns are consistent with an iron pool in the top 10 cm of the sediment. In late summer 1994 total iron concentrations in the Sacca di Goro ranged from 116 (station 8) to 202 (station 4) $\mu$mol $cm^{-3}$. Comparable quantities were also measured at station C1 in the Certes fishponds, Bassin d'Arcachon (134 $\mu$mol $cm^{-3}$). In contrast the minimum value found in the Étang du Prévost (42 $\mu$mol $cm^{-3}$) was at least 3 times less than those measured in the Sacca di Goro.

Differences in iron concentration at the different sampling sites suggest that SRP fluxes are controlled by the iron hydroxide-phosphate-sulphide system (Golterman, 1995). This hypothesis is supported by the fact that at station 11, in the Étang du Prévost, iron limitation is coupled to large AVS concentrations (up to 35 $\mu$mol $cm^{-3}$) indicating that the iron hydroxide-sulphide system is principally regulated by sulphide. Thus the probability that SRP might be mobilised from the sediment to the overlaying water should be higher at station 11 than at stations A and B, where AVS does not accumulate (Giordani et al., 1996). In addition, at the intertidal sites A and B at low tide, the sediment surface is exposed to the air, thus leading to AVS reoxidation. In the Sacca di Goro, due to the availability of iron, AVS production can be buffered thereby maintaining a more dynamic iron hydroxide-sulphide system than in the Étang du Prévost. At station C1, in the Bassin d'Arcachon, the situation is different, due to the high organic matter content in the surface sediment and slow water renewal. Nevertheless, SRP fluxes were always negligible probably due to the presence of additional sinks, such as calcium and carbonate as well as clays (Golterman, 1995).

## Concluding remarks

The importance of N relative to P fluxes appears to be regulated mainly by exchangeable $NH_4^+$ to exchangeable SRP ratios, while SRP flux appears to be related to iron availability in the surface sediment. This leads to a more general mechanism to explain eutrophication and dystrophy in these shallow lagoon systems. The Étang du Prévost, which is the most eutrophic lagoon, has internal pools of both total nitrogen and phosphorus which are on average lower than those in the Bassin d'Arcachon while pore water SRP is higher than that at the other sites. In an analogous manner, dystrophic events are unusual in the latter system which has sediment pools of iron two-three times higher than those in the Prévost lagoon. Thus, the conclusion from this study is that the Bassin d'Arcachon appears to be protected against dystrophic events and the sediments act as a sink for nutrients.

In previous years even allowing for the high iron concentrations the Sacca di Goro has undergone severe dystrophic crisis and very high SRP fluxes (Viaroli et al., 1993). Thus the iron hypothesis must be modified to take into account additional factors, i.e. organic matter accumulation and biodegradability as well as the hydrodynamics of the system. At present many of these questions remain unanswered, even though we have been able to identify two major factors: primary producers and biogeochemical buffers. The ultimate question is how do these interact to maintain ecosystem stability.

## Acknowledgements

This study was financed by the joint EU project 'Coastal Lagoon Eutrophication and ANaerobic processes (CLEAN)', contract number EV5V-CT92-0080.

## References

Alvisi, A., D. Marzocchi & S. Edelwais, 1993. Indagine quali-quantitativa delle acque superficiali dei bacini Burana-Volano e Canal Bianco. Dimensione Ambiente, Amm. Prov. Ferrara, Servizio Ambiente, 207 pp.

A.P.H.A., A.W.W.A., W.P.C.F., 1975. Standard Methods for the Examination of Water and Wastewater. 14th edn. APHA, Washington, 1193 pp.

Aspila, K. I., H. Agemian & A. S. Y. Chau, 1976. A semiautomated method for the determination of inorganic, organic and total phosphate in sediments. Analyst 101: 187–197.

Bachelet, G., X. de Montaudouin, I. Auby & P. J. Labourg, 1994. A comparative study of the seasonal changes in macrophytes and macrozoobenthos assemblages in three coastal lagoons under varying degrees of eutrophication. In P. Caumette (Coord.) C.L.E.A.N. Progress Report 1994. EU Environment Programme DG XII, Brussels: 353–367.

Barbanti, A., M. C. Bergamini, F. Frascari, S. Miserocchi & G. Rosso, 1993. Investigations on some critical aspects of sedimentary phosphorus chemical fractionation. J. envir. Qual. 23: 1093–1102.

Barbanti, A., F. Frascari, D. Paltrinieri & G. Rosso, 1992. Transport of nutrients in rivers: investigation on the Po river (Italy). Sci. Total. Envir. Suppl. 92: 337–344.

Bartoli, M., M. Cattadori, G. Giordani & P. Viaroli, 1994. Oxygen, sulphide and nutrient fluxes in shallow eutrophic lagoons with different primary producer communities. II. Superficial sediment profiles of nutrients, iron and reduced sulfur pools. In P. Caumette (Coord.) C.L.E.A.N. Progress Report 1994. EU Environment Programme DG XII, Brussels: 353–367.

Blackburn, T. H. & N. D. Blackburn, 1993. Rates of microbial processes in sediments. Phil. Trans. r. Soc., Lond. 344: 49–58.

Boynton, W. R. & W. M. Kemp, 1985. Nutrient regeneration and oxygen consumption by sediments along an estuarine salinity gradient. Mar. Ecol. Prog. Ser. 23: 45–55.

Buchsbaum, R., I. Valiela, T. Swain, M. Dzierzesky & S. Allen, 1991. Available and refractory nitrogen in detritus of vascular plants and macroalgae. Mar. Ecol. Prog. Ser. 72: 131–143.

Castel, J., P. Caumette & R. Herbert, 1996. Eutrophication gradients in coastal lagoons as exemplified by the Bassin d'Arcachon and the Étang du Prévost. Hydrobiologia 329 (Dev. Hydrobiol. 117): xi–xxx.

Caumette, P., 1986. Phototrophic sulphur bacteria and sulphate reducing bacteria causing red waters in a shallow brackish coastal lagoon (Prévost Lagoon, France). FEMS Microbiol. Ecol. 38: 113–124.

Dal Cin, R. & P. Pambianchi, 1991. I sedimenti della Sacca di Goro (Delta del Po). In S. Bencivelli & N. Castaldi (eds), Studio integrato dell'ecologia della Sacca di Goro, Francoangeli, Milano: 253–263.

Enoksson, V., 1993. Nutrient recycling by coastal sediments: effects of added algal material. Mar. Ecol. Prog. Ser. 92: 245–254.

Escaravage, V., 1990. Daily cycles of dissolvd oxygen and nutrient content in a shallow fishpond: the impact of water renewal. Hydrobiologia 207: 131–136.

Giordani, G., M. Bartoli, M. Cattadori & P. Viaroli, 1996. Sulphide release from anoxic sediments in relation to iron availability and organic matter recalcitrance and its effects on inorganic phosphorus recycling. Hydrobiologia 329 (Dev. Hydrobiol. 117): 205–216.

Golterman, H. L. & A. Booman, 1988. Sequential extraction of iron-phosphate and calcium-phosphate from sediments by chelating agents. Verh. int. Ver. Limnol. 23: 904–909.

Golterman, H. L., 1995. The role of iron hydroxide-phosphate-sulphide system in the phosphate exchange between sediments and overlying water. Hydrobiologia 297: 43–54.

Hansen, L. S. & T. H. Blackburn, 1991. Aerobic and anaerobic mineralization of organic material in marine sediment microcosms. Mar. Ecol. Prog. Ser. 75: 283–291.

Hopkinson, C. S. Jr., 1987. Nutrient regeneration in shallow-water sediments of the estuarine plume region of the nearshore Georgia Bight, USA. Mar. Biol. 94: 127–142.

Johnston, C. A., 1991. Sediment and nutrient retention by freshwater wetlands: effects on surface water quality. Critical Reviews in Environmental Control 21: 491–565.

Koroleff, F., 1970. Direct determination of ammonia in natural waters as indophenol blue. Information on techniques and methods for seawater analysis. I.C.E.S. Interlaboratory. Rep No. 3: 19–32.

Labourg, P. J., 1975. Contribution à l'hydrologie des étangs saumâtres de la région d'Arcachon: description des phénomènes d'eaux blanches. Bull. soc. linn. Bordeaux 5: 3–8.

Lapointe, B., M. M. Littler & D. S. Littler, 1992. Nutrient availability to marine macroalgae in siliciclastic versus carbonate-rich coastal waters. Estuaries 15: 75–82.

Lavery, P. S. & J. A. Mc Comb, 1991. Macroalgal-Sediment nutrient interactions and their importance to macroalgal nutrition in a eutrophic estuary. Estuar. coast. Shelf Sci. 32: 281–295.

Mann, K. H., 1988. Production and use of detritus in various freshwater, estuarine, and coastal marine ecosystems. Limnol. Oceanogr. 33: 894–910.

Mesnage, V. & B. Picot, 1995. The distribution of phosphate in sediment and its relation with eutrophication of a Mediterranean coastal lagoon. Hydrobiologia 297: 29–41.

Miller-Way, T., G. S. Boland, G. T. Rowe & R. R. Twilley, 1994. Sediment oxygen consumption and benthic nutrient fluxes on the Louisiana continental shelf: a methodological comparison. Estuaries 17: 809–815.

Nienhuis, P. H., 1992. Eutrophication, water management, and the functioning of Dutch estuaries and coastal lagoons. Estuaries 15: 538–548.

O'Kane, J. P., M. Suppo, E. Todini & J. Turner, 1992. Physical intervention in the lagoon of Sacca di Goro. An examination using a 3-D numerical model. Sci. Total Envir. suppl. 92: 489–510.

Pugnetti, A., P. Viaroli & I. Ferrari, 1992. Processes leading to dystrophy in a Po River Delta lagoon (Sacca di Goro): phytoplancton-macroalgae interactions. Sci. Total Envir. suppl.: 445–456.

Rizzo, W. M., G. J. Lackey & R. R. Christian, 1992. Significance of euphotic, subtidal sediments to oxygen and nutrient cycling in a temperate estuary. Mar. Ecol. Prog. Ser. 86: 51–61.

Sand-Jensen, K. & J. Borum, 1991. Interactions among phytoplankton, periphyton, and macrophytes in temperate freshwater and estuaries. Aquat. Bot. 41: 137–175.

Sfriso, A., B. Pavoni, A. Marcomini & A. A. Orio, 1992. Macroalgae, nutrient cycles, and pollutants in the lagoon of Venice. Estuaries 15: 517–528.

Sloth, N. P., N. Risgaard-Petersen, S. Rysgaard, S. Pedret Pelegri, 1993. Nitrification, denitrification and nitrate ammonification in sediments of two coastal lagoons in southern France. In P. Caumette (Coord.), CLEAN Progress Report, European Commission. EU Environment Programme DG XII, Brussels: 159–185.

Valderrama, J. C., 1981. The simultaneous analysis of total nitrogen and total phosphorus in natural waters. Mar. Chem. 10: 109–122.

Val Klump, J. & C. Martens, 1981. Biogeochemical cycling in an organic rich coastal marine basin-II. Nutrient sediment–water exchange processes. Geochim. Cosmochim. Acta 45: 101–121.

Viaroli, P., A. Pugnetti & I. Ferrari, 1992. *Ulva rigida* growth and decomposition processes and related effects on nitrogen and phosphorus cycles in a coastal lagoon (Sacca di Goro, Po River Delta). In G. Colombo, I. Ferrari, V. U. Ceccherelli & R. Rossi (eds), Marine eutrophication and population dynamics. Olsen & Olsen, Fredensborg: 77–84.

Viaroli, P. & M. Naldi, 1992. Ricerche sui cicli di azoto e fosforo in una laguna costiera eutrofizzata (Sacca di Goro, Delta del Po). S.IT.E ATTI 15: 95–115.

Viaroli, P., M. Naldi, R. R. Christian & I. Fumagalli, 1993. The role of macroalgae and detritus in the nutrient cycles in a shallow water dystrophic lagoon. Verh. int. Ver. Limnol. 25: 1048–1051.
Viaroli, P., M. Bartoli, C. Bondavalli, R. R. Christian, G. Giordani & M. Naldi, 1996. Macrophyte communities and their impact on benthic fluxes of oxygen, sulpide and nutrients in shallow eutrophic environments. Hydrobiologia 329 (Dev. Hydrobiol. 117): 105–119.

Hydrobiologia **329**: 161–174, 1996.
P. Caumette, J. Castel & R. Herbert (eds), Coastal Lagoon Eutrophication and ANaerobic Processes (C.L.E.AN.).



# Seasonal variation in rates of heterotrophic nitrogen fixation (acetylene reduction) in *Zostera noltii* meadows and uncolonised sediments of the Bassin d'Arcachon, south-west France

David T. Welsh[1,2*], Sophie Bourguès, Rutger de Wit[2] & Rodney A. Herbert[1]
[1]*Department of Biological Sciences, University of Dundee, Miller's Wynd, Dundee DD1 4HN, Scotland*
[2]*Laboratoire d'Océanographie Biologique, Centre d'Océanographie et de Biologie Marine, Université Bordeaux I, 2 rue Professeur Jolyet, 33120 Arcachon, France*
* *Present address. Fax: 33-56-83 51 04*
*Key words:* Acetylene reduction, nitrogen fixation, sulphate reduction, rhizosphere, *Zostera noltii*, root exudates

**Abstract**

Nitrogen fixation (acetylene reduction) rates were measured over an annual cycle in meadows of the seagrass *Z. noltii* and uncolonised sediments of the Bassin d'Arcachon, south-west France, using both slurry and whole core techniques. Measured rates using the slurry technique in *Z. noltii* colonised sediments were consistently higher than those determined in isolated cores. This was probably due to the release of labile organic carbon sources during preparation of the slurries. Thus, in colonised sediments the whole core technique may provide a more accurate estimate of *in situ* activity. Acetylene reduction rates measured by the whole core technique in colonised sediments were 1.8 to 4-fold greater, dependent upon the season, in the light compared with those measured in the dark, indicating that organic carbon released by the plant roots during photosynthesis was an important factor regulating nitrogen fixation. In contrast acetylene reduction rates in uncolonised sediments were independent of light.

Addition of sodium molybdate, a specific inhibitor of sulphate reduction inhibited acetylene reduction activity in *Z. noltii* colonised sediments by >80% as measured by both slurry and whole core techniques irrespective of the light regime, throughout the year inferring that sulphate reducing bacteria (SRB) were the dominant component of the nitrogen fixing microflora. A mutualistic relationship between *Z. noltii* and nitrogen fixing SRB in the rhizosphere, based on the exchange of organic carbon and fixed nitrogen is proposed. In uncolonised sediments sodium molybdate initially severely inhibited acetylene reduction rates, but the level of this inhibition declined over the course of the year. These data indicate that the nitrogen fixing SRB associated with the *Zostera* roots and rhizomes were progressively replaced by an aerobic population of nitrogen fixers associated with the decomposition of this recalcitrant high C:N ratio organic matter.

Acetylene and sulphate reduction rates in the seagrass beds showed distinct summer maxima which correlated with a reduced availability of $NH_4^+$ in the sediment and the growth cycle of *Z. noltii* in the Bassin. Overall, these data indicate that acetylene reduction (nitrogen fixation) activity in the rhizosphere of *Z. noltii* was regulated both by release of organic carbon from the plant roots and maintenance of low ammonium concentrations in the root zone due to efficient ammonium assimilation.

Nitrogen fixation rates determined from acetylene reduction rates measured by the whole core technique ranged from 0.1 to 7.3 mg N $m^{-2}$ $d^{-1}$ in the *Z. noltii* beds and between 0.02 and 3.7 mg N $m^{-2}$ $d^{-1}$ in uncolonised sediments, dependent upon the season. Nitrogen fixation in the rhizosphere of *Z. noltii* was calculated to contribute between 0.4 and 1.1 g N $m^{-2}$ $y^{-1}$ or between 6.3 and 12% of the annual fixed nitrogen requirement of the plants. Heterotrophic nitrogen fixation therefore represents a substantial local input of fixed nitrogen to the sediments of this shallow coastal lagoon and contributes to the overall productivity of *Z. noltii* in this ecosystem.

## Introduction

It is generally considered that nitrogen availability is one of the major factors regulating primary productivity in coastal marine environments (Ryther & Dunstan, 1971; Eppley et al., 1979). Inshore coastal waters are often characterised by high primary production of plankton and rooted macrophytes (sea and salt marsh grasses) (McRoy & McMillan, 1977; Nixon & Pilson, 1983; Moriarty et al., 1990). In order to sustain these high levels of primary production substantial inputs of fixed nitrogen are required (Patriquin, 1972). Whilst, efficient recycling of organic nitrogen in the sediment can supply a large proportion of this fixed nitrogen (Iizumi et al., 1982; Dennison et al., 1987; Caffrey & Kemp, 1992), several investigations have demonstrated that porewater concentrations of inorganic N are insufficient to meet the growth requirements of the plant communities (Patriquin, 1972; Short, 1983; Moriarty et al., 1985). Heterotrophic nitrogen fixation in the phylosphere and rhizosphere of seagrasses may therefore play an important role in regulating primary production in these ecosystems. High rates of heterotrophic nitrogen fixation have been reported in seagrass colonised sediments and estimated to supply upto 50% of the nitrogen requirement of the plant communities (Patriquin & Knowles, 1972; McRoy et al., 1973; Capone et al., 1979; Capone & Taylor, 1980; Wolfenden & Jones, 1987; O'Donohue et al., 1991a; Moriarty & O'Donohue, 1993). However, there is a substantial energy cost associated with nitrogen fixation, estimated to be equivalent to 16 ATP per molecule of $N_2$ fixed (Postgate, 1982) and thus rates of heterotrophic nitrogen fixation in natural environments are generally considered to be limited by the availability of suitable organic carbon substrates (Herbert, 1975; Zuberer & Silver, 1978; Nedwell & Aziz, 1980; Jones, 1982). The high rates of heterotrophic nitrogen fixation reported in seagrass and salt marsh grass sediments have been demonstrated to be associated with the excretion of organic compounds from the plant roots and closely coupled to the photosynthetic activity of the plants (Capone et al., 1979; Boyle & Patriquin, 1981; Whiting et al., 1986; O'Donohue et al., 1991a).

Previous studies on the rates of nitrogen fixation in seagrass meadows and the relationship between the plants and the heterotrophic nitrogen fixing microflora in the rhizosphere have been undertaken primarily in tropical or sub-tropical areas and thus these relationships are much less well characterised in temperate areas. In this study we have investigated the seasonal variation in nitrogen fixation (acetylene reduction) rates in *Zostera noltii* Hornem. colonised and uncolonised sediments in the Bassin d'Arcachon, South-West, France. The dependency of nitrogen fixation on plant photosynthesis and the potential role played by nitrogen fixing sulphate reducing bacteria (SRB) in the rhizosphere was investigated. The ability to fix nitrogen is widely distributed amongst SRB (Riederer-Henderson & Wilson, 1970; Postgate et al., 1985; Postgate et al., 1988), which have previously been proposed as potentially the most important heterotrophic nitrogen fixers in coastal marine sediments (Herbert, 1975; Nedwell & Aziz, 1980). Where, sulphate reduction is the dominant metabolic process, accounting for up to 50% of all organic matter mineralisation (Jørgensen, 1977; Jørgensen, 1982; Canfield, 1989).

## Materials and methods

### *Sampling site*

The sampling site used in this study is described by Welsh et al., (1996) and corresponds to the CLEAN sampling Station A. Samples were collected between March 1994 and February 1995, both from an area within the *Z. noltii* beds and an adjacent area where the *Zostera* had recently died back but decaying roots and rhizomes were still present.

### *Determination of acetylene reduction (nitrogen fixation) rates*

Nitrogen fixation rates were measured using the acetylene reduction technique of Stewart et al., (1967), using both slurry and whole core techniques as described below.

#### *Slurry experiments*

Acetylene reduction rates in slurry experiments were determined as described by Welsh et al. (1996).

#### *Whole core measurements*

Large sediment cores were collected by inserting 5 × 25 cm (internal diameter) grey plastic core tubes into the sediment until the sediment surface was level with the rim of the core tube. The surrounding sediment was removed, the sediment below the core tube was sliced using a steel wire and the core carefully transferred onto a perspex sheet for transportation to the laboratory. In the laboratory cores were stored under natural light conditions in a 50 cm (water depth) × 3 m (internal diameter) water bath (volume approx 600 litres), circulated with aerated seawater from the Bassin for a maximum of 2 days before use. After equilibration the cores were transferred to small water baths and the water level lowered to the level of the sediment surface. Plexiglass core tubes (20 × 5 cm, internal diameter) were inserted into the sediment, taking care not to damage the *Zostera* leaves, when present. The core tubes were sealed with rubber bungs and 10% of the headspace volume was replaced with acetylene via a Suba-seal sampling port on the side of the core tube. The cores were incubated under either natural light or in the dark by covering the core tube with a double layer of aluminium foil. Molybdate treated cores were pre-incubated for 12–16 hours with an overlying water column of natural seawater supplemented with 25 mmol $l^{-1}$ sodium molybdate. During the incubation period, triplicate 1 ml samples of the headspace gas were collected at 2-hour intervals over a 12-hour period and stored by inserting the needle into a butyl rubber bung.

Gas samples were analysed for ethylene and acetylene within 1–2 hours of sampling by gas chromatography, using a Perkin Elmer Autosystem Gas Chromatograph fitted with a 3 m × 2.2 mm (internal diameter) Chromosorb 101 (80/100 mesh) column with $N_2$ as carrier gas and flame ionisation detection. Flow rates for $N_2$, air and $H_2$ were 30, 450 and 50 ml $min^{-1}$ respectively and the oven temperature was 30 °C. Ethylene concentrations were calculated by reference to known standards and all the data were corrected for the small quantities of ethylene present as a contaminant in the acetylene used in this study.

*Enumeration of sulphate reducing bacteria*

Populations of viable SRB were enumerated using the Most Probable Number (MPN) technique. The growth medium consisted of filtered (0.22 $\mu$m pore size) seawater (1000 ml), 5 mM $NH_4Cl$, 1 mM $KH_2PO_4$, 20 mM $NaHCO_3$, 100 $\mu$M $Na_2S_2O_4$, 200 $\mu$M $FeSO_4$, 0.1 g yeast extract, 0.5 ml SL 12B trace elements solution without EDTA (Pfennig & Trüper, 1992), 1.0 ml vitamin V7 solution (Pfennig et al., 1981), 10 mM sodium lactate and 5 mM sodium acetate. The vitamin solution, sodium dithionite and iron solutions were filter sterilised (0.22 $\mu$m pore size) and aseptically added together with the autoclaved carbonate buffer to the bulk autoclaved medium when cool. The medium was prepared according to the procedure described by Widdel & Bak, (1992) and the pH adjusted to pH 7.2 using sterile 1 M HCl and NaOH before aseptically dispensing 3 ml volumes into sterile 4 ml Venoject tubes.

Populations of viable SRB were determined for 4 replicate sediment cores (3 cm internal diameter), the 0–2 cm depth horizon of each core was sectioned, transferred to a sterile petri dish and homogenised using a sterile spatula. A 1 ml aliquot was transferred to the first dilution tube and gently sonicated for 1 min to release attached bacteria, decimal dilution series were prepared using anaerobic autoclaved filtered seawater supplemented with $Na_2S \cdot 9\,H_2O$ (final concentration 0.8 mM) in Hungate tubes previously flushed with $N_2/CO_2$ 90/10% v/v. Aliquot volumes (0.25 ml) of each dilution were used to inoculate 8 replicate Venoject tubes. The tubes were incubated in the dark at room temperature (20–25 °C) and regularly checked for the formation of black FeS precipitates which were scored as positive for the growth of SRB. MPN estimates and their 95% confidence intervals were calculated using the computer programme developed by Clarke & Owens, (1983).

*Determination of sulphate reduction rates and sediment exchangeable $NH_4^+$ concentration*

Sulphate reduction rates and sediment exchangeable ammonium concentrations were determined as described by Welsh et al. (1996), except that sulphate reduction was determined only for the 0–2 and 2–5 cm depth horizons.

## Results

Data presented in Figure 1. shows depth profiles of acetylene reduction activity (ARA) in slurries of *Z. noltii* colonised sediments. ARA was detectable throughout the 0–5 cm depth horizon during the sampling periods in March, July and October 1994 and January 1995. The highest activities were recorded during the summer and autumn sampling programmes, with the peak of activity occurring in the 0–2 cm depth horizon (Figure 1). In winter and spring ARA was reduced and the activity maximum occurred at greater depth (Figure 1), presumably in response to changes in the depth of $O_2$ penetration and carbon availability. The addition of 20 mmol $l^{-1}$ sodium molybdate a specific inhibitor of bacterial sulphate reduction (Taylor & Oremland, 1979; Smith & Klug, 1981; Oremland & Capone, 1988) to sediment slurries severely inhibited ARA by between 75–95% throughout the sediment profiles in all seasons (Figure 1).

Acetylene reduction activity was also detectable throughout the 0–5 cm depth horizon in slurries prepared from uncolonised sediments (Figure 2). Rates of acetylene reduction were always lower than those recorded in sediments colonised by *Z. noltii* and showed a much lower seasonal variation, although a

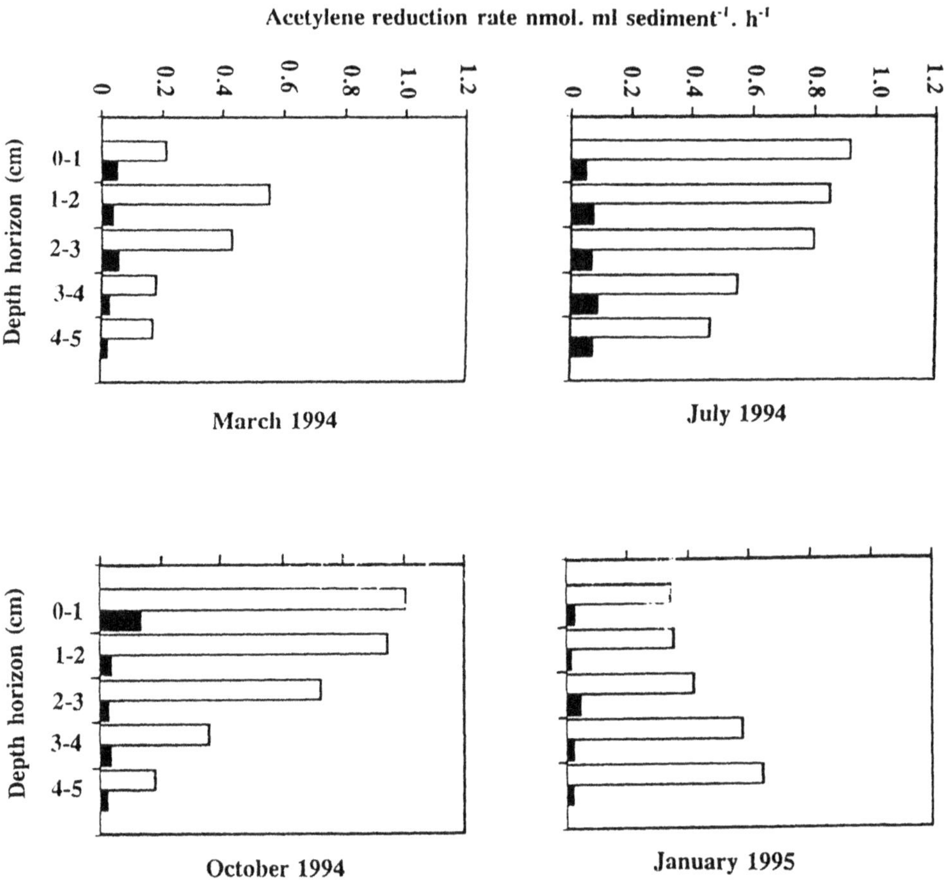


*Figure 1.* Seasonal variation in depth profiles of acetylene reduction activity (nitrogen fixation) recorded in *Zostera noltii* colonised sediments in March, July and October 1994 and January 1995, using the slurry technique in the absence (open bars) and presence (solid bars) of 20 mmol $l^{-1}$ sodium molybdate. Data points represent the means of 5 replicate determinations, standard deviations were generally less than 5%.

summer maximum was apparent (Figure 2.). Sodium molybdate additions inhibited ARA by >50% (Figure 2), indicating that SRB were also the dominant component of the nitrogen fixing microflora present in the uncolonised sediments.

A typical data set obtained using the whole core technique in July 1994 in *Z. noltii* colonised sediments are shown in Figure 3. ARA was significantly influenced by light with the rate in light incubated cores being 4-fold greater than that recorded in the dark. These data demonstrate that the photosynthetic activity of the *Zostera* was a major factor influencing nitrogen fixation (ARA) in the rhizosphere, since, nitrogen fixing cyanobacteria were not observed in the sediments or as epiphytes on the *Zostera* leaves. Additionally, data from the slurry experiments demonstrate that a large proportion of ARA occurred at depths where light was not available for photosynthetic nitrogen fixers. Rates recorded by the whole core technique for cores pre-incubated with sodium molybdate were 15.5 and 28-fold lower during light and dark incubations respectively than those recorded in the absence of sodium molybdate (Figure 3), indicating that SRB were responsible for the bulk of the recorded activity. In molybdate treated cores the acetylene reduction rate during light incubations was 6.7-fold greater than during dark incubations (Figure 3), inferring that the molybdate resistant component of the nitrogen fixing microflora were also dependent upon the photosynthetic activity of the *Zostera*. In contrast, little difference was recorded between the acetylene reduction rates during light and dark incubations in uncolonised sediments indicating that the stimulation observed by light in colonised sediments was a result of photosynthetic inputs by the *Zostera* rather than an effect of light *per se*.

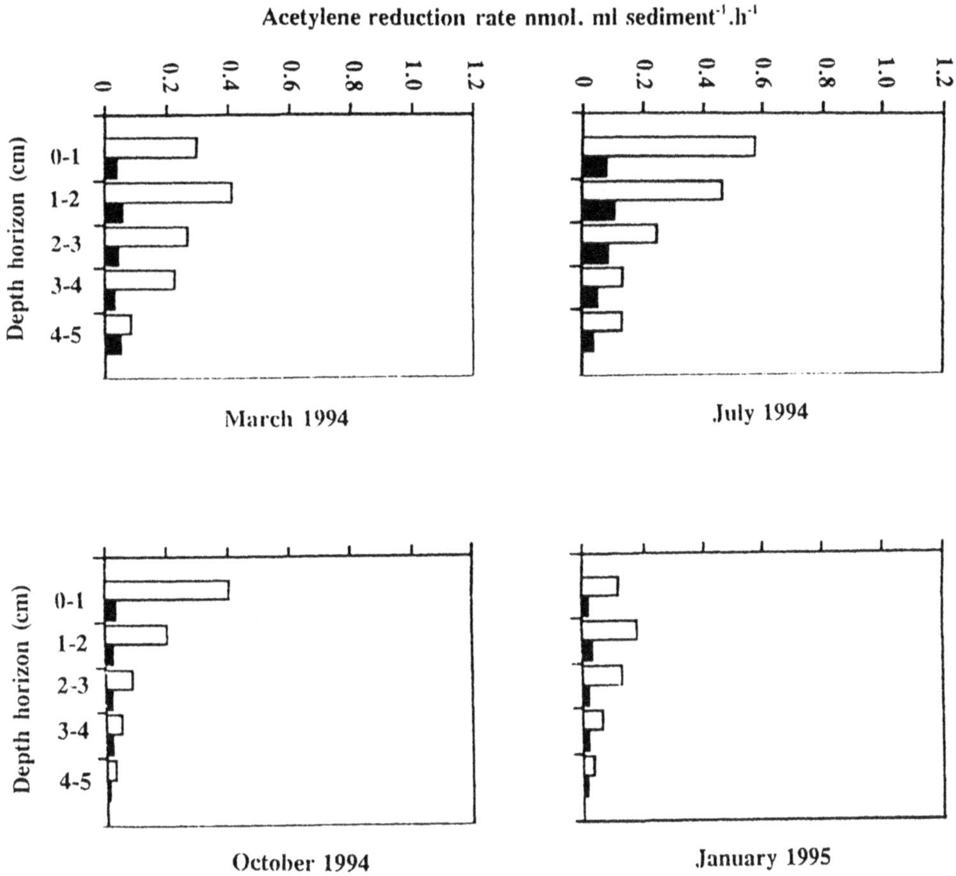


*Figure 2.* Seasonal variation in depth profiles of acetylene reduction activity (nitrogen fixation) recorded in uncolonised sediments in March, July and October 1994 and January 1995, using the slurry technique in the absence (open bars) and presence (solid bars) of 20 mmol $l^{-1}$ sodium molybdate. Data points represent the means of 5 replicate determinations, standard deviations were generally less than 5%.

Seasonal patterns of ARA measured in *Z. noltii* colonised sediments using both the slurry and whole core techniques are presented in Figure 4. Acetylene reduction activity showed a mid-summer peak with highest rates recorded in July independently of which technique was used. However, the degree of seasonal variation observed differed considerably dependent on which technique was used. Acetylene reduction rates recorded by the slurry technique were always higher than those recorded by the whole core technique irrespective of light regimes. These differences between the two techniques varied considerably with season with rates measured by the slurry technique being 9.4 and 23-fold greater than those recorded using the whole core technique (light incubation) and 19.2 and 41.8-fold greater than whole core dark incubations in March and January respectively (Figure 4.). In contrast little difference was evident between values obtained by the two techniques during the summer and autumn sampling periods with rates measured in slurries being only 1.1 and 1.6-fold greater than those measured during light incubations of whole cores in July and October respectively (Figure 4A & B).

In *Z. noltii* colonised sediments the acetylene reduction rates recorded during light incubations of whole cores were always higher than those measured in the dark (Figure 4B & C). The degree of stimulation observed, however, varied between 1.8 and 4-fold dependent on the season, being greatest in July and lowest in January, indicating that the influence of *Z. noltii* photosynthesis on ARA varied with the growth phase of the plants. Addition of sodium molybdate severely inhibited ARA activity throughout the annual cycle by between 85 and 95%, irrespective of which technique was used (Figure 4), inferring that ARA was primarily mediated by SRB.

Rates of acetylene reduction recorded in uncolonised sediments were substantially lower than

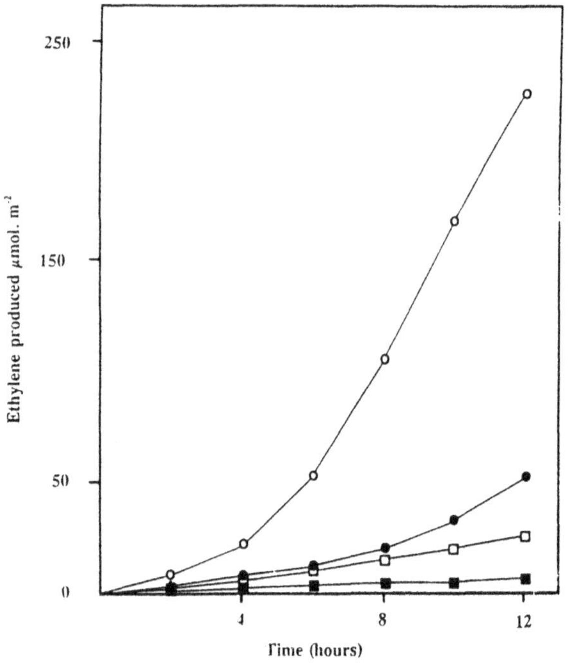


*Figure 3.* A typical data set of ARA measured using the whole core technique under natural light conditions (open circles), dark incubation (solid circles), natural light conditions in the presence of 20 mmol $l^{-1}$ sodium molybdate (open squares) and dark incubation in the presence of 20 mmol $l^{-1}$ sodium molybdate (solid squares). Data points represent the means of triplicate determinations, standard deviations have been omitted for the sake of clarity, but were less than 10% of the mean values.

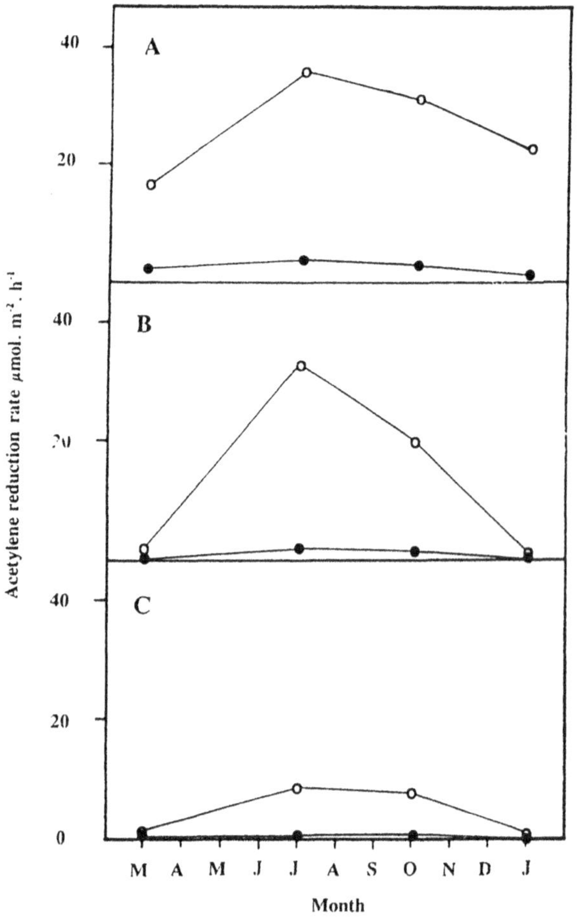


*Figure 4.* Seasonal variation in the rates of acetylene reduction measured in *Zostera noltii* colonised sediments between March 1994 and January 1995. Rates were measured in the absence (open circles) and presence (solid circles) of 20 mmol $l^{-1}$ sodium molybdate.

A. Integrated rates for the 0-5 cm depth horizon determined using the slurry technique.
B. Rates determined under natural light conditions using the whole core technique.
C. Rates determined during dark incubations using the whole core technique.

those recorded in *Z. noltii* colonised sediments (Figure 5). Rates of acetylene reduction recorded using the slurry technique showed only a small annual variation and were always inhibited by >75% following the addition of sodium molybdate (Figure 5A), inferring that SRB were responsible for the bulk of this activity. Little difference was recorded between light and dark incubations of the whole core technique (Figure 5B & C). Initially, sodium molybdate additions severely inhibited ARA by approximately 80–90% during both light and dark incubations in March and July respectively (Figure 5B & C). However, this level of inhibition decreased to 12 and 36% during light and dark incubations respectively in October and similarly decreased in January (Figure 5B & C). These data indicate a shift in the composition of the heterotrophic nitrogen fixing microflora had occurred over this period.

The mid-summer peak in nitrogen fixation (acetylene reduction) rates in *Z. noltii* colonised sediments correlated well with the availability of $NH_4^+$ in these sediments (Figure 6). Sediment exchangeable $NH_4^+$ concentrations were highest in winter with a concentration of approximately 290 $\mu$mol per litre of sediment recorded in January 1995. The exchangeable $NH_4^+$ concentration declined rapidly during spring and early summer attaining a minimal value of approximately 190 $\mu$mol 1 sediment$^{-1}$ in May which was maintained throughout the summer. The $NH_4^+$ pool slowly regenerated during the autumn and winter period (Figure 6).

This annual cycling of sediment exchangeable $NH_4^+$ pool is consistent with the life cycle of *Z. noltii* in the Bassin, which has a minimal standing crop dur-

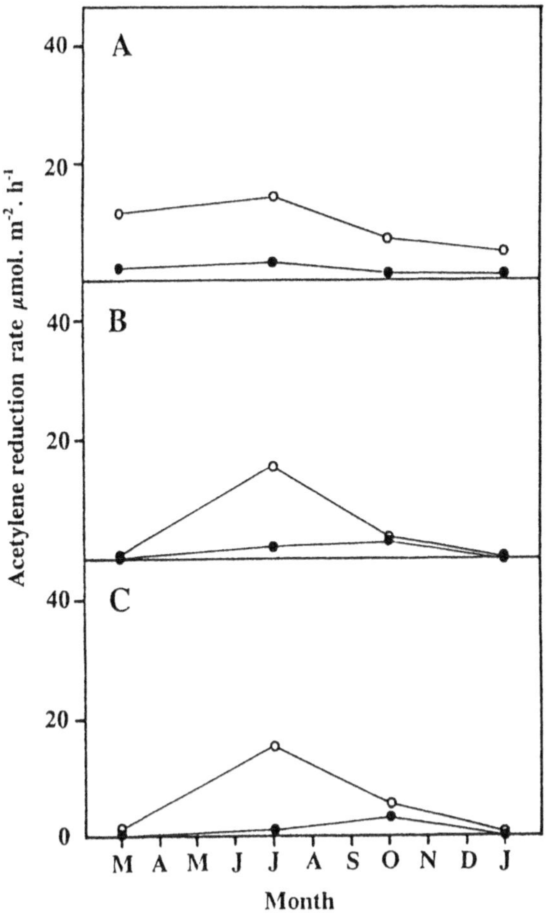


*Figure 5.* Seasonal variation in the rates of acetylene reduction measured in uncolonised sediments in the absence (open circles) and presence (solid circles) of 20 mmol $l^{-1}$ sodium molybdate.

A. Integrated rates for the 0–5 cm depth horizon determined using the slurry technique.
B. Rates determined under natural light conditions using the whole core technique.
C. Rates determined during dark incubations using the whole core technique.

ing the winter. Seagrass growth commences in February and continues throughout the spring and summer period achieving maximal biomass concentrations in June (root and rhizome biomass) and August (shoot biomass). Thereafter the standing crop slowly declines during autumn and winter (Auby, 1991). In contrast to the *Z. noltii* colonised sediments, the sediment exchangeable $NH_4^+$ concentration in uncolonised sediments remained at a relatively constant value of 280 to 300 $\mu$mol 1 sediment$^{-1}$ throughout the annual cycle (Figure 6).

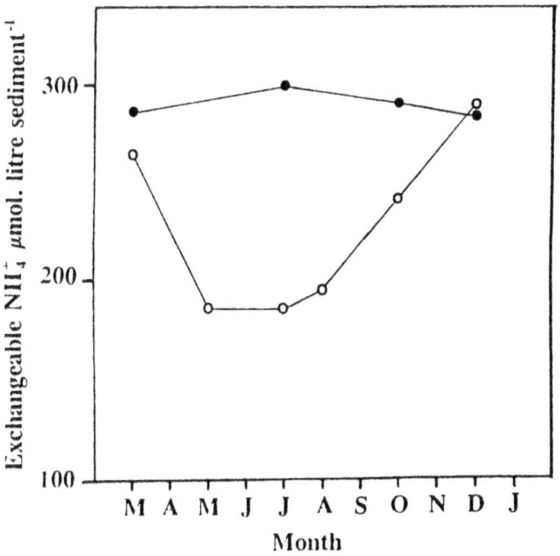


*Figure 6.* Seasonal variation in the sediment exchangeable ammonium concentration between March 1994 and January 1995 in *Zostera noltii* colonised sediments (open circles) and uncolonised sediments (solid circles). Data points represent the means of triplicate determinations, standard deviations have been omitted for the sake of clarity, but were generally 15% (colonised sediments) and 5% (uncolonised sediments) of the mean values.

Rates of sulphate reduction determined in *Z. noltii* colonised sediments showed a relationship with seasonal changes in both the seagrass community biomass and acetylene reduction rates, with a maximal rate of $1.36 \pm 0.44$ mmol m$^{-2}$ h$^{-1}$ recorded in July 1994 (Figure 7). In contrast, in March the sulphate reduction rate was $0.68 \pm 0.13$ mmol m$^{-2}$ h$^{-1}$. These changes in sulphate reduction rate were primarily due to an increase in the rate recorded in the surficial 0–2 cm depth horizon, with the activity in this horizon representing 87% of the total activity in July 1994. In contrast in March the activity in this zone was >3-fold lower than in July and represented only 56% of the total activity (data not shown). Thus, the increase in sulphate reduction activity occurred concurrently with the observed shift in the acetylene reduction activity peak towards the sediment surface of *Zostera noltii* colonised sediments in summer (Figure 2).

Populations of SRB in the surficial 0–2 cm depth horizon of *Z. noltii* colonised sediments determined by the most probable number technique ranged from $4.6 \times 10^5$ to $3.8 \times 10^6$ ml sediment$^{-1}$ (Figure 8). However, seasonal trends in these population densities were small and hardly significant (ANOVA analysis; $F(6,21)$: 2.902; $p \sim 0.05$) inferring that the observed changes in sulphate reduction rate in *Z. noltii* colonised sediments (Figure 7) were due to changes in the activ-

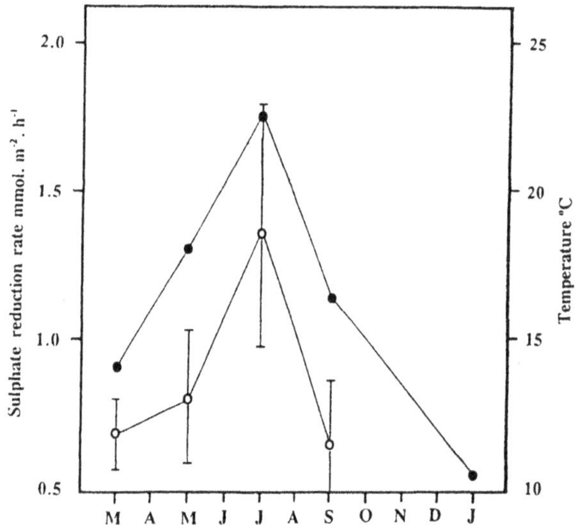


*Figure 7.* Seasonal variation of surficial sediment temperature (solid circles) and the integrated sulphate reduction rate (open circles) for the 0–5 cm depth horizon determined under natural light conditions in *Zostera noltii* colonised sediments. Data points for sulphate reduction rates represent the means of triplicate determinations ± standard deviation. The rate of sulphate reduction was not determined in January 1995.

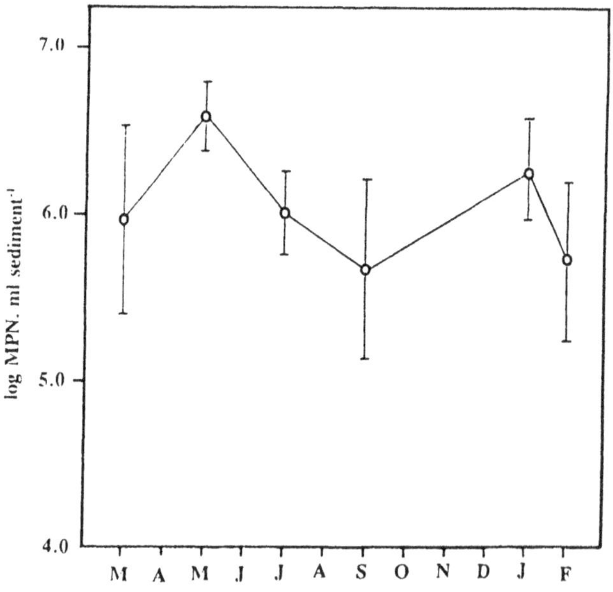


*Figure 8.* Seasonal variation of the most probable number of sulphate reducing bacteria in the 0–2 cm depth horizon of *Zostera noltii* colonised sediments determined between March 1994 and February 1995. Data points represent the means of 4 replicate determinations ± standard deviation.

ity per cell rather than shifts in the population density of SRB.

## Discussion

ARA was detectable throughout the year in both *Z. noltii* colonised sediments and uncolonised sediments in the Bassin d'Arcachon (Figures 1–5). Rates of acetylene reduction were substantially higher in colonised sediments as compared to uncolonised sediments and are in agreement with previous studies which have demonstrated high nitrogen fixation (acetylene reduction) rates in the rhizosphere of seagrasses (Capone et al., 1979; Capone, 1988; O'Donohue et al., 1991a; Moriarty & O'Donohue, 1993; Pereg et al., 1994). However, the recorded acetylene reduction rates varied considerably dependent upon the assay method used, with rates measured using the slurry technique overestimating those recorded in isolated cores, especially for *Z. noltii* colonised sediments during the spring and autumn sampling programmes (Figures 4 & 5). This over-estimation may be an inherent artefact of the slurry technique due to the release of carbon sources during preparation of the slurries. In *Z. noltii* colonised sediments internal pools of dissolved organic carbon in the plant roots may have been released due to root damage. It has previously been demonstrated that under oxygen limiting conditions *Zostera marina* L. roots progressively switch to fermentative metabolism producing ethanol, $CO_2$ and lactate as the major metabolic endproducts (Smith et al., 1988). Internal pools of lactate sequestered in the root tissues therefore represent a substantial pool of organic carbon, which may be liberated if the plant roots are inadvertently damaged. It has previously been reported that physical disturbance of vegetated salt marsh sediments can significantly alter sediment porewater dissolved organic carbon concentrations. For example, porewater acetate concentrations measured following sediment slicing and centrifugation or sediment squeezing extraction techniques can be as much as 100-fold greater than those determined by the non-destructive 'sediment sipper' technique (Howes et al., 1985; Hines et al., 1994). These authors attributed these differences to the degree of root damage and associated carbon release during porewater sampling. Similarly, the carbon status of slurries prepared from uncolonised sediments may also have been enhanced by the release of carbon due to the physical break-up and redistribution of the decaying *Z. noltii* roots and rhizomes present in these sediments.

Thus, acetylene reduction rates measured by the slurry technique are more indicative of the potential ARA due to the elevated carbon status of the slurries, whereas *in situ* rates will be lower due to the limited availability of suitable carbon sources. This hypothesis is supported by the fact that the addition of a range of different carbon sources to slurries of *Z. noltii* colonised sediments at a final concentration of 15 mmol $l^{-1}$ resulted in only a minor 15–30% stimulation of ARA (Welsh et al., 1996).

In contrast the intact core technique may provide a more realistic estimate of the *in situ* rate of acetylene reduction, since, using this technique the disturbance of physicochemical gradients of $O_2$ sulphide and nutrients and the relationship between the plant roots and the rhizosphere microflora is minimal. Additionally, the use of the intact core technique in seagrass colonised sediments may preferentially measure ARA associated with the plant roots, since diffusion of acetylene to the rhizosphere will be facilitated by transfer via lacunal air spaces within the plant leaves and roots (O'Donohue et al., 1991b). This method may however, lead to a slight underestimate of the overall rate of acetylene reduction, since nitrogen fixing bacteria in zones remote from the plant roots may not be subject to a saturating acetylene concentration.

Acetylene reduction rates in *Z. noltii* colonised sediments measured by the whole core technique in the light were always higher than those measured in the dark (Figures 3 & 4). In contrast, in uncolonised sediments the rates measured during light and dark incubations were similar (Figure 5) and comparable with rates measured during dark incubations of colonised sediments (Figure 4). These data demonstrate that heterotrophic nitrogen fixation (acetylene reduction) in the rhizosphere of *Z. noltii* was fuelled by inputs of organic carbon from the plant roots during photosynthesis, whereas, during dark incubations or in uncolonised sediments activity is linked to the degradation of moribund organic matter in the sediment and thus independent of light. These results are in agreement with previous studies of sea and saltmarsh grasses which have demonstrated that nitrogen fixation/acetylene reduction rates are closely coupled to the photosynthetic activity of the plants and stimulated by release of labile carbon from the plant roots (Boyle & Patriquin, 1981; Whiting et al., 1986; O'Donohue et al., 1991a; Moriarty & O'Donohue, 1993). In this study the level of stimulation of ARA by light was variable with season, ranging from a 1.8-fold stimulation in January to a 4-fold stimulation in July, with the overall ARA increasing from 1.02 (light incubation) and 0.56 $\mu$mol $m^{-2}$ $h^{-1}$ (dark incubation) in January to 32.41 (light incubation) and 8.70 $\mu$mol $m^{-2}$ $h^{-1}$ (dark incubation) in July (Figure 4b & c). These differences in both the overall rates of ARA and the level of stimulation by light are consistent with the annual variation of the *Z. noltii* root and shoot biomass which is minimal in winter and maximal in summer (Auby, 1991). Therefore, potential photosynthetic inputs to the rhizosphere would be greatest in mid-summer, whereas in winter these would be substantially reduced due to the lower standing crop and shorter day length. Thus, seasonal variations in plant biomass and rate of carbon release by the plant roots, may be one of the major factors regulating heterotrophic nitrogen fixation in the rhizosphere. The hypothesis that the input of labile organic carbon to the rhizosphere via the *Z. noltii* roots is a major determinant of ARA in the rhizosphere is supported, if as discussed earlier, we consider that the slurry technique is a measure of the potential ARA, whereas, intact core measurements approximate to the *in situ* rate. Thus, in July, ARA measured by the whole core technique during light incubations was equivalent to >90% of the potential (integrated slurry) activity (Figure 4a & b) whereas, in January rates measured by the whole core technique in the light represented <5% of the potential activity in the slurries. These data indicate that in January ARA was severely limited by availability of suitable organic carbon sources, whereas in July when the potential photosynthetic inputs of organic carbon from the *Zostera* roots was greatest the rate measured during light incubations of whole cores approximates to the potential rate as measured by the slurry technique.

In contrast to colonised sediments acetylene reduction rates measured in uncolonised sediments, using the whole core technique were not light dependent (Figure 5b & c) and were similar to those recorded during dark incubations of *Z. noltii* colonised sediments (Figure 4c). In these sediments ARA appears to be linked to the degradation of high C:N ratio (~38:1, Welsh, unpublished data) moribund root and rhizome biomass which was present. Previous, studies have similarly recorded heterotrophic nitrogen fixation associated with decomposing seagrass roots (Kenworthy et al., 1987; O'Donohue et al., 1991a). Rates of acetylene reduction in these sediments also showed a mid-summer peak (Figure 5) and are most probably regulated by the quantity and quality of the residual organic matter.

The availability of fixed nitrogen sources in the sediment would also be expected to exert a major influence on nitrogen fixation rates as both the synthesis and activity of nitrogenase are sensitive to feedback inhibition (Zumft & Castillo, 1978; Postgate, 1982; Dixon, 1984; Postgate & Kent, 1984). Additions of $NH_4^+$ have previously been demonstrated to inhibit *in situ* ARA in the rhizosphere of the salt marsh grass *Spartina alterniflora* Loisel (Yoch & Whiting, 1986). Sediment exchangeable $NH_4^+$ concentrations in *Z. noltii* colonised sediments measured in this study showed a distinct annual cycle (Figure 6), with the $NH_4^+$ pool exhibiting an inverse relationship to ARA (Figure 4). The exchangeable $NH_4^+$ pool decreased during spring reaching a plateau concentration in May of approximately 190 $\mu$mol l sediment$^{-1}$ which was maintained until August, corresponding to the period of maximal ARA, thereafter the $NH_4^+$ pool slowly regenerated during the autumn and winter period and this was accompanied by a corresponding reduction in ARA. In contrast the sediment exchangeable $NH_4^+$ pool in uncolonised sediments maintained a stable value of between 280 and 300 $\mu$mol l sediment$^{-1}$ throughout the year (Figure 6), inferring that the changes observed in colonised sediments were a result of $NH_4^+$ assimilation by the plants. Thus nitrogen fixation in the rhizosphere of *Z. noltii* appears to be regulated by the activity of the plant roots due both to the supply of organic carbon as root exudates during photosynthesis and the maintenance of low $NH_4^+$ concentration in the rhizosphere during the growth season.

ARA in *Z. noltii* colonised sediments was severely inhibited (>75%) in the presence of sodium molybdate, in both slurries and whole cores irrespective of light regime throughout the year (Figures 1, 3 & 4). Sodium molybdate is a potent inhibitor of sulphate reduction (Taylor & Oremland, 1979; Smith & Klug, 1981; Oremland & Capone, 1988) and these data provide strong evidence that SRB were the dominant component of the nitrogen fixing microflora in the rhizosphere. In a previous study, O'Donohue et al., (1991a), estimated that approximately 90% of nitrogen fixation in meadows of the seagrass *Zostera capricornia* in Moreton Bay, Australia was due to the activity of anaerobic or microaerophilic bacteria in the rhizosphere, which would have included SRB, although no attempts were made to further identify the bacteria responsible. Additionally, numerous studies have reported a stimulation of nitrogen fixation activity under anaerobic or microaerophilic conditions in seagrass sediments or isolated roots and rhizomes (Patriquin & McClung, 1978; Capone & Taylor, 1980; Capone & Budin, 1982), conditions which would also stimulate SRB.

Sulphate reduction rates measured in *Z. noltii* colonised sediments were high ranging from 15.6 to 32.6 mmol $SO_4^{2-}$ m$^{-2}$ d$^{-1}$ and are comparable with those determined in other organically rich sediments, including seagrass (Blackburn et al., 1994), salt marsh (Hines et al., 1989; Hines et al., 1994) and tropical mangrove sediments (Hines & Lyons, 1982; Kristensen et al., 1994; Nedwell et al., 1994). Sulphate reduction rates were highest in summer and approximately double those measured in spring and autumn (Figure 7). The seasonal changes in sulphate reduction rate correlated well with changes in sediment temperature (Figure 7) and acetylene reduction rates (Figure 4). This increase in sulphate reduction rate during summer was primarily due to increased activity in the surficial 0–2 cm depth horizon (data not shown) and thus correlates with the shift in ARA towards the sediment surface in summer observed in *Z. noltii* colonised sediments (Figure 1). However, the relationship between the sulphate reduction rate and sediment temperature may be coincidental, since these changes also correlate with the changes in the *Z. noltii* standing crop (Auby, 1991) and thus potential carbon inputs from the plant roots. It has previously been demonstrated in salt marsh sediments, colonised by *Spartina alterniflora*, that sulphate reduction rates in the rhizosphere were closely coupled to the growth phase of the plants and driven by the release of dissolved organic carbon exudates from the plant roots during active growth (Hines et al., 1989). Similar seasonal variations in bacterial productivity in the rhizosphere of seagrasses and plant growth have also been observed and linked to the exudation of organic carbon from the plant roots (Moriarty & Boon, 1989). A similar relationship between the growth of *Z. noltii* and the release of organic carbon by the plant roots could account for the observed changes in sulphate reduction rate in these sediments. Thus, a mutualistic relationship may exist between the *Z. noltii* and SRB in the rhizosphere.

In uncolonised sediments sodium molybdate additions also inhibited ARA (Figure 5) indicating that nitrogen fixing SRB were present. However, the degree of inhibition measured in whole core incubations (Figure 5b & c) decreased over the year indicating a population change in the nitrogen fixing microflora, with SRB being replaced by aerobic or possibly microaerophilic bacteria, since sodium molybdate consistently inhibited ARA by approximately 80 to 90% throughout the

*Table 1.* Comparison of nitrogen fixation rates recorded in the Bassin d'Arcachon with rates recorded in other seagrass meadows. Data for the Bassin d'Arcachon are based on acetylene reduction rates from whole core experiments only and were calculated using the theoretical 3:1 ratio between acetylene reduction and nitrogen fixation.

| Seagrass species | Locality | Nitrogen fixation rate (mg N $m^{-2}$ $d^{-1}$) | Reference |
|---|---|---|---|
| None | Bassin d'Arcachon | 0.02–3.7 | This study |
| *Zostera noltii* | Bassin d'Arcachon (spring) | 0.2–0.4 | This study |
| *Zostera noltii* | Bassin d'Arcachon (summer) | 2.0–7.3 | This study |
| *Zostera noltii* | Bassin d'Arcachon (autumn) | 1.8–4.4 | This study |
| *Zostera noltii* | Bassin d'Arcachon (winter) | 0.1–0.2 | This study |
| *Syringodium isoetifolium* & *Cymodocea serrulala* | Gulf of Carpentaria Australia (summer) | 16–47 | Moriarty & O'Donohue, 1993 |
| *Thalassia hemprichii* | Gulf of Carpentaria (summer) | 16 | Moriarty & O'Donohue, 1993 |
| *Enhalus acoroides* | Gulf of Carpentaria (summer) | 25 | Moriarty & O'Donohue, 1993 |
| *Zostera capricornia* | Australia (summer) | 25–40 | O'Donohue et al., 1991 a |
| *Zostera capricornia* | Australia (winter) | 10 | O'Donohue et al., 1991 a |
| *Thalassia testudinium* | Barbados | 27–140 | Patriquin & Knowles, 1972 |
| *Thalassia testudinium* | Florida | 0.03 | McRoy et al., 1973 |
| *Thalassia testudinium* | Southern Florida | 5–24 | Capone & Taylor, 1980 |
| *Thalassia testudinium* | Bahamas | 6–9 | Capone et al., 1979 |
| *Thalassia testudinium* | Bahamas | 14–41 | Oremland et al., 1976 |
| *Hadodule beaudetti* | Jamaica | 28 | Blackburn et al., 1994 |

year in anaerobic slurries (Figures 2 & 5a). These data indicate that in this area where the *Zostera* had recently died back at the beginning of the experimental period, the initial population of nitrogen fixing SRB associated with the plant roots was progressively replaced by an aerobic or microaerophilic nitrogen fixing population associated with the degradation of residual root and rhizome biomass.

Rates of acetylene reduction can be used to calculate rates of nitrogen fixation using the theoretical ratio of 3:1 based on the provision of reducing equivalents as shown in equations 1 and 2.

$$C_2H_2 + 2H^+ + 2e^- \rightarrow C_2H_4 \quad (1)$$

$$N_2 + 6H^+ + 6e^- \rightarrow 2NH_3 \quad (2)$$

However, *in vitro* studies of nitrogenase have demonstrated that $e^-$ are also shunted to $H_2$ production during nitrogen fixation as shown in equation 3, but this does not occur during acetylene reduction (Postgate, 1982; Capone, 1988).

$$N_2 + 8H^+ + 8e^- \rightarrow 2NH_3 + H_2 \quad (3)$$

This equation indicates that the ratio of acetylene reduction to nitrogen fixation should be 4:1, if $e^-$ which are shunted to $H_2$ production are accounted for. However, SRB in common with many other nitrogen fixing microorganisms possess uptake hydrogenases (Adams & Mortensen, 1984; Houchins, 1984; Voordouw et al., 1990) and thus *in vivo* $e^-$ lost due to hydrogen evolution during nitrogen fixation may be efficiently recycled to nitrogenase. Additionally, there is some controversy over the use of the theoretical 3:1 ratio since acetylene is not a natural substrate for nitrogenase and inhibits other sedimentary processes (Payne, 1984; Oremland & Capone, 1987; Capone, 1988). Nevertheless, studies directly comparing acetylene reduction rates and $^{15}N_2$ fixation rates in seagrass meadows or isolated seagrass roots have shown reasonable agreement with the theoretical 3:1 ratio. (Patriquin & Knowles, 1972; Capone & Budin, 1982; O'Donohue et al., 1991 a) and we believe it is therefore appropriate to use this ratio in this study.

Acetylene reduction rates measured in this study using the whole core technique, yield rates equivalent to nitrogen fixation rates of between 0.1 and 7.3 mg N $m^{-2}$ $d^{-1}$ for *Z. noltii* colonised sediments and between 0.02 and 3.7 mg N $m^{-2}$ $d^{-1}$ in uncolonised sediments, dependent on the season if the 3:1 ratio is used. The rates recorded in the *Z. noltii* beds fall within the lower region of those determined in previous studies and show a much higher degree of temporal variation (Table 1). However, the majority of previous studies reported in the literature have focused on the role played by heterotrophic nitrogen fixation in tropical and sub-tropical seagrass ecosystems, generally in oligotrophic coastal areas (Table 1). In contrast the Bassin d'Arcachon is a semi-enclosed lagoon, subject to much greater seasonal variations in both temperature and light regimes, which receives substantial anthropogenic inputs of fixed nitrogen, calculated to be equivalent to 1000 tons N $y^{-1}$ (Auby et al., 1994). It is therefore not surprising that the nitrogen fixation rates measured in this ecosystem are somewhat lower and seasonally more variable than those previously reported for tropical seagrass ecosystems.

Auby, (1991) has previously calculated that the net productivity of *Z. noltii* in the Bassin d'Arcachon is equivalent to between 74 and 111 g carbon $m^{-2}$ $y^{-1}$ for shoot biomass and between 53 and 70 g carbon $m^{-2}$ $y^{-1}$ for root and rhizome biomass. This biomass has carbon to nitrogen ratios of 15.2:1 and 37.8:1 for shoot and root and rhizome biomass respectively (Welsh, unpublished data, determined in October 1994). Thus, by calculation *Z. noltii* in the Bassin d'Arcachon requires a fixed nitrogen input of between 6.3 and 9.2 g nitrogen $m^{-2}$ $y^{-1}$. Based on the acetylene reduction rates measured in this study by the whole core technique, heterotrophic nitrogen fixation in the rhizosphere can provide between 0.4 and 1.1 g nitrogen $m^{-2}$ $y^{-1}$ or between 6.3 and 12.0% of the fixed nitrogen requirement of the plants. Heterotrophic nitrogen fixation in the rhizosphere of *Z. noltii*, therefore represents a substantial local input of fixed nitrogen to the surficial sediments of this shallow coastal lagoon and contributes to the overall productivity of *Z. noltii* community in this ecosystem.

## Acknowledgments

This research was carried out as part of the E.U. funded Coastal Lagoon Eutrophication and ANaerobic processes (CLEAN) project, grant No EVSV-CT92-0080. Sophie Bourguès was supported by a grant from the French Ministry of Research (MESR). The authors would like to thank all the participants in the CLEAN project for their helpful discussions on the ecology of this ecosystem and especially Isabelle Auby for invaluable local knowledge on the life cycle and ecology of *Zostera noltii* in the Bassin d'Arcachon. We would also like to thank Prof. Pierre Caumette for the use of laboratory facilities in his Institute and all the staff and students of the Laboratoire d'Océanographie, Arcachon for their kindness and assistance during the course of this study.

## References

Adams, M. W. W. & L. E. Mortensen, 1984. The physical and catalytical properties of hydrogenase II of *Clostridium pasteurianum*. J. Biol. Chem. 259: 7045–7055.

Auby, I., 1991. Contribution à l'étude des herbiers de *Zostera noltii* dans le Bassin d'Arcachon: dynamique, production et dégradation, macrofaune associée. Ph.D. thesis, Université de Bordeaux I.

Auby, I., F. Manaud, D. Maurer & G. Trut, 1994 . Etude de la prolifération des algues vertes dans le Bassin d'Arcachon. Rapport IFREMER–CEMAGREF–SSA–SABARC. 163 pp, IFREMER, Arcachon, France.

Blackburn, T. H., D. B. Nedwell & W. J. Wiebe, 1994. Active mineral cycling in a Jamaican seagrass sediment. Mar. Ecol. Prog. Ser. 110: 233–239.

Boyle, C. D. & D. G. Patriquin, 1981. Carbon metabolism of *Spartina alterniflora* Loisel in relation to that of associated nitrogen-fixing bacteria. New Phytol. 89: 275–288.

Caffrey, J. M. & W. M. Kemp, 1992. Influence of the submersed plant, *Potamogeton perfoliatus*, on nitrogen cycling in estuarine sediments. Limnol. Oceanogr. 37: 1483–1495.

Canfield, D. E., 1989. Sulfate reduction and oxic respiration in marine sediments: implications for organic carbon preservation in euxinic environments. Deep Sea Res. 36: 121–138.

Capone, D. G., 1988. Benthic nitrogen fixation. In T. H. Blackburn & J. Sørensen (eds), Nitrogen cycling in coastal marine environments. John Wiley & Sons Ltd, Chichester. 85–123.

Capone, D. G. & J. M. Budin, 1982. Nitrogen fixation associated with rinsed roots and rhizomes of the eelgrass *Zostera marina*. Pl. Physiol. 70: 1601–1604.

Capone, D. G., P. A. Penhale, R. S. Oremland & B. F. Taylor, 1979. Relationship between productivity and $N_2$ ($C_2H_2$) fixation in the rhizosphere of *Thalassia testudinium*. Limnol. Oceanogr. 24: 117–125.

Capone, D. G. & B. F. Taylor, 1980. $N_2$ fixation in the rhizosphere of *Thalassia testudinium*. Can. J. Microbiol. 26: 998–1005.

Clarke, K. R. & N. J. P. Owens, 1983. A simple and versatile microcomputer program for the determination of 'most probable number'. J. Microbiol. Meth. 1: 133–137.

Dennison, W. C, R. C. Aller & R. S. Alberte, 1987. Sediment ammonium availability and eelgrass (*Zostera marina*) growth. Mar. Biol. 94: 469–477.

Dixon, R. A., 1984. The genetic complexity of nitrogen fixation. J. gen. Microbiol. 130: 2745–2755.

Eppley, R. W., E. H. Renger & W. G. Harrison, 1979. Nitrate and phytoplankton production in southern California waters. Limnol. Oceanogr. 24: 483–494.

Herbert, R. A., 1975. Heterotrophic nitrogen fixation in shallow estuarine sediments. J. exp. mar. Biol. Ecol. 18: 215–225.

Hines, M. E. & W. B. Lyons, 1982. Biogeochemistry of nearshore Bermudan sediments. I. Sulfate reduction rates and nutrient regeneration. Mar. Ecol. Prog. Ser. 8: 87–94.

Hines, M. E., S. L. Knollmeyer & J. B. Tugel, 1989. Sulfate reduction and other sedimentary biogeochemistry in a northern New England salt marsh. Limnol. Oceanogr. 34: 578–590.

Hines, M. E., G. T. Banta, A. E. Giblin & J. E. Hobbie, 1994. Acetate concentrations and oxidation in salt-marsh sediments. Limnol. Oceanogr. 39: 140–148.

Houchins, J. P., 1984. The physiology and biochemistry of hydrogen metabolism in cyanobacteria. Biochim. Biophys. Acta. 768: 227–255.

Howes, B. L., J. W. H. Dacey & S. G. Wakeham, 1985. Effects of sampling technique on measurements of porewater constituents in salt marsh sediments. Limnol. Oceanogr. 30: 221–227.

Iizumi, H., A. Hattori & C. P. McRoy, 1982. Ammonium regeneration and assimilation in eelgrass (*Zostera marina*) beds. Mar. Biol. 66: 59–65.

Jones, K., 1982. Nitrogen fixation in the temperate estuarine intertidal salt marsh sediments of the river Lune. Limnol. Oceanogr. 22: 814–832.

Jørgensen, B. B., 1977. The sulfur cycle of a coastal marine sediment (Limfjorden, Denmark). Limnol. Oceanogr. 22: 814–832.

Jørgensen, B. B., 1982. Mineralization of organic matter in the sea bed: the role of sulphate reduction. Nature 296: 643–645.

Kenworthy, W. J., C. Currin, G. Smith & G. Thayer, 1987. The abundance biomass and acetylene reduction activity of bacteria associated with decomposing rhizomes of two seagrasses, *Zostera marina* and *Thalassia testudinium*. Aquat. Bot. 27: 97–119.

Kristensen, E., G. M. King, M. Holmer, G. T. Banta, M. H. Jensen, K. Hansen & N. Bussarawit, 1994. Sulfate reduction, acetate turnover and carbon metabolism in sediments of the Ao Nam Bor mangrove, Phuket, Thailand. Mar. Ecol. Prog. Ser. 109: 245–255.

McRoy, C. P. & C. McMillan, 1977. Production, ecology and physiology of seagrasses. In C. P. McRoy & C. Helfferich (eds). Seagrass ecology. Marcel Dekker, New York. 53–87.

McRoy, C. P., J. J. Goering & B. Chaney, 1973. Nitrogen fixation associated with seagrasses. Limnol. Oceanogr. 18: 998–1002.

Moriarty, D. J. W. & P. I. Boon, 1989. Interactions of seagrasses with sediment and water. In A. W. D. Larkum, A. J. McComb & S. A. Shepherd (eds), Biology of seagrasses. Elsevier, Amsterdam: 500–535.

Moriarty, D. J. W. & M. J. O'Donohue, 1993. Nitrogen fixation in seagrass communities during summer in the Gulf of Carpentaria, Australia. Aust. J. mar. Freshwat. Res. 44: 117–125.

Moriarty, D. J. W., P. I. Boon, J. A. Hansen, W. G. Hunt, I. R. Poiner, P. C. Pollard, G. W. Skyring & D. C. White, 1985. Microbial biomass and productivity in seagrass beds. Geomicrobiol. J. 4: 21–51.

Moriarty, D. J. W., D. G. Roberts & P. C. Polard, 1990. Primary and bacterial productivity of tropical seagrass communities in the Gulf of Carpentaria, Australia. Mar. ecol. Prog. Ser. 61: 145–157.

Nazina, T. N., E. P. Rozanova & T. A. Kalininskaya, 1979. Fixation of molecular nitrogen by sulfate-reducing bacteria from oil strata. Mikrobiologiya 48: 133–136.

Nedwell, D. & S. Aziz, 1980. Heterotrophic nitrogen fixation in an intertidal salt marsh sediment. Estuar. coast. mar. Sci. 10: 699–702.

Nedwell, D. B., T. H. Blackburn & W. J. Wiebe, 1994. Dynamic nature of the turnover of organic carbon, nitrogen and sulphur in the sediments of a Jamacaican mangrove forest. Mar. Ecol. Prog. Ser. 110: 223–231.

Nixon, S. W. & M. E. Q. Pilson, 1983. Nitrogen in estuarine and coastal marine ecosystems. In J. E. Carpenter & D. G. Capone (eds), Nitrogen in the marine environment. Academic Press, New York. 565–647.

O'Donohue, M. J., D. J. W. Moriarty & I. C. MacRae, 1991a. Nitrogen fixation in sediments and the rhizosphere of the seagrass *Zostera capricornia*. Microb. Ecol. 22: 53–64.

O'Donohue, M. J., D. J. W. Moriarty & I. C. MacRae, 1991b. A comparison of methods for determining rates of acetylene reduction (nitrogen fixation) by heterotrophic bacteria in seagrass sediment. J. Microbiol. Meths. 13: 171–183.

Oremland, R. S. & D. G. Capone, 1988. Use of 'specific inhibitors' in biogeochemistry and microbial ecology. Adv. Microbiol. Ecol. 10: 285–383.

Oremland, R. S., J. W. Gotto & B. F. Taylor, 1976. $N_2$ ($C_2H_2$) fixation associated with the rhizosphere communities of the seagrass *Thalassia testudinium*. Abstracts of the annual meeting of the American Society of Microbiology. abstr. 171.

Patriquin, D. G., 1972. The origin of nitrogen and phosphorus for the growth of the marine angiosperm *Thalassia testudinium*. Mar. Biol. 15: 25–46.

Patriquin, D. G. & R. Knowles, 1972. Nitrogen fixation in the rhizosphere of marine angiosperms. Mar. Biol. 16: 49–58.

Patriquin, D. G. & C. McClung, 1978. Nitrogen accretion, and the nature and possible significance of $N_2$ fixation (acetylene reduction) in a Nova Scotian *Spartina alterniflora* stand. Mar. Biol. 47: 227–242.

Payne, W. J., 1984. Influence of acetylene on microbial and enzymatic assays. J. Microbiol. Methods. 2: 117–133.

Pereg, L. L., Y. Lipkin & N. Sar., 1994. Different niches of the *Halophila stipulacea* seagrass bed harbor distinct populations of nitrogen fixing bacteria. Mar. Biol. 119: 327–333.

Pfennig, N., F. Widdel & H. G. Trüper, 1981. The dissimilatory sulfate reducing bacteria. In M. P. Starr, H. Stolp, H. G. Trüper, A. Balows & H. G. Schlegel (eds), The Prokaryotes vol 1. Springer-Verlag, New York. 926–940.

Pfennig, N., H. G. Trüper, 1992. The family *Chromatiaceae*. In A. Balows, H. G. Trüper, M. Dworkin & K. H. Schleifer (eds), The Prokaryotes, vol 4. Springer-Verlag, New York. 3200–3331.

Postgate, J. R., 1982. The fundamentals of nitrogen fixation. Cambridge University Press, London.

Postgate, J. R. & H. M. Kent, 1984. Derepression of nitrogen fixation in *Desulfovibrio gigas* and its stability to ammonia or oxygen stress *in vivo*. J. gen. Microbiol. 130: 2824-1.

Postgate, J. R., H. M. Kent, S. Hill & H. Blackburn, 1985. Nitrogen fixation by *Desulfovibrio gigas* and other species of *Desulfovibrio*. In P. W. Ludden & J. E. Burris (eds). Nitrogen fixation and $CO_2$ metabolism. Elsevier, Amsterdam. 225–234.

Postgate, J. R., H. M. Kent & R. L. Robson, 1988. Nitrogen fixation by *Desulfovibrio*. In J. A. Cole & S. J. Ferguson (eds). The nitrogen and sulphur cycles. Society of General Microbiology Symposium 42. Cambridge University Press, Cambridge. 457–471.

Riederer-Henderson, M. A. & P. W. Wilson, 1970. Nitrogen fixation by sulphate reducing bacteria. J. gen. Microbiol. 61: 27–31.

Ryther, J. H. & W. M. Dunstan, 1971. Nitrogen, phosphorus and eutrophication in the coastal marine environment. Science 171: 1008–1013.

Short, F. T., 1983. The response of interstitial ammonium in eelgrass (*Zostera marina*) beds to environmental perturbations. J. exp. mar. Biol. Ecol. 68: 195–208.

Smith, R. D., A. M. Pregnall & R. S. Alberte, 1988. Effects of anaerobiosis on root metabolism of *Zostera marina* (eelgrass): implications for survival in reducing sediments. Mar. Biol. 98: 131–141.

Smith, R. I. & M. J. Klug, 1981. Electron donors utilised by sulfate-reducing bacteria in eutrophic lake sediments. Apl. envir. Microbiol. 42: 116–121.

Stewart, W. D. P., G. P. Fitzgerald & R. H. Burris, 1967. *In situ* studies on $N_2$ fixation using the acetylene reduction technique. Proc. nat. Acad. Sci. U.S.A. 58: 2071–2078.

Taylor, B. F. & R. S. Oremland, 1979. Depletion of adenosine triphosphate in *Desulfovibrio* by oxyanions of Group IV elements. Curr. Microbiol. 3: 101–103.

Voordouw, G., V. Niviere, F. G. Ferris, P. M. Fedorak & D. W. S. Westlake, 1990. Distribution of hydrogenase genes in *Desulfovibrio* spp. and their use in identification of species from the oilfield environment. Apl. envir. Microbiol. 56: 3748–3754.

Welsh, D. T., R. Wellsbury, S. Bourgues, R. de Wit & R. Herbert, 1996. Relationship between porewater organic carbon intent, sulphate reduction and nitrogen fixation (acetylene reduction) in the rhizosphere of *Zostera noltii*. Hydrobiologia 329 (Dev. Hydrobiol. 117): 169–177.

Whiting, G. J., E. L. Gandy & D. C. Yoch, 1986. Tight coupling of root-associated nitrogen fixation and plant photosynthesis in the salt marsh grass *Spartina alterniflora* and carbon dioxide enhancement of nitrogenase activity. Apl. envir. Microbiol. 52: 108–113.

Widdel, F. & F. Bak, 1992. Gram-negative mesophilic sulfate-reducing bacteria. In A. Balows, H. G. Trüper, M. Dworkin, W. Harder & K. H. Schleifer (eds), The Prokaryotes vol 4. Springer-Verlag, New York. 3352–3378.

Wolfenden, J. & K. Jones, 1987. Seasonal variation of *in situ* nitrogen fixation ($C_2H_2$ reduction) in an expanding marsh of *Spartina anglica*. J. Ecol. 75: 1011–1021.

Yoch, D. C. & G. J. Whiting, 1986. Evidence for the $NH_4^+$ switch-off regulation of nitrogenase activity by bacteria in salt marsh sediments and roots of the grass *Spartina alterniflora*. Apl. envir. Microbiol. 51: 143–149.

Zuberer, D. A. & W. S. Silver, 1978. Biological dinitrogen fixation (acetylene reduction) associated with Florida mangroves. Apl. envir. Microbiol. 35: 567–575.

Zumft, W. G. & F. Castillo, 1978. Regulatory properties of the nitrogenase from *Rhodopseudomonas palustris*. Arch. Microbiol. 117: 53–60.

*Hydrobiologia* **329**: 175–183, 1996.
*P. Caumette, J. Castel & R. Herbert (eds), Coastal Lagoon Eutrophication and ANaerobic Processes (C.L.E.AN.).*



# Relationship between porewater organic carbon content, sulphate reduction and nitrogen fixation (acetylene reduction) in the rhizosphere of *Zostera noltii*

David T. Welsh[1,3*], Peter Wellsbury[2], Sophie Bourguès[3], Rutger de Wit[3] & Rodney A. Herbert[1]
[1] *Department of Biological Sciences, University of Dundee, Miller's Wynd, Dundee DD1 4HN, Scotland*
[2] *Department of Geology, Wills Memorial Building, University of Bristol, Queen's Road, Bristol BS8 1RJ England*
[3] *Laboratoire d'Océanographie Biologique et de Biologie Marine, Université Bordeaux I, 2 rue Professeur Jolyet, 33120 Arcachon, France*
**Present address. Fax: 33-56-83 51 04*

*Key words:* Acetylene reduction, nitrogen fixation, sulphate reduction, acetate, seagrasses

## Abstract

Depth profiles of nitrogen fixation (acetylene reduction), sulphate reduction, $NH_4^+$ concentration and porewater volatile fatty acids concentrations were measured in *Zostera noltii* colonised sediments in the Bassin d'Arcachon, France in March 1994. Acetylene reduction activity (ARA) was detectable throughout sediment profiles. Addition of sodium molybdate (20 mmol $l^{-1}$) a specific inhibitor of sulphate reduction to slurries inhibited ARA by >75% inferring that sulphate-reducing bacteria (SRB) were the dominant component of the nitrogen fixing microflora. The peak of ARA was coincident with that of sulphate reduction and a relatively constant relationship of 40 mole sulphate reduced per mole acetylene reduced was recorded throughout the profiles. From this ratio it was calculated that at least 17% of the ATP yield from sulphate reduction would be required to support the measured rates of nitrogen fixation (acetylene reduction).

Acetate was the dominant constituent of the porewater volatile fatty acids pool, accounting for >90% of the total pool as measured by HPLC. Concentrations of porewater acetate recorded by HPLC were compared with those measured using an enzymatic technique and these data indicate that approximately 10% of the total porewater acetate pool was not available to microbial metabolism. Profiles of porewater acetate concentrations measured by both techniques were similar to those recorded for both ARA and sulphate reduction and thus acetate oxidation may fuel these activities.

## Introduction

Seagrass and saltmarsh communities are amongst the most productive coastal marine ecosystems (McRoy & McMillan, 1977; Moriarty et al., 1990; Pollard & Moriarty, 1991). In order to achieve and maintain these high levels of productivity, such communities require substantial inputs of fixed nitrogen (Patriquin, 1972). Whilst, efficient recycling of organic nitrogen within the sediment can supply a large proportion of this requirement (Iizumi et al., 1982; Dennison et al., 1987; Caffrey & Kemp, 1992), several studies have clearly demonstrated that porewater dissolved nitrogen concentrations are insufficient to meet the growth requirements of the plants (Patriquin, 1972; Short, 1983; Moriarty et al., 1985). Heterotrophic nitrogen fixation in the rhizosphere of seagrasses therefore represents a potentially significant source of 'new' nitrogen for plant growth, which may compensate for the fixed nitrogen deficit in the sediment and thereby influence the productivity of the plant community. Rates of nitrogen fixation ranging from 0.02 to 140 mg N $m^{-2}$ $d^{-1}$ have been measured in seagrass sediments and calculated to supply upto 50% of their nitrogen requirements (Patriquin & Knowles, 1972; McRoy et al., 1973; Capone et al., 1979; Capone & Taylor, 1980; Wolfden & Jones, 1987; O'Donohue et al., 1991; Moriarty & O'Donohue, 1993). However, nitrogen fixation has a substantial metabolic cost, which has been estimated to be equivalent to 16 ATP per molecule of $N_2$ fixed (Postgate, 1982) and thus rates of heterotrophic nitrogen fixation are generally considered to be limited by the availability of organic carbon (Herbert, 1975; Nedwell & Aziz, 1980; Jones, 1982). Heterotrophic nitrogen fixation in the rhizosphere of seagrasses has been shown to be closely coupled to plant photosynthesis and dependent upon exudates of labile organic carbon from the plant roots (Capone et al., 1979; Boyle & Patriquin, 1981; Whiting et al., 1986; O'Donohue et al., 1991).

In this study we have investigated nitrogen fixation (acetylene reduction) activity in the seagrass meadows in the Bassin d'Arcachon, south-west France. The dependency of nitrogen fixation on porewater ammonium and dissolved organic carbon concentration and the potential role of sulphate-reducing bacteria (SRB) as nitrogen fixers has been investigated.

## Materials and methods

### *Description of the sampling station*

The Bassin d'Arcachon is a large mesotidal coastal lagoon situated on the French Atlantic coast (Figure 1). The Bassin has a tidal range of 2–3 m and a total surface area of 155 $km^2$ of which 40 $km^2$ is covered at low tide. Seventy percent of the Bassin is composed of intertidal mudflats, consisting of clayish muds and muddy sands which are dominated by the seagrass *Zostera noltii* Hornem. These seagrass meadows cover an area greater than 70 $km^2$ and are the largest in Western Europe (Auby, 1991). The salinity of the Bassin normally ranges between 30–33‰, but can decrease to 20‰ following periods of heavy rainfall, whilst the water temperature ranges from 4–6 °C in mid-winter to 21–22 °C in August. A full description of the hydrology of the Bassin is given by Robert et al. (1987). The sampling station used during this study was situated in a large seagrass bed on the Ile aux oiseaux, a large mudflat situated in the central region of the Bassin (Figure 1) and corresponds to the CLEAN project Station A. All the sediment samples were collected on the $8^{th}$ March 1994.

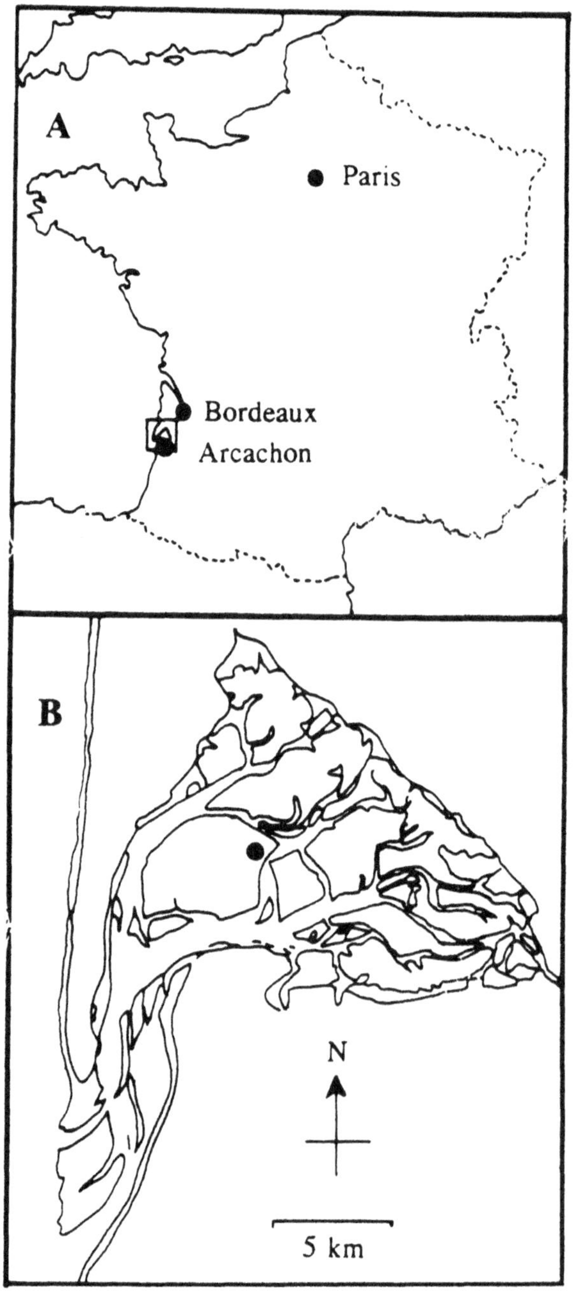


*Figure 1.* A. Location of the Bassin d'Arcachon on the French Atlantic coast. B. Site diagram of the Bassin d'Arcachon at low tide showing the location of the study site (●) in the central region of the Bassin and the major mudflats and channels.

### *Nitrogen fixation (acetylene reduction) activity*

Rates of nitrogen fixation were measured using the acetylene reduction technique of Stewart et al. (1967). Intact sediment cores were collected in 25 × 5 cm (i.d.)

Plexiglass core tubes and transported to the laboratory on ice. In the laboratory the cores were carefully extruded, the *Z. noltii* leaves were removed and the upper 5 cm of sediment were sliced into 1 cm thick sections. The sediment slices were immediately transferred into 360 ml volume glass bottles and 20 ml of deoxygenated filter-sterilised (0.2 $\mu$m pore size) seawater was added. The bottles were sealed with screw tops fitted with rubber-sealed injection ports and immediately gassed out with oxygen-free nitrogen for 10 mins. Following degassing the sediment slurries were thoroughly mixed by vigorously shaking the bottles. Experiments were initiated by replacing 10% v/v of the headspace gas with acetylene and the slurries were incubated for 24 hours at *in situ* temperature in the dark. Sodium molybdate was added when required prior to degassing to give a final concentration of 20 mmol $l^{-1}$ and the slurries were pre-incubated at the *in situ* temperature for 1 hour following degassing prior to the addition of acetylene. Acetate when added was added prior to the addition of acetylene by injection of a filter sterilised stock solution to give a final concentration of 5 mmol $l^{-1}$.

In tests to determine the effect of carbon source additions on acetylene reduction activity a 1:1 (sediment: filter-sterilised seawater) slurry was prepared under nitrogen atmosphere from the pooled 0–3 cm depth horizons of 8 cores. Aliquot volumes (50 ml) of slurry were dispensed into 360 ml glass bottles and the slurries were thereafter treated as described previously. Carbon sources were added by injection to give a final concentration of 15 mmol $l^{-1}$ prior to the addition of acetylene.

Following incubation the assays were terminated by the injection of 1 M trichloroacetic acid to give a final concentration of 5% w/v. Aliquot 1 ml volumes of the headspace gas were analysed for both acetylene and ethylene by gas chromatography using a Perkin Elmer Autosystem Gas Chromatograph fitted with a 3 m × 2.2 mm (internal diameter) Chromosorb 101 (80/100 mesh) column with $N_2$ as carrier gas and flame ionisation detection. Flow rates for $N_2$, air and $H_2$ were 30, 450 and 50 ml min$^{-1}$ respectively and the oven temperature was 30 °C. Ethylene concentrations were determined by reference to known standards and all the data were corrected for the small quantities of ethylene present as a contaminant in the acetylene used in this study.

*Determination of sulphate reduction rates*

Sulphate reduction rates were determined using the core injection method of Jørgensen & Fenchel (1974). Sediment cores were collected in 20 × 4 cm (i.d.) Plexiglass core tubes fitted with pre-drilled silicon rubber filled injection ports. Cores were injected at 2 cm depth intervals with 10 $\mu$l of carrier free $^{35}SO_4^{2-}$ [370 kbq] (Isotopchim, France). The core tubes were covered in a double layer of aluminium foil to the level of the sediment surface and incubated for 6 hours at *in situ* temperature under natural light conditions. Following, incubation the sediment was extruded and sliced into 1 cm thick sections which were immediately fixed by vigorous shaking with an equal volume of 2% w/v zinc acetate to trap the sulphides and inhibit further activity.

The production of radiolabelled sulphides was determined by the single step chromium reduction method of Fossing & Jørgensen (1989) and the released sulphides trapped in 2% w/v zinc acetate. The specific activities of $^{35}S^{2-}$ and $^{35}SO_4^{2-}$ were determined in aliquot volumes (1 ml) of the zinc acetate traps and the supernatant (following the initial centrifugation of the sediment respectively) by liquid scintillation counting (Intertechnique SL-3ccm liquid scintillation counter). Sulphate concentrations were determined as previously described by Tabatabai (1974) and the sulphate reduction rate calculated according to the method of Fossing & Jørgensen (1989).

*Determination of porewater and sediment exchangeable ammonium concentrations*

Sediment cores were collected in 20 × 5 cm (i.d.) Plexiglass core tubes and transported to the laboratory on ice. Cores for porewater ammonium content were sectioned into 1 cm slices and immediately centrifuged at 5000 × ***g*** for 10 min at 5 °C to extract the sediment porewater. The collected porewater was passed through a 25 mm Whatman GF/F filter and frozen at −40 °C before analysis. Cores for sediment exchangeable ammonium concentration were sectioned into 1 cm slices and the sediment extracted for 1 hour by shaking in an equal volume of 1 M KCl. The resultant sediment slurry was centrifuged at 5000 × ***g*** for 10 min at 5 °C, the supernatant was decanted, passed through a 25 mm Whatman GF/F filter and frozen at −40 °C. Thawed samples were suitably diluted with distilled deionised water and analysed for ammonium content using a Technicon continuous flow autoanalyser (Chemlab Ltd, England).

*Determination of porewater volatile fatty acid and bioavailable acetate concentrations*

Sediment cores were collected using 25 × 5 cm (i.d.) Plexiglass core tubes and transported to the laboratory on ice. The cores were extruded and sectioned into 1 cm slices and the sediment centrifuged at 5000 × ***g*** for 10 min at 5 °C. The porewater was decanted and passed through 25 mm Whatman GF/F filters and frozen at −40 °C.

Porewater samples were analysed for total volatile fatty acid concentrations using the 2-nitrophenylhydrazine hydrochloride (2 NPH) derivitisation method of Parkes et al. (1989) as modified by Wellsbury and Parkes (1995). Porewater bioavailable acetate concentrations were determined using the enzymic method of King (1991), using HPLC separation and quantification of AMP and ATP as described by Wellsbury and Parkes (1995).

## Results

Data presented in Figure 2 show profiles of acetylene reduction activity (ARA) in the surficial sediments of the *Z. noltii* beds. ARA was detectable at all depth horizons tested with peak activities of 0.55 and 0.43 nmol $ml^{-1}$ $h^{-1}$ recorded in the 1–2 and 2–3 cm depth horizons respectively. Ethylene accumulation was never observed in control slurries to which no acetylene was added (Data not shown). Additions of 20 mmol $l^{-1}$ sodium molybdate (final concentration) a specific inhibitor of bacterial sulphate reduction to sediment slurries severely inhibited ARA by 75 to 85% (Figure 2) throughout the profile inferring that SRB were responsible for the bulk of the recorded activity. Addition of sodium acetate to sediment slurries to give a final concentration of 5 mmol $l^{-1}$ had little effect on ARA with a maximal stimulation of 33% observed in the 1–2 cm depth horizon (Figure 2). Similarly, addition of a range of carbon sources at a concentration of 15 mmol $l^{-1}$ only stimulated ARA between 18 and 35% when added to the slurries (Table 1).

The depth profile of sulphate reduction activity was similar to that recorded for acetylene reduction with maximal activities of 20 and 17.5 nmol sulphate reduced ml sediment$^{-1}$ $h^{-1}$ recorded in the 1–2 and 2–3 cm depth horizons respectively (Figure 3) and 85% of total sulphate reduction occurred within the 0 – 5 cm depth horizon (Data not shown). A relatively constant ratio between sulphate and acetylene reduction of approximately 40 (range 36 – 48) nmol sulphate reduced per nmol acetylene reduced was recorded throughout the depth profile with the exception of the 0–1 cm depth horizon, where the ratio was 11.8:1 (Table 2).

Acetate was the predominant volatile fatty acid present in the sediment porewaters, representing >90% of the total volatile fatty acid pool. Concentrations of total (measured by 2NPH method) and bioavailable acetate (enzyme method) were similar

*Table 1*. Effect of a range of organic carbon sources on the rate of acetylene reduction recorded in slurries of the 0–3 cm depth horizon of *Zostera noltii* colonised sediments. All carbon sources were present at a final concentration of 15 mmol $l^{-1}$.

| Carbon source | Acetylene reduction rate (mmol ml sed$^{-1}$ $h^{-1}$) | Control |
|---|---|---|
| No additions | 0.786 | 100 |
| Ethanol | 0.988 | 126 |
| Pyruvate | 1.058 | 135 |
| Lactate | 0.929 | 118 |
| Glucose | 1.031 | 131 |
| Galactose | 0.991 | 126 |

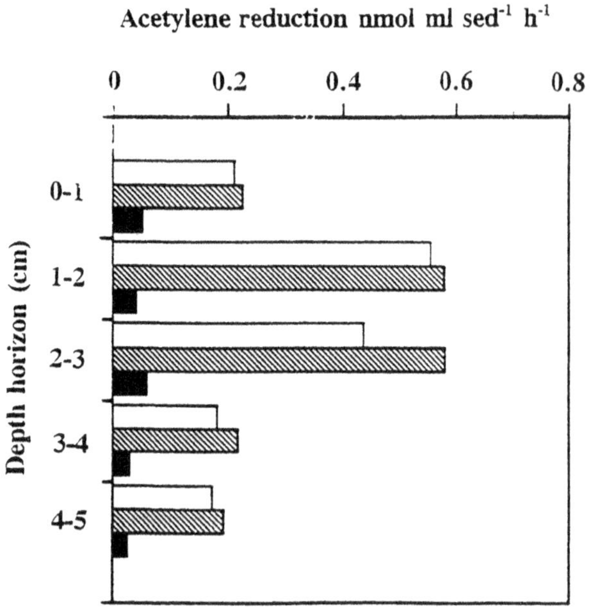


*Figure 2.* Profiles of acetylene reduction activity in the surficial sediments determined in slurry experiments with no additions (open bars), supplemented with 5 mmol $l^{-1}$ sodium acetate (hatched bars) and supplemented with 20 mmol $l^{-1}$ sodium molybdate (solid bars). Data represent the means of 5 determinations, standard deviations were less than 5%.

*Table 2.* Relationship between sulphate and acetylene reduction activity with depth in *Zostera noltii* colonised sediments.

| Depth horizon (cm) | Acetylene reduction rate (nmol $ml^{-1}$ $h^{-1}$) | Sulphate reduction rate (nmol $ml^{-1}$ $h^{-1}$) | Ratio of acetylene reduction: sulphate reduction activity |
|---|---|---|---|
| 0–1 | 0.211 | 2.5 | 11.8 |
| 1–2 | 0.553 | 20.0 | 36.2 |
| 2–3 | 0.434 | 17.5 | 40.3 |
| 3–4 | 0.181 | 8.7 | 48.1 |
| 4–5 | 0.172 | 6.7 | 39.0 |

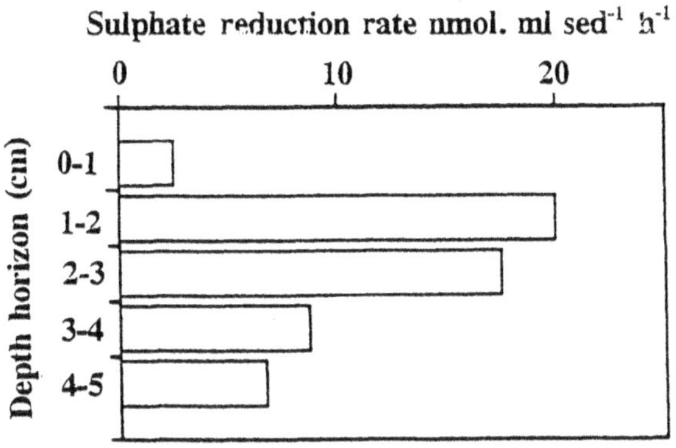


*Figure 3.* Depth profile of sulphate reduction activity in *Zostera noltii* colonised sediments.

throughout the profiles (Figure 4) with approximately 90% of the total acetate pool available for microbial metabolism. Profiles were similar to those recorded for both acetylene reduction and sulphate reduction with peak concentrations of 76 and 52 $\mu$M (total acetate) and 73 and 48 $\mu$M (bioavailable acetate) recorded in the 1–2 and 2–3 cm depth horizons respectively (Figure 4). Concentrations of propionate were considerably lower than those recorded for acetate, ranging from 0.03 to 5.21 $\mu$M with a maximum in the 1–2 cm depth horizon (Figure 4). Concentrations of other volatile fatty acids were below the detection limit of the HPLC system.

Profiles of porewater and sediment exchangeable $NH_4^+$ showed little variation with depth (Figure 5). Porewater $NH_4^+$ concentrations varied from 65 to 106 $\mu$M with the maximum concentration recorded in the 1–2 cm depth horizon and the minimum in the 0–1 cm horizon (Figure 5). Whereas, sediment exchangeable ammonium concentrations ranged from 219 to 303 $\mu$mol 1 sediment$^{-1}$ with the maximal concentration recorded in the 3–4 cm depth horizon and the minimum in the 4–5 cm horizon (Figure 5).

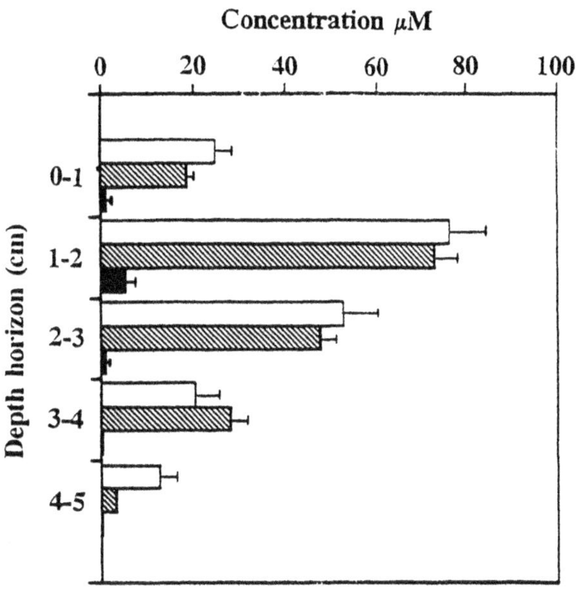


*Figure 4.* Depth profiles of porewater total acetate concentrations measured by HPLC (open bars), bioavailable acetate measured by the enzyme method (hatched bars) and propionate measured by HPLC (solid bars), error bars indicate standard deviations (n = 3).

## Discussion

ARA was detectable throughout the depth profiles of sediments from *Z. noltii* beds with maximal activities recorded in the root zone of the plants (Figure 2). These data are in agreement with several previous studies which have demonstrated high levels of nitrogen fixation/acetylene reduction in the rhizosphere of sea and salt marsh grasses (Patriquin, 1972; Short, 1983; Moriarty et al.,1985; O'Donohue et al.,1991; Moriarty & O'Donohue, 1993). Addition of sodium molybdate at a final concentration of 20 mmol $l^{-1}$ severely inhibited ARA by >75% throughout the depth profile (Figure 2). Sodium molybdate is a specific inhibitor of sulphate

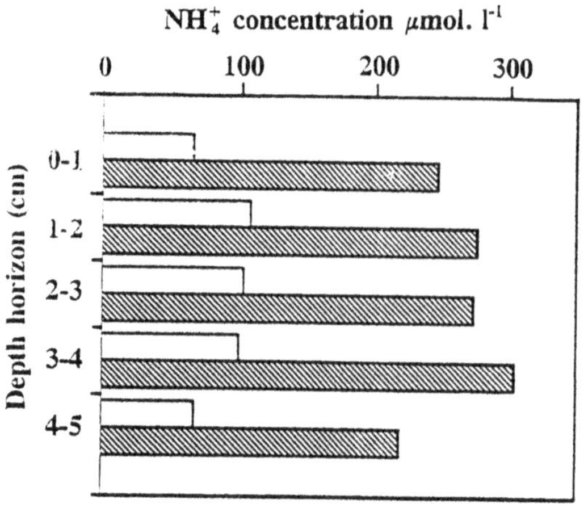


*Figure 5.* Depth profiles of porewater $NH_4^+$ (open bars) and sediment exchangeable $NH_4^+$ (hatched bars) concentrations.

reduction (Taylor & Oremland, 1979; Smith & Klug, 1981; Oremland & Capone, 1988). These data indicate that SRB were responsible for the bulk of the recorded ARA. The ability to fix molecular nitrogen has long been recognised amongst SRB (Sisler & Zobell, 1951; Le Gall et al.,1959; Postgate et al., 1988) and these bacteria have previously been proposed as potentially the most important heterotrophic fixers of nitrogen in coastal marine sediments (Herbert, 1975; Nedwell & Aziz, 1980).

Depth profiles of sulphate reduction (Figure 3) showed a similar overall pattern to those recorded for ARA and with the exception of the 0–1 cm depth horizon, a relatively constant molar ratio of approximately 40:1 was recorded between ARA and sulphate reduction rates at all depths (Table 2). These data provide further evidence to support the hypothesis that SRB are principally responsible for the observed acetylene reduction activity (nitrogen fixation) in *the Zostera* root zone. Deviation from the 40:1 ratio in the 0–1 cm depth horizon is probably due to an underestimation of the sulphate reduction rate in this zone due to the reoxidation of reduced $^{35}S$ by molecular oxygen during the 6 hour incubation period. Jørgensen (1994) recently reported an 8-fold decrease in the calculated sulphate reduction rates of oxic microbial mat sediments incubated for 30 min and 3 hours, which he attributed to the reoxidation of radiolabelled sulphides. Oxygen penetration in *Z. noltii* colonised sediments in the Bassin d'Arcachon is spatially highly variable due to the presence of plant roots but is typically limited to a depth of 6–13 mm (de Wit, unpublished data). It is therefore not surprising that sulphate reduction rates in this zone would be underestimated.

It has been estimated that the net ATP yield for *Desulfobacter postgatei* growing on acetate and sulphate is 0.8 mol ATP per mol sulphate reduced (See Widdel & Hansen, 1991). Thus, using the recorded ratio of 40:1 between sulphate reduction and acetylene reduction and the theoretical ratio of 3:1 (see Welsh et al., 1996) to convert acetylene reduction rates to nitrogen fixation rates, the energy requirement for this process in the *Zostera noltii* colonised sediments can be determined. Postgate, (1982) calculated that 16 mol ATP are required to fix 1 mol of $N_2$. On this basis approximately 17% of the ATP generated by sulphate reduction would be required to fuel the recorded nitrogen fixation (acetylene reduction) rates in the rhizosphere. This figure, however, represents a minimum value since, the theoretical energy requirement for nitrogen fixation does not take into account additional costs such as the synthesis of nitrogenase and the electron carriers. Reported ATP:$N_2$ ratios for pure cultures vary considerably e.g. 30 in *Klebsiella pneumoniae* (Hill, 1976) and 42 for *Rhizobium* ORS 571 (Stam et al., 1984) but are generally much greater than 16. Thus, it is probable that the true energy demand of nitrogen fixation in these sediments is substantially greater than 17% of the total ATP produced by sulphate reduction.

The addition of exogenous carbon sources to slurries had little effect on ARA of *Z. noltii* colonised sediments (Figure 2, Table 1). These results are at variance with other studies of heterotrophic nitrogen fixation in marine sediments, where carbon source additions were found to significantly stimulate nitrogen fixation/acetylene reduction rates (e.g. Herbert, 1975; Nedwell & Aziz, 1980). However, this may be due to an inherent artifact of the slurry technique in vegetated sediments due to the release of organic carbon from the plant roots during the preparation of the slurries. Internal pools of labile carbon in the roots of *Z. noltii* represent a large pool of potentially available organic carbon within the sediment, since it has been demonstrated for *Zostera marina* L. that under conditions of oxygen limitation the roots switch to a fermentative metabolism producing $CO_2$, ethanol and lactate as metabolic end products, the latter accumulating in the root tissue (Smith et al., 1988). Such internal pools of labile carbon may then be released during the preparation of sediment slurries, since, previous studies have shown that physical disturbance of salt marsh sediments can substantially alter the carbon composition of the sediment porewater (Hines et al., 1994). Thus, organic carbon may be more readily available in slurries of *Z. noltii* colonised sediments due to root damage and perhaps it is therefore not surprising that carbon source additions result in only a minor stimulation of ARA. This explanation is supported by the observation that acetylene reduction rates recorded at the same time by a whole core technique were substantially lower than those recorded in slurries, indicating activity was limited by carbon source availability *in situ* (Welsh et al., 1996).

Porewater acetate concentrations were determined by both HPLC following 2-NPH derivitisation and enzyme assay methods. The HPLC method measures the total acetate pool and involves the efficient extraction and derivitisation of both bioavailable and non-bioavailable pools. In contrast the enzyme assay method potentially only measures the bioavailable acetate pool, which is converted to acetyl CoA by acetyl CoA synthetase according to the following equation:

$$\text{acetate} + \text{ATP} + \text{CoA} \xrightarrow{\text{acetyl CoA synthetase}} \text{acetyl CoA} + \text{AMP} + \text{PP}_i$$

This reaction represents the first step in the uptake and metabolism of acetate by SRB (Widdel & Hansen, 1991) and methanogens (Whitman et al., 1991) and is thus thought to yield a more accurate estimate of the bioavailable acetate pool in sediment porewaters (King, 1991; Kristensen et al., 1994). However, the recent study by Wellsbury & Parkes (1995), indicates that this method may also overestimate the true bioavailability of acetate in marine sediment porewaters.

Depth profiles of porewater acetate concentrations measured by both techniques were similar (Figure 4) with maximal concentrations measured in the 1–2 and 2–3 cm depth horizons. However, concentrations measured by the enzyme assay technique were generally lower than those measured by HPLC, indicating that approximately 10% of the porewater acetate pool measured using the HPLC (2-NPH) technique was not available to the indigenous microflora. Propionate was the only other volatile fatty acid present in measurable concentrations but was present at much lower concentration than acetate which represented >90% of the total volatile fatty acid pool (Figure 5). The depth profiles of porewater acetate pools were similar to those recorded for ARA and sulphate reduction activity (Figures 3 & 4). Acetate could fuel these activities, since this volatile fatty acid has been demonstrated to be the principal substrate for sulphate reduction in marine sediments (Sørensen et al., 1981; Parkes et al., 1989) and to support nitrogen fixation in pure cultures of SRB (Widdel, 1987). This relationship is however difficult to analyse, since depth profiles of acetate also correlate with the distribution of the *Z. noltii* roots and thus may have been influenced by the release of organic carbon from the plant roots during porewater extractions as discussed previously. Additionally, it cannot be completely excluded that the recorded porewater pools of acetate are the result of incomplete lactate oxidation. However, this seems unlikely, since, in molybdate inhibition studies of sediments from the same sampling station, where, carbon sources normally consumed during sulphate reduction would accumulate, acetate was the major VFA which accumulated and lactate was never detected (Finster et al., 1994). These authors also calculated that VFA oxidation rates, based on accumulation rates during molybdate inhibition and assuming steady state conditions could account for approximately 90% of the recorded sulphate reduction rate at this station, indicating that other potential carbon and energy sources do not significantly contribute to sulphate reduction and thus to nitrogen fixation by SRB.

Nitrogen fixation/acetylene reduction rates are likely to be influenced by the availability of $NH_4^+$, since, nitrogenase is subject to feed-back inhibition by this source of fixed nitrogen (Postgate, 1982; Dixon, 1984; Capone, 1988). Porewater $NH_4^+$ concentrations in the *Z. noltii* beds ranged from 65 to 106 $\mu$M (Figure 5), concentrations which would totally suppress nitrogenase activity of pure cultures growing in liquid media, due to both repression of nitrogenase synthesis and reversable 'switch-off' of preformed nitrogenase (Postgate & Kent, 1984; Postgate et al., 1985). However, a number of previous studies have reported nitrogen fixation/acetylene reduction activity in marine sediments at several-fold higher porewater $NH_4^+$ concentrations (see Capone, 1988 for review). Additionally, in these sediments, zones of low $NH_4^+$ concentration may exist around the plant roots due to $NH_4^+$ assimilation by the plants and such microenvironments would allow nitrogen fixation to continue due to the absence of feedback inhibition.

In conclusion data presented in this paper indicate that an association exists between nitrogen fixation (acetylene reduction) rates and SRB in the rhizosphere of *Z. noltii* and that acetate may represent the principal carbon source fuelling nitrogen fixation by these bacteria.

## Acknowledgments

This Research was carried out as part of the E.U. funded Coastal Lagoon Eutrophication and ANaerobic processes (CLEAN) project, grant No EV5V-CT92-0080. P. Wellsbury was supported by a grant from the E.U. (Grant No. EV5V-CT92-0065) and S. Bourguès by a grant from the French Ministry of Research and Education (MESR). D. Welsh and R. A. Herbert would like to thank Prof. P. Caumette for the use of laboratory space in his Institute and all the staff and students of the Laboratoire d'Océanographie, Arcachon for their assistance during the course of this study.

## References

Auby, I., 1991. Contribution à l'étude des herbiers de *Zostera noltii* dans le Bassin d'Arcachon: dynamique, production et dégradation, macrofaune associée. Ph.D. thesis, Université de Bordeaux I.

Boyle, C. D. & D. G. Patriquin, 1981. Carbon metabolism of *Spartina alterniflora* Loisel in relation to that of associated nitrogen-fixing bacteria. New Phytol. 89: 275–288.

Caffrey, J. M. & W. M. Kemp, 1992. Influence of the submersed plant, *Potamogeton perfoliatus*, on nitrogen cycling in estuarine sediments. Limnol. Oceanogr. 37: 1483–1495.

Capone, D. G., 1988. Benthic nitrogen fixation. In: T. H. Blackburn & J. Sørensen (eds), Nitrogen cycling in coastal marine environments. John Wiley & Sons Ltd, Chichester. 85–123.

Capone, D. G. & B. F. Taylor, 1980. $N_2$ fixation in the rhizosphere of *Thalassia testudinium*. Can. J. Microbiol. 26: 998–1005.

Capone, D. G., P. A. Penhale, R. S. Oremland & B. F. Taylor, 1979. Relationship between productivity and $N_2$ ($C_2H_2$) fixation in a Thalassia testudinium community. Limnol. Oceanogr. 24: 117–125.

Dennison, W. C., R. C. Aller & R. S. Alberte, 1987. Sediment ammonium availability and eelgrass (Zostera marina) growth. Mar. Biol. 94: 469–494.

Dixon, R. A., 1984. The genetic complexity of nitrogen fixation. J. gen. Microbiol. 130: 2745–2755.

Finster, K., A. Gammelgaard & L. S. Hansen, 1994. Sulfate reduction rates, inorganic sulfur pools and the accumulation of volatile fatty acids in the Bassin of Arcachon and the Prévost lagoon, southern France: a comparative study. C.L.E.A.N. (Coastal Lagoon Eutrophication and ANaerobic processes) progress report: 223–255.

Fossing, H. & B. B. Jørgensen, 1989. Measurements of bacterial sulphate reduction in sediments: Evaluation of a single step chromium reduction method. Biogeochemistry 8: 205–222.

Hill, S., 1976. The apparent ATP requirement for nitrogen fixation in growing *Klebsiella pneumoniae*. J. gen. Microbiol. 95: 297–312.

Herbert, R. A., 1975. Heterotrophic nitrogen fixation in shallow estuarine sediments. J. exp. mar. Biol. Ecol. 18: 215–225.

Hines, M. E., G. T. Banta, A. E. Giblin & J. E. Hobbie, 1994. Acetate concentrations and oxidation in salt-marsh sediments. Limnol. Oceanogr. 39: 140–148.

Iizumi, H., A. Hattori & C. P. McRoy, 1982. Ammonium regeneration and assimilation in Eelgrass (*Zostera marina*) beds. Mar. Biol. 66: 59–65.

Jones, K., 1982. Nitrogen fixation in the temperate estuarine intertidal salt marsh sediments of the river Lune. Limnol. Oceanogr. 27: 455–460.

Jørgensen, B. B. & T. Fenchel, 1974. The sulfur cycle of a marine sediment model system. Mar. Biol. 24: 189–201.

King, G. M., 1991. Measurement of acetate concentrations in marine pore water by using an enzymatic approach. Appl. envir. Microbiol. 57: 3476–3481.

Kristensen, E., G. M. King, M. Holmer, G. T. Banta, M. H. Jensen, K. Hansen & N. Bussarawit, 1994. Sulfate reduction, acetate turnover and carbon metabolism in sediments of the Ao Nam Bor mangrove, Phuket, Thailand. Mar. Biol. Prog. Ser. 109: 245–255.

Le Gall, J., J. C. Senez & F. Pichinoty, 1959. Fixation de l'azote par les bactéries sulfato-réductrices: isolement et caracterisation de souches actives. Annales Institute Pasteur 96: 223–230.

McRoy, C. P., J. J. Goering & B. Chaney, 1973. Nitrogen fixation associated with seagrasses. Limnol. Oceanogr. 18: 998–1002.

McRoy, C. P. & C. McMillan, 1977. Production, ecology and physiology of seagrasses. In C. P. McRoy & C. Helfferich (eds), Seagrass Ecology. Marcel Dekker, New York. 53–87.

Moriarty, D. J. W. & M. J. O'Donohue, 1993. Nitrogen fixation in seagrass communities during summer in the Gulf of Carpentaria, Australia. Aust. J. mar. Freshwat. Res. 44: 117–125.

Moriarty, D. J. W., P. I. Boon, J. A. Hansen, W. G. Hunt, I. R. Poiner, P. C. Pollard, G. W. Skyring & D. C. White, 1985. Microbial biomass and productivity in seagrass beds. Geomicrobiol. J. 4: 21–51.

Moriarty, D. J. W., D. G. Roberts & P. C. Pollard, 1990. Primary and bacterial productivity of tropical seagrass communities in the Gulf of Carpentaria, Australia. Mar. Ecol. Prog. Ser. 61: 145–157.

Nedwell, D. B. & S. Aziz, 1980. Heterotrophic nitrogen fixation in an intertidal salt marsh sediment. Estuar. coast. mar Sci. 10: 699–702.

O'Donohue, M. J., D. J. W. Moriarty & I. C. McRae, 1991. Nitrogen fixation in sediments and the rhizosphere of the seagrass *Zostera capricornia*. Microb. Ecol. 22: 53–64.

Oremland, R. S. & D. G. Capone, 1988. Use of 'specific inhibitors' in biogeochemistry and marine microbiology. Adv. Microbiol. Ecol. 10: 285–383.

Parkes, R. J., G. R. Gibson, I. Mueller-Harvey, W. J. Buckingham & R. A. Herbert, 1989. Determination of the substrates for sulphate reducing bacteria within marine and estuarine sediments with different rates of sulphate reduction. J. gen. Microbiol. 135: 175–187.

Patriquin, D. G., 1972. The origin of nitrogen and phosphorus for the growth of the marine angiosperm *Thalassia testudinium*. Mar. Biol. 16: 25–46.

Patriquin, D. G. & R. Knowles, 1972. Nitrogen fixation in the rhizosphere of marine angiosperms. Mar. Biol. 16: 49–58.

Pollard, P. C. & D. J. W. Moriarty, 1991. Organic carbon decomposition, primary and bacterial productivity and sulphate reduction in tropical seagrass beds of the Gulf of Carpentaria, Australia. Mar. Ecol. Prog. Ser. 69: 149–159.

Postgate, J. R., 1982. The fundamentals of nitrogen fixation. Cambridge University Press, London.

Postgate, J. R. & H. M. Kent, 1984. Derepression of nitrogen fixation in *Desulfovibrio gigas* and its stability to ammonia or oxygen stress *in vivo*. J. gen. Microbiol. 130:2824–2831.

Postgate, J. R., H. M. Kent, S. Hill & H. Blackburn, 1985. Nitrogen fixation by *Desulfovibrio gigas* and other species of *Desulfovib-*

*rio*. In W. Ludden & J. E. Burris (eds), Nitrogen fixation and $CO_2$ metabolism. Elsevier Science Publishing, Amsterdam. 225–234.

Postgate, J. R., H. M. Kent & R. L. Robson, 1988. Nitrogen fixation by *Desulfovibrio*. In J. A. Cole & S. J. Ferguson (eds). The nitrogen and sulphur cycles. Society for General Microbiology Symposium 42. Cambridge University Press, Cambridge: 457–471.

Robert, R., N. Guillocheau & Y. Collos, 1987. Hydrological parameters during an annual cycle in the Arcachon Bassin. Mar. Biol. 95: 631–640.

Short, F. T., 1983. The response of interstitial ammonium in eelgrass (*Zostera marina*) beds to environmental pertubations. J. exp. mar. Biol. Ecol. 68: 195–208.

Sisler, F. D. & C. E. Zobell, 1951. Nitrogen fixation by sulphate-reducing bacteria indicated by nitrogen/argon ratios. Science 262: 209–210.

Stam, H., H. W. van Versveld, W. de Vries & A. H. Stouthamer, 1984. Hydrogen oxidation and efficiency of nitrogen fixation in succinate-limited chemostat cultures of *Rhizobium* ORS 571 Arch. Microbiol. 139: 53–60

Smith, R. D., A. M. Pregnall & R. S. Alberte, 1988. Effects of anaerobiosis on root metabolism of *Zostera marina* (eelgrass): implications for survival in reducing sediments. Mar. Biol. 98: 131–141.

Smith, R. I. & M. J. Klug, 1981. Electron donors utilised by sulfate-reducing bacteria in eutrophic lake sediments. Appl. envir. Microbiol. 42: 116–121.

Sørensen, J., D. Christensen & B. B. Jørgensen, 1981. Volatile fatty acids and hydrogen as substrates for sulfate-reducing bacteria in anaerobic marine sediment. Appl. envir. Microbiol. 42: 5–11.

Stewart, W. D. P., G. P. Fitzgerald & R. H. Burris, 1967. *In situ* studies on $N_2$ fixation using the acetylene reduction technique. Proc. natn. Acad. Sci. U.S.A. 58: 2071–2078.

Tabatabai, M. A., 1974 Determination of $SO_4^{2-}$ in water samples. Sulphur Inst. J. 10: 11–14.

Taylor, B. F. & R. S. Oremland, 1979. Depletion of adenosine triphosphate in *Desulfovibrio* by oxyanions of Group IV elements. Curr. Microbiol. 3: 101–103.

Wellsbury, P. & R. J. Parkes, 1995. Acetate turnover and bioavailability in an estuarine sediment. FEMS Microbiol. Ecol. 17: 85–94.

Welsh, D. T., S. Bourguès, R. de Wit & R. A. Herbert, 1996. Seasonal variation in rates of heterotrophic nitrogen fixation (acetylene reduction) in *Zostera noltii* meadows and uncolonised sediments of the Bassin d'Arcachon, South-west France. Hydrobiologia 329 (Dev. Hydrobiol. 117): 161–174.

Whiting, G. J., E. L. Gandy & D. C. Yoch, 1986. Tight coupling of root-associated nitrogen fixation and plant photosynthesis in the salt marsh grass *Spartina alterniflora* and carbon dioxide enhancement of nitrogenase activity. Appl. envir. Microbiol. 52: 108–113.

Whitman, W. B., T. L. Bowen & D. R. Boone, 1991. The methanogenic bacteria. In A. Balows, H. G. Trüper, M. Dworkin, W. Harder & K. H. Schleifer (eds), The Prokaryotes. Springer-Verlag, New York: 719–767.

Widdel, F., 1987. New types of acetate-oxidising, sulfate-reducing *Desulfobacter* species, *D. hydrogenophilus* sp. nov., *D. latus* sp. nov., and *D. curvatus* sp. nov. Arch. Microbiol.

Widdel, F. & T. A. Hansen, 1991. The dissimilatory sulfate- and sulfur-reducing bacteria. In A. Balows, H. G. Trüper, M. Dworkin, W. Harder & K. H. Schleifer (eds), The Prokaryotes. Springer-Verlag, New York: 583–625.

Wolfden, J. & K. Jones, 1987. Seasonal variation of *in situ* nitrogen fixation ($C_2H_2$ reduction) in an expanding marsh of *Spartina anglica*. J. Ecol. 75: 1011–1021.

*Hydrobiologia* **329:** 185–198, 1996.
*P. Caumette, J. Castel & R. Herbert (eds), Coastal Lagoon Eutrophication and ANaerobic Processes (C.L.E.AN.).*



# The biogeochemistry of two eutrophic marine lagoons and its effect on microphytobenthic communities

Lucas J. Stal, Simone B. Behrens, Marlies Villbrandt, Stef van Bergeijk & Finn Kruyning
*Laboratory for Microbiology, ARISE, University of Amsterdam, Nieuwe Achtergracht 127, NL-1018 WS Amsterdam, The Netherlands*

*Key words:* Coastal lagoon, cyanobacteria, iron, phosphate, sulfide, nitrogen fixation

## Abstract

This paper summarizes the results of a study on the biogeochemistry of two eutrophic marine lagoons. The lagoons investigated were the Bassin d'Arcachon situated on the Atlantic coast, and Étang du Prévost on the Mediterranean coast. The sites chosen for this study were characterized by the presence of dense communities of microphytobenthos. Both lagoons receive a large input of nutrients but they differ in several aspects. The Bassin d'Arcachon receives a large amount of iron. Iron is of great importance in reducing the effects of eutrophication. Ferric iron is an efficient scavenger of phosphate and it has been proposed that this is one of the mechanisms that controls primary productivity and algal growth in this lagoon. The mechanisms of phosphate mobilization were studied by using sediment slurries. These experiments demonstrated that not only ferric iron but presumably also calcium was responsible for phosphate binding. Another effect of the high iron content in the Bassin d'Arcachon was the precipitation of sulfide as iron sulfide or pyrite. In the Étang du Prévost sulfate reduction resulted in the accumulation of free sulfide. The relative low content of iron in Étang du Prévost not only allowed the formation of free sulfide but may also have limited the binding capacity of phosphate in the sediment. On the other hand sulfate reduction was not important for the release of phosphate from the sediment. In Étang du Prévost primary productivity is nitrogen rather than phosphorus limited. In contrast in the Bassin d'Arcachon primary productivity was presumably mostly phosphate limited. In Étang du Prévost the non-heterocystous cyanobacterium *Oscillatoria* sp. was the dominant nitrogen-fixing species. Heterocystous species were excluded from this lagoon as a result of the presence of free sulfide. It was demonstrated that heterocystous cyanobacteria are more sensitive towards sulfide than non-heterocystous species. The absence of free sulfide explained the presence of the heterocystous cyanobacterium *Anabaena* sp. in Bassin d'Arcachon.

## Introduction

Increased input of nutrients such as phosphate and nitrogen compounds in water bodies usually leads to excess algal growth and primary productivity (Pugnetti et al., 1992, Reynolds & Walsby, 1975, Riegman et al., 1992). The enhanced level of organic matter that results from eutrophication may lead to serious problems when microbiological activity causes anoxia of the water column. Coastal lagoons are special in this regard. The high concentration of sulfate in seawater selects for sulfate reducing bacteria that are responsible for end oxidation of organic matter, resulting in the formation of highly toxic sulfide (Howarth & Merkel, 1984; Oremland & Polcin, 1982; Skyring, 1987). In the worst case, dystrophy may cause complete anoxia of coastal lagoons and such waters may turn purple as a result of blooms of anoxygenic phototrophic bacteria that live at the expense of sulfide (Caumette, 1986). It will be evident that, apart from a number of microorganisms, all other life is eliminated in waters that suffer from dystrophy. Increasing eutrophication of coastal lagoons may change the communities from seagrass beds (e.g. *Zostera* spp.) in non-polluted or moderately eutrophic waters to excessive growth of green macroalgae (*Ulva, Monostroma, Enteromorpha*) to blooms of planktonic cyanobacteria where the turbidity of the water is such that green algae can not proliferate.

In this investigation two coastal lagoons in France were compared. The Étang du Prévost is a highly eutrophic lagoon situated on the Mediterranean coast, close to the city of Montpellier, whilst the Bassin d'Arcachon is situated on the Atlantic coast, 60 km south-west of the city of Bordeaux. Whereas the Mediterranean lagoon may experience dystrophic crisis, this phenomenon has never been recorded in the Bassin d'Arcachon. The processes that are involved in causing dystrophy were investigated by studying one comparable site in each of these lagoons as a model system. Shallow areas in both lagoons are characterized by the occurrence of microphytobenthic communities. The role of iron in these sediments was investigated. Iron reacts with sulfide to produce the virtually insoluble precipitate FeS. It was postulated that iron might function as an efficient scavenger of sulfide what would prevent it from reaching toxic concentrations. In contrast, oxygen production by the microphytobenthos may result in iron oxidation. Ferric iron is known to form an insoluble complex with phosphate. Thus, iron may also play a role in regulating the availability of phosphate for algal growth.

Nitrogen fixation may be important in providing an additional source of nitrogen. Among diazotrophic organisms cyanobacteria are particularly important (Fay, 1992). As oxygenic photoautotrophic organisms they can provide the enzyme nitrogenase with sufficient energy and reducing equivalents. In both lagoons benthic communities of cyanobacteria were present (microbial mats) in which nitrogen fixation was important. In the Bassin d'Arcachon heterocystous cyanobacteria appeared to be the predominant diazotrophs, whereas non-heterocystous species were found in Étang du Prévost. Nitrogen fixation is particularly sensitive to oxygen and the oxygenic phototrophic cyanobacteria have evolved special adaptations (Gallon, 1992). Heterocystous cyanobacteria develop special cells, the heterocysts, that have lost the capacity for oxygenic photosynthesis and in which the nitrogen fixation takes place (Haselkorn, 1978). Such cyanobacteria can be considered to be the best adapted for diazotrophic growth (Stal et al., 1994). It has been shown that sulfide is important for the selection of cyanobacteria (Howsley & Pearson, 1979).

## Dynamics of iron content of the sediments of Bassin d'Arcachon and Étang du Prévost

In Table 1 the iron content of the sediments of the stations C (Bassin d'Arcachon) and 12 (Étang du Prévost) are shown. Both stations were sampled in September 1993 and in May 1994. Two methods were used to assay iron. Total reactive iron was extracted with 10 M HCl and assayed with phenanthroline. Bioavailable iron was extracted from the sediment with 0.5 M HCl and assayed using the ferrozine reagent (Lovley & Phillips, 1987a). The iron content was highest in Station C of Bassin d'Arcachon. This is in accordance with the observation that the input of iron in the Bassin d'Arcachon via river discharge is much higher than in the Étang d Prévost (P. J. Labourg, pers. comm.). In September 1993 the iron content was particularly high whereas in May 1994 the levels were much lower. This was the case in both stations but particularly evident in Station C. The September:May iron ratios in Station C were 3.2 and 3.1 for total reactive iron and bioavailable iron, respectively. For Station 12 these ratios were 1.5 and 1.1. It is conceived that iron accumulated during the spring and summer was mobilized during the autumn and winter. The water of Bassin d'Arcachon is subject to frequent exchange with the Atlantic Ocean because of the tides and the relatively large entrance of the lagoon to the open sea. In contrast the Étang du Prévost has only a narrow connection with the Mediterranean Sea and the tidal range is small ($\sim$20 cm). Therefore, the water in Étang du Prévost has a much longer residence time than in Bassin d'Arcachon. This explains the smaller differences in iron content between spring and summer in the Étang du Prévost.

The mechanisms of iron immobilization and mobilization are not precisely known. The bulk of iron is present as ferrous sulfide which is virtually insoluble. Considerable iron is fixed as pyrite. Pyrite is a very stable compound and pyrite iron is not included in the iron measurements presented in Table 1. Oxidation of ferrous iron results in the formation of insoluble ferric compounds (Phillips et al., 1993). Benthic microorganisms such as cyanobacteria may also bind and accumulate iron (Stal, 1994). Thus, immobilization of iron may be associated with microbial activities such as (i) sulfate- and sulfur reduction which produce the sulfide that will precipitate as FeS, (ii) oxygenic photosynthesis by microphytobenthos that produce the oxygen that will oxidize iron and form insoluble iron hydroxides either chemically or biologically,

*Table 1.* Summary of the results of iron determinations in the upper 5 cm of the sediment of Station C, the Bassin d'Arcachon and Station 12, Étang du Prévost. Samples were collected in September 1993 and in May 1994. Two methods of iron determination were used. Total reactive iron was extracted with 10 M HCl and assayed with phenanthrolinechloride. Bioavailable iron was extracted in 0.5 M HCl and assayed with ferrozine reagent. Iron is expressed either as $\mu$mol cm$^{-3}$ or as molar ratios.

| | | $Fe^{2+} + Fe^{3+}$ | ($\mu$mol cm$^{-3}$) | ratio |
|---|---|---|---|---|
| Sample Station | Fe Assay | September | May | September:May |
| Arcachon C | total reactive Fe | 125 | 39 | 3.2 |
| Prévost 12 | total reactive Fe | 44 | 29 | 1.5 |
| Arcachon C | bioavailable Fe | 104 | 34 | 3.1 |
| Prévost 12 | bioavailable Fe | 20 | 19 | 1.1 |
| Arcachon C | ratio reactive:bio | 1.2 | 1.1 | |
| Prévost 12 | ratio reactive:bio | 2.2 | 1.5 | |
| ratio C:12 | total reactive Fe | 2.8 | 1.3 | |
| ratio C:12 | bioavailable Fe | 5.2 | 1.8 | |

*Table 2.* Ferrous and ferric iron in the upper 5 cm of the sediments of the sampling stations C (Bassin d'Arcachon) and 12 (Étang du Prévost) in September 1993 and May 1994. Two methods of iron determination were used. Total reactive iron was extracted with 10 M HCl and assayed with phenanthrolinechloride. Bioavailable iron was extracted in 0.5 M HCl and assayed with ferrozine reagent.

| | | $Fe^{2+}$ or $Fe^{3+}$ ($\mu$mol cm$^{-3}$) | | | | ratio $Fe^{2+}$:$Fe^{3+}$ | |
|---|---|---|---|---|---|---|---|
| Sample Station | Fe Assay | September | | May | | September | May |
| | | $Fe^{2+}$ | $Fe^{3+}$ | $Fe^{2+}$ | $Fe^{3+}$ | | |
| Arcachon C | total reactive Fe | 55 | 70 | 4 | 35 | 0.8 | 0.1 |
| Prévost 12 | total reactive Fe | 7 | 37 | 9 | 21 | 0.2 | 0.4 |
| Arcachon C | bioavailable Fe | 97 | 11 | 33 | 1 | 8.8 | 33 |
| Prévost 12 | bioavailable Fe | 17 | 4 | 13 | 6 | 4.3 | 2.2 |

(iii) binding to extracellular polysaccharide sheaths of benthic microorganisms (e.g. cyanobacteria) (Decho, 1990). Mobilization and liberation of the iron from the sediments is therefore possible when the activities described above cease. When acid volatile sulfide is subsequently oxidized (chemically and/or biologically) iron will be liberated.

In September 1993 the sediment at Station C contained almost three times (ratio Station C: Station 12 = 2.8) as much total reactive iron on a volume basis than Station 12. In May 1994 this difference was much smaller (ratio Station C: Station 12 = 1.3). Another remarkable difference between the two stations was the ratio of total reactive iron and bioavailable iron. The ratio of bioavailable iron Station C: Station 12 was 5.2 and 1.8 in September 1993 and May 1994, respectively. In Station C virtually all iron was present as bioavailable iron (ratio 1.2 and 1.1 in September and May, respectively). In Station 12 only half of the total reactive iron was determined as bioavailable (ratio 2.2 and 1.5 in September and May, respectively). The relatively low content of bioavailable iron in Station 12 in September coincided with a high content of ferric iron, many forms of which are not readily available for organisms (Phillips et al., 1993).

Data presented in Table 2 summarizes the amounts of ferrous and ferric iron as well as their ratios at the two sampling stations in September 1993 and May 1994. It is notable that the two methods applied give totally different results with regard to the differentiation between ferrous and ferric iron. These data indicate that the methods used cannot readily compared when ferrous and ferric iron are considered separately. Because it is more likely that the bulk of the iron is in the reduced form, only the ferrozine method will be considered here since this clearly yielded the highest amounts

*Table 3.* Ratios of ferric:ferrous iron in sediment slurries from the Bassin d'Arcachon (Station C) and Étang du Prévost (Station 12) with and without the addition of 25 mM $Fe^{3+}$. The slurries were incubated in the dark for up to 90 h. Experiments were carried out in May 1994.

| Station | Fe addition | Incubation time (h) | | | | |
|---|---|---|---|---|---|---|
| | | 0 | 1 | 24 | 48 | 90 |
| | | ratio ferric:ferrous iron ($Fe^{3+}$:$Fe^{2+}$) | | | | |
| Arcachon C | no Fe add. | 0.09 | 0.09 | 0.10 | 0.01 | nd |
| Arcachon C | 25 mM $Fe^{3+}$ | 0.79 | 0.07 | 0.06 | 0.07 | nd |
| Prévost 12 | no Fe add. | 0.02 | 0.13 | 0.13 | 0.02 | 0.08 |
| Prévost 12 | 25 mM $Fe^{3+}$ | 1.76 | 2.50 | 0.31 | 0.57 | 0.30 |

*Table 4.* The amounts of acid volatile sulfide (AVS) and pyrite in the upper 5 cm of the sediments of the sampling stations C (Bassin d'Arcachon) and 12 (Étang du Prévost), compared with the iron content (amounts in $\mu$mol $cm^{-3}$).

| Sample Station | Month | AVS | Total reactive Fe | Total bioavailable Fe | Pyrite |
|---|---|---|---|---|---|
| Arcachon C | September | 47 | 125 | 104 | 1.99 |
| Arcachon C | May | 8 | 39 | 34 | 0.23 |
| Prévost 12 | September | 35 | 44 | 20 | 0.83 |
| Prévost 12 | May | 1 | 29 | 19 | 0.07 |

of ferrous iron. Station 12 shows the highest relative amounts of ferric iron (lowest ratios $Fe^{2+}$:$Fe^{3+}$). The sediment of Station 12 was covered by a dense layer (mat) of cyanobacteria. The oxygen produced by this community may have been responsible for the oxidation of the iron. Because the cyanobacterial mat was most active in spring, the May ratio $Fe^{2+}$:$Fe^{3+}$ was lower than in September when cyanobacterial activity was low, presumably because evaporation resulted in hypersaline conditions. In Station C the situation was totally different. The relative contribution of ferrous to total iron was much larger than in Station 12. In Station C the microphytobenthic community (cyanobacteria and diatoms) developed during summer and photosynthetic oxygen production may therefore result in a higher ferric iron content in September compared with May. However, the highly reduced conditions in this station ensured that ferrous iron was quantitatively the most important species present both in September and in May.

Depth profiles of iron revealed similar patterns regardless whether total reactive iron or bioavailable iron was considered. The profile of iron ($Fe^{2+} + Fe^{3+}$) in Station C showed an increase in content with depth (Figure 1a, b). This gradient was more pronounced in May 1994 (Figure 1a) compared with September 1993 (Figure 1b). The iron content in May 1994 was lower than in September 1993, particularly in the top layer of the sediment. As stated previously, virtually all iron was in the ferrous form. Ferrous iron (when not precipitated as metal sulfide) is very soluble and this might explain the mobilization and dissolution of iron from the sediment, which would be most pronounced in the surface layer. The situation was different in Station 12 (Figure 1c, d). The iron profile in September 1993 showed a sharp increase with depth (Figure 1d). In May highest concentrations were found in the top layers but the gradient was less clearly defined (Figure 1c). Although the highest iron concentrations were measured in September 1993, the depth profile of iron at Station 12 indicated that it was in the process of being mobilized and dissolved. This was not only evident from the gradient but also from the greater proportion of ferrous iron. In contrast, in May iron was apparently being immobilized. This is evident from the gradient and from the lower proportion of ferrous iron. Moreover, the immobilization of iron was also coincident with the presence of an active cyanobacterial mat at this station during the spring. Oxygen produced by the cyanobacteria may precipitate ferric iron and these organisms have also been shown to accumulate iron in their extracellular polysaccharide sheath (Stal, 1994).

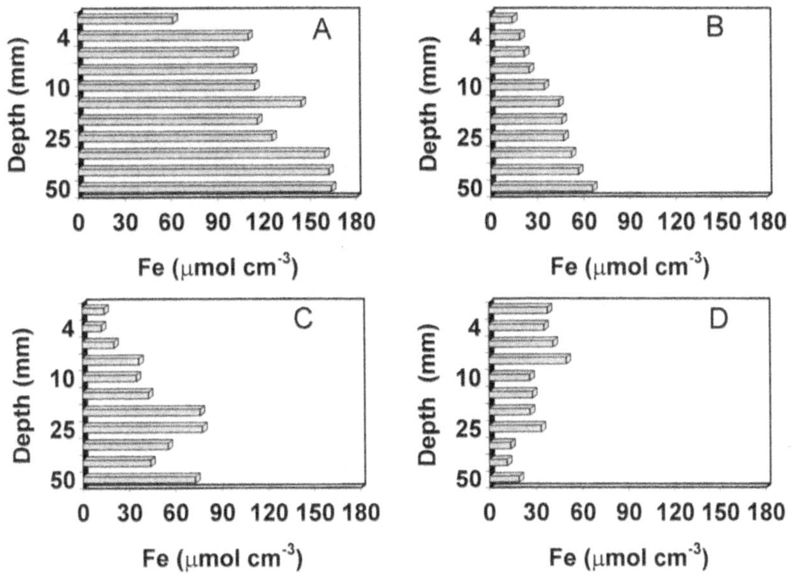


*Figure 1.* Depth profiles of total ($Fe^{2+} + Fe^{3+}$) reactive iron in the sediments of the Bassin d'Arcachon (Station C) (a,b) and Étang du Prévost (Station 12) (c,d) in May 1994 (a,c) and September 1993 (b,d).

Experiments in which sediment slurries were used demonstrated that ferric iron was rapidly reduced. The results of these experiments are summarized in Table 3. The sediment slurries were incubated in the dark and the ratio of ferric:ferrous iron measured immediately after the addition of 25 mM $Fe^{3+}$ and after 1, 24, 48 and 90 hours. Sediment slurries without added iron were used as controls. In the control slurries, most of the iron was in the reduced state and very little change in the ferric:ferrous iron ratio was recorded in sediment samples from the Étang du Prévost (Station 12) as well as from Bassin d'Arcachon (Station C). The addition of 25 mM $Fe^{3+}$ resulted in a substantial increase in the ferric:ferrous iron ratio. At Station C this increase was less dramatic since the iron content at this site was much higher than at Station 12 (Table 1). At Station C ferric iron was already reduced after 1 h incubation, after which the ratio ferric:ferrous iron returned to its original value. These results demonstrate the enormous potential for iron reduction at this site. Sediment slurries from Station 12 behaved slightly differently. Due to the lower iron content at this site the addition of 25 mM ferric iron resulted in a much higher ferric:ferrous iron ratio. This ratio increased somewhat after 1 h incubation, probably due to oxidation by oxygen introduced in the slurry during preparation. Thereafter the ferric:ferrous iron ratio decreased but remained above the original value. Although it was evident that this site also possessed a high potential for iron reduction, it was clearly less dramatic than at Station C. Sediment slurries were also amended with glucose (20 mM), molybdate (20 mM) and a combination of glucose and molybdate. By increasing the organic matter content the reducing potential should increase (Lovley & Phillips, 1986). The addition of molybdate inhibits sulfate reduction (Oremland & Capone, 1988) and it was expected that sulfide produced by sulfate reducing bacteria would contribute significantly to iron reduction (see equations p. 11). However, the results show that the iron-reducing potential of the sediment slurry was already maximal and could not be significantly increased by the addition of glucose. Equally, inhibition of sulfate reduction did not decrease this potential (Figure 2). These data indicate that processes other than sulfate reduction were responsible for iron reduction. Ferric iron may serve as terminal electron acceptor in the anaerobic degradation of organic matter (Lovley, 1987, 1991). Clearly, the organic matter content of the sediments was sufficient to sustain these processes since the addition of glucose had no effect.

## Acid volatile sulfide in the sediments of Bassin d'Arcachon and Étang du Prévost

The concentrations of acid volatile sulfide (AVS) in the sediments of the Stations C (Bassin d'Arcachon) and 12 (Étang du Prévost) were highest in September.

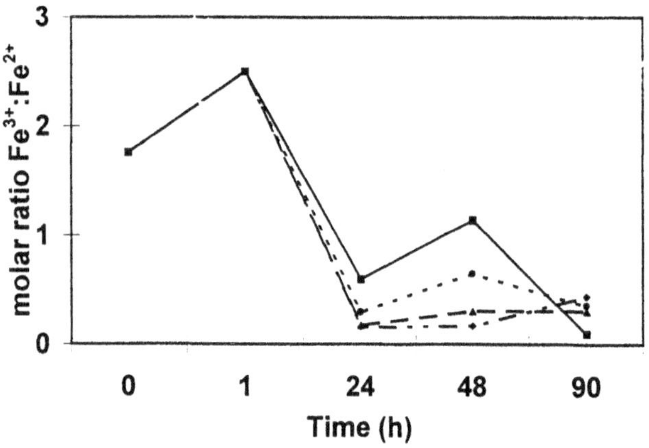


*Figure 2.* Ratio ferric ($Fe^{3+}$) to ferrous ($Fe^{2+}$) iron after the addition of 25 mM of $Fe^{3+}$ at t = 0 to a sediment slurry from the Étang du Prévost (Station 12). Slurries were incubated for up to 90 h in the dark at ambient temperature. —■— : no further additions; —●— : glucose added; —△—: molybdate added; —◇—: glucose and molybdate added.

*Table 5.* Comparison of total extractable phosphate and ferric iron (bio-available) in the upper 5 cm of the sediments of the sampling stations C (Bassin d'Arcachon) and 12 (Étang du Prévost). Amounts represent $\mu$mol cm$^{-3}$ of sediment. Samples were collected in September 1993 and May 1994.

| Month | Station | $Fe^{3+}$ | $PO_4^{3-}$ |
|---|---|---|---|
| September | Arcachon C | 11.0 | 0.7 |
| | Prévost 12 | 4.3 | 4.1 |
| May | Arcachon C | 1.2 | 1.2 |
| | Prévost 12 | 5.7 | 1.2 |

Acid volatile sulfide in the top 5 cm of Station was 47 and 8 $\mu$mol cm$^{-3}$ in September and May, respectively. In Station 12 these numbers were respectively 35 and 1 (Table 4). Sulfide is produced by sulfate- and

*Table 6.* Sediment slurries collected from the Bassin d'Arcachon (Station C) and Étang du Prévost (Station 12) amended with 100 $\mu$M phosphate. The effect of iron was measured by adding 25 mM $Fe^{3+}$. Incubation was for 1 h. Experiments were done in May 1994.

| Treatment | Station C | Station 12 |
|---|---|---|
| | $\mu$M phosphate in water | |
| no iron addition | 10 | 24 |
| 25 mM $Fe^{3+}$ added | 0.3 | 4.2 |

*Table 7.* Effects of sulfide on nitrogenase activity in a non-heterocystous cyanobacterium (*Oscillatoria* sp.) and a heterocystous species (*Anabaena* sp.). The control (0 mM) was set to 100. Experiments were performed at a pH of 8.

| Sulfide concentration (mM) | *Oscillatoria* sp. | *Anabaena* sp. |
|---|---|---|
| 0 | 100 | 100 |
| 0.5 | 154 | 46 |
| 1 | 104 | 27 |
| 5 | 94 | 13 |
| 10 | 118 | nd |

sulfur reducing microorganisms. These processes are only of importance after sufficient organic matter has been produced, i.e. in summer or autumn, depending on the geographical location. It is evident that during winter and spring most of the sulfide in the sediment is oxidized due to either biotic or abiotic processes. In September 1993 the sulfide concentrations were slightly higher in Station C as compared to Station 12, but in both stations acid volatile sulfide concentrations were very high. If these concentrations are compared with the total iron content of these sampling stations it was evident that at Station C iron was present in considerable excess and therefore any sulfide produced at this station would be immediately precipitated as FeS. Free sulfide ($HS^-$) would not be expected to be present at this site. The situation was different in Station 12 where in September the concentration of total reactive iron was only slightly higher than acid volatile sulfide (Table 4). Moreover, when bioavailable iron was considered acid volatile sulfide was in excess. Under these conditions free sulfide was expected to be present. Pyrite was also present at both sampling sites but was quantitatively unimportant. The dynamics of pyrite followed that of acid volatile sulfide and iron, i.e. it was also oxidized and leached during winter.

The profiles of acid volatile sulfide in the sediment of the Stations C and 12 are shown in Figure 3. In Station 12 in May sulfide was apparently produced mainly at a depth of 4–8 mm (Figure 1c), while in September this had moved down to 12–25 mm (Figure 1d). In Station C a different pattern was recorded. In May AVS concentrations increased with depth which suggests that sulfide oxidation was occurring at the sediment surface (Figure 1a). In September virtually no gradient was present and AVS was equally high at all depths except the surface layer (Figure 1b). In the surface layer presumably little sulfide was produced.

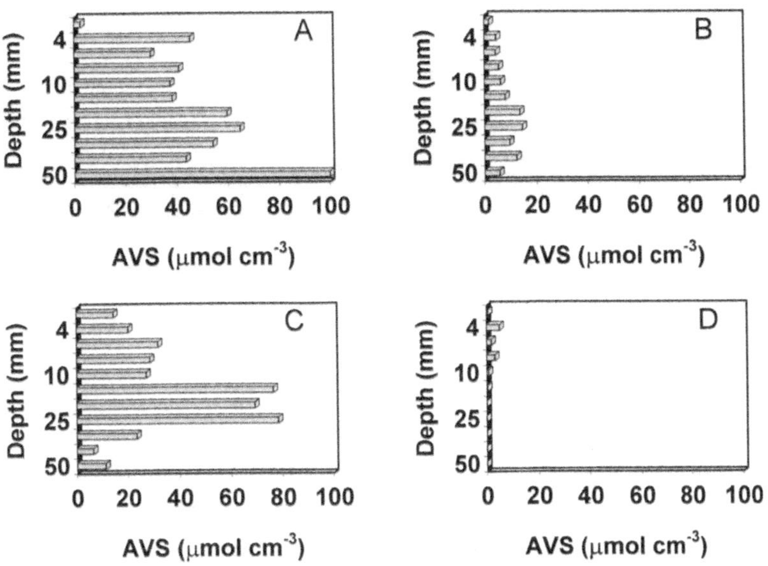


*Figure 3.* Depth profiles of acid volatile sulfide (AVS) in the sediments of the Bassin d'Arcachon (Station C) (a,b) and Étang du Prévost (Station 12) (c,d) in May 1994 (a,c) and September 1993 (b,d)

Profiles of pyrite are shown in Figure 4. In Station 12 in May highest concentrations of pyrite were found in the top layer indicating that pyrite formation was occurring in this layer. In May Station 12 also had a relatively high ferric iron content and it is probable that this was important either directly or indirectly in pyrite formation (see equation). Thus sulfide production and pyrite formation appear to proceed in different horizons in the sediment. In September sediment concentrations of pyrite was much higher but the depth profile was reversed, indicating a possible oxidation of pyrite at the surface (Figure 4d). At Station C the sediment content of pyrite was totally different. The pyrite concentrations were higher at Station C compared with Station 12. Depth profiles of pyrite showed highest concentrations in the deeper layers of the sediment both in September and May. This suggested that pyrite was formed in the deeper layers of the sediment, concurrent with sulfide production. The large excess of ferrous iron probably prevented the formation of elemental sulfur and thus pyrite formation, as was probably the case at Station 12. Under these conditions pyrite may have been produced here as a direct consequence of sulfate reduction (Howarth & Merkel, 1984).

$$2Fe^{3+}+S^{2-} \rightarrow 2Fe^{2+}+S^0$$
$$2Fe^{2+}+2S^{2-} \rightarrow 2FeS$$
$$FeS+S^0 \rightarrow FeS_2$$
$$\overline{2Fe^{3+}+3S^{2-} \rightarrow FeS+FeS_2}$$

## Dynamics of phosphate in the sediments of Bassin d'Arcachon and Étang du Prévost

In Table 5 the concentration of total extractable phosphate in the sediments of the sampling stations of Bassin d'Arcachon and Étang du Prévost are shown. At Station C highest phosphate concentrations were detected in May, while at Station 12 highest values were recorded in September. These differences may be attributed to differences in growth season. In the Étang du Prévost the cyanobacterial mats were fully developed in spring and early summer, while in the Bassin d'Arcachon they did not appear until late summer. The active growth of the microphytobenthos may also result in the lowering of phosphate concentration. In addition there are significant differences in the phosphate load that the two lagoons receive. Highest phosphate concentrations were measured in the sediments of Étang du Prévost. This lagoon clearly received a much higher phosphate load. The phosphate that was extracted from the sediments might not have

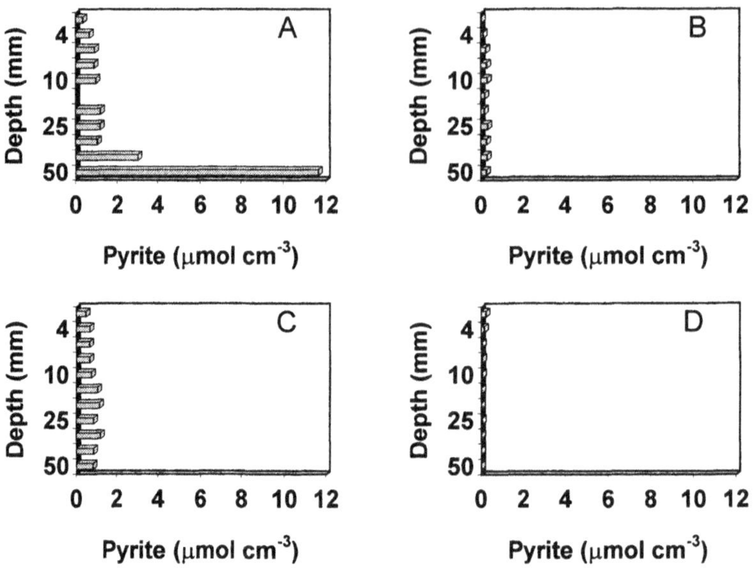


*Figure 4.* Depth profiles of pyrite ($FeS_2$) in the sediments of the Bassin d'Arcachon (Station C) (a,b) and Étang du Prévost (Station 12) (c,d) in May 1994 (a,c) and September 1993 (b,d).

been available for growth of microphytobenthos. Phosphate forms a complex with ferric iron (Ehrlich, 1990, Danenlouwerse et al., 1993). In Table 5 the phosphate concentrations are compared with the concentrations of ferric iron in the sediments. It is clearly evident that in all cases ferric iron equalled or exceeded the phosphate concentrations. At Station C in May phosphate and iron concentrations were equal which suggests that the microphytobenthos was supplied with sufficient phosphate. In September, however, ferric iron exceeded the sediment phosphate content by a factor of 15, which is likely to have resulted in phosphate limitation. At Station 12 the situation again was reversed. In September ferric iron and phosphate showed equal concentrations, indicating that phosphate supply probably was not limiting primary productivity. However, in May ferric iron was in large excess (almost 5-fold).

In order to study binding and mobilization of phosphate at both sampling sites, sediment slurry experiments were performed. Phosphate was added to sediment slurries was rapidly immobilized but the amount that was bound differed between the two sampling sites. At Station C 90% of the phosphate added to the slurry (100 $\mu$M) was bound while at Station 12 this was only 76%. If the sediment slurries were enriched by the addition of 25 mM ferric iron (estimated final concentrations of ferric iron in the slurries of Stations C and 12 are respectively 26 and 30 mM) 99.7% and 95.8% of the phosphate was immobilized at stations C and 12, respectively (Table 6).

Because phosphate concentrations at both sites in May 1994 were comparable (1.2 $\mu$mol cm$^{-3}$, Table 5), differences in phosphate immobilization between the stations C and 12 cannot be explained in terms of sediment phosphate content. Although total iron content in Station C was greater than at Station 12, more ferric iron was present at the latter station. Since phosphate complexes with ferric rather than with ferrous iron it might be expected that more phosphate would be bound at Station 12. Furthermore, the addition of 25 mM ferric iron would be expected to completely bind phosphate. Notwithstanding the fact that, in contrast to Station 12, this ferric iron was almost instantaneously reduced in Station C (Table 2), still more phosphate was bound at Station C. Thus, although iron was responsible for part of the phosphate binding, the results cannot be solely explained in terms of iron concentration. It is possible that the differences in physico-chemical conditions between the two stations were responsible for the observed differences (Robertson & Alexander, 1992). Considerable differences have been observed in temperature, salinity and pH between these sampling sites. Moreover, phosphate can also be bound by calcium (Ehrlich, 1990, Hirschler et al., 1990; Sukenik et al., 1985).

In order to understand the mechanisms of phosphate release from the sediments, experiments were carried out with sediment slurries. It was hypothesized that phosphate was mainly complexed with ferric iron. Consequently, sulfide production by sulfate reducing bacteria would reduce iron, thereby mobilizing phosphate (Caraco et al., 1989). In order to stimulate sulfate reduction sediment slurries were enriched with glucose. As a control, slurries were incubated with 20 mM molybdate which inhibited sulfate reduction. As shown in Figure 5 the addition of 100 $\mu$M phosphate to a slurry of Station 12 resulted first in the immediate immobilization of a large portion of the phosphate. However, after 24 hours phosphate was released from the sediment. This release was clearly stimulated by glucose. The presence of glucose led to a greater phosphate release than was added to the sediment, indicating that phosphate already present in the sediment was mobilized. Sulfate reduction was apparently not required for the release of phosphate. This confirmed the observations on iron reduction. Iron may also be reduced by microorganisms in a dissimilatory process (Lovley & Phillips, 1988, Lovley et al., 1992). Even sulfate reducing bacteria have been shown to reduce ferric iron (Coleman et al., 1993). In the zone of ferric iron reduction sulfate reduction and methanogenesis are inhibited (Lovley & Phillips, 1987b). In fact, highest phosphate release was observed in glucose amended sediments in the presence of 20 mM molybdate. In further experiments undertaken with sediments from the Bassin d'Arcachon (Station C) and Étang du Prévost (Station 12) in May 1994, molybdate appeared to be required for phosphate mobilization (Figure 6). As proposed previously, phosphate may well be bound to calcium which is present in large amounts in these sediments. While sulfate reduction is a process which leads to slightly alkaline conditions, decomposition of organic matter by iron reduction would lead to acid production. This decrease in pH would cause the dissolution of calcium phosphate (Petterson et al., 1988). Fractionated extraction of phosphate from the sediments also confirmed that a large portion of phosphate may be bound as calcium phosphate (Stal, unpublished results). Thus the mobilization of phosphate from sediments is a complex process in which the contribution of iron and sulfate reduction in the decomposition of organic matter plays a central role.

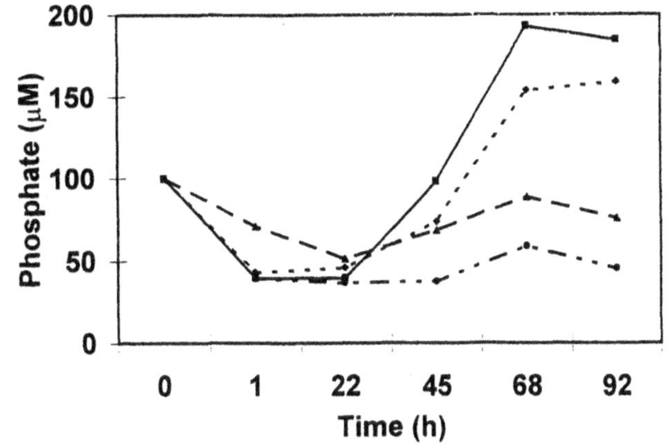


*Figure 5.* Binding and mobilization of phosphate in the sediment of the Étang du Prévost (station 12) in September 1994. At t=0 100 $\mu$M phosphate was added to a sediment slurry which was subsequently incubated for over 90 h and phosphate was determined in the water.: —△— no further additions; —◆—: glucose added; —●— molybdate added; —■— glucose and molybdate added.

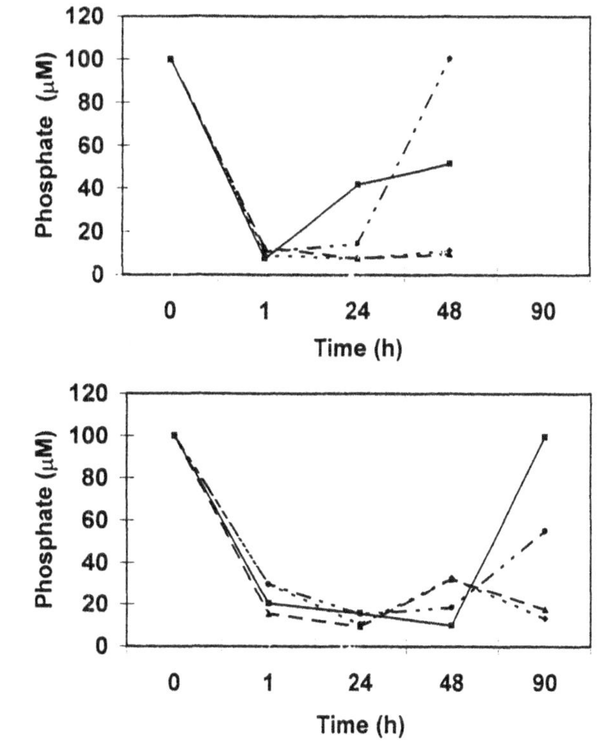


*Figure 6.* As Figure 5. but measured in May 1994. A. Bassin d'Arcachon (Station C). B. Étang du Prévost (Station 12).: —△— no further additions; —◆—: glucose added; —●— molybdate added; —■— glucose and molybdate added.

## Nitrogen fixation

As shown in the preceeding sections, the sediment at Station C in the Bassin d'Arcachon possesses a high phosphate binding capacity. Therefore, primary productivity in this system may be phosphate limited at particular times of the year. In contrast, Station 12 in Étang du Prévost had a significantly lower phosphate binding potential and particularly in summer a much higher phosphate content. This would almost certainly result in nitrogen limitation. The sediments of both stations were characterized by benthic communities of phototrophic microorganisms. In the Étang du Prévost the sediment at Station 12 was covered by a microbial mat composed of *Oscillatoria* sp., a filamentous, non-heterocystous cyanobacterium. In the Bassin d'Arcachon the sediment of Station C was frequently colonized by benthic diatoms but sometimes mats of the heterocystous cyanobacterium *Anabaena* sp. developed. Because environmental conditions at Station C undergo considerable fluctuations the benthic cyanobacterial communities were not particularly stabile. The occurrence of benthic communities of diatoms in Station C suggested phosphate rather than nitrogen limited primary productivity. However, massive developments of benthic diatoms may successively lead to nitrogen limitation. It was considered that under such conditions nitrogen-fixing cyanobacteria may develop.

Cyanobacteria are oxygenic photoautotrophic prokaryotes and many species are capable of diazotrophic growth (Fay, 1992). However since nitrogenase is extremely oxygen sensitive, nitrogen-fixing cyanobacteria have evolved different strategies in order to protect the enzyme from inactivation (Gallon, 1992). Nitrogen-fixing cyanobacteria can be subdivided in 3 groups. (i) Heterocystous cyanobacteria differentiate special cells, heterocysts, that have lost the capacity of oxygenic photosynthesis and in which nitrogen fixation takes place. Heterocysts possess special cell walls that provide a diffusion barrier that limit the entrance of oxygen. The interior of heterocysts is therefore virtually anoxic. Among the cyanobacteria heterocystous species can be regarded as best adapted for diazotrophic growth. (ii) Non-heterocystous, filamentous or unicellular species fix nitrogen only under anoxic or extremely low-oxygen conditions. The strategy of these species is clearly the avoidance of oxygen since an adequate protection mechanism is absent. Although many cyanobacteria (may be up to 50% of all species) may belong to this group, their ecological relevance has not been satisfactory elucidated. (iii) Non-heterocystous, filamentous or unicellular species that fix nitrogen under fully oxic conditions. The mechanisms by which these organisms protect nitrogenase from oxygen inactivation are not precisely understood (Gallon, 1992). It has been proposed that such cyanobacteria separate nitrogen fixation temporally from oxygenic photosynthesis, i.e. nitrogen fixation is confined to the dark (Gallon & Stal, 1992). However, nitrogen fixation in the non-heterocystous, planktonic cyanobacterium *Trichodesmium* spp. is strictly light-dependent. Recent discoveries indicate that confinement of nitrogen fixation to the dark in many non-heterocystous cyanobacteria is determined by oxygen dynamics which cause supersaturation in the light (Stal, 1995). Aerobic nitrogen-fixing non-heterocystous cyanobacteria are particularly known from marine microbial mats which are characterized by oxygen supersaturation during the day and anoxic conditions during the night. It must be emphasized that aerobic nitrogen-fixing, non-heterocystous cyanobacteria are not optimal adapted for diazotrophic growth (Stal et al., 1994). Nitrogenase in these organisms suffer clearly from oxygen inactivation which is counteracted by a high rate of *de novo* synthesis of the enzyme. Therefore, notwithstanding their capability of diazotrophy, growth of these organisms may still be nitrogen limited. In contrast, heterocystous species can provide all nitrogen necessary and are rarely nitrogen-limited.

Diel variations of nitrogenase activity at Station C (Bassin d'Arcachon) showed a strict light dependence. The cyanobacterial community in this station was composed of the heterocystous species *Anabaena* sp. in which nitrogen fixation was light dependent. In Station 12 (Étang du Prévost) nitrogenase activity was usually much higher and a clear diel variation was not observed. The cyanobacterial community was composed of the filamentous non-heterocystous species *Oscillatoria* sp. In Figure 7 the effects of light and oxygenic photosynthesis on nitrogenase activity of both communities are depicted. The *Anabaena* community of Bassin d'Arcachon showed highest activity in the light (Figure 7a). In the dark activity was much decreased while the addition of $10^{-5}$ M 3-(3,4-dichlorophenyl)1,1- dimethylurea (DCMU), an inhibitor of oxygenic photosynthesis (PS II), resulted in a slight decrease of activity. This result was expected with heterocystous species since heterocysts do not contain photosystem II and therefore DCMU has no effect. The slightly lower activity in the presence of DCMU can be explained by an overall inhibitory effect on cyanobacterial metabolism or the limited supply with reducing equivalents from the vegetative cells. The inhibition of photosystem II in the *Oscillatoria* sp. community of Étang du Prévost had a stimulatory effect (Figure 7b). This demonstrated that oxygenic photosynthesis in this community seriously impaired nitrogen fixation. Light incubation resulted in slightly higher nitrogenase activity as compared to dark incubation. In aerobic nitrogen fixing non-heterocystous cyanobacteria dark activity is often higher than light incubations. This highlights the problems these organisms have with photosynthetic oxygen production and explains why nitrogen fixation in natural populations is usually confined to the dark. The different behaviour of the Étang du Prévost population of *Oscillatoria* can be explained by the fact that this organism is not an aerobic nitrogen-fixing species. This cyanobacterium was isolated and subsequently shown to fix nitrogen only under anoxic conditions. Experiments were carried out with the two populations of cyanobacteria in order to further investigate the effects of additions on nitrogenase activity. The addition of glucose (Figure 8a), sulfide (Figure 9a) or ferrous iron (Figure 10a) all stimulated nitrogenase activity in samples from Étang du Prévost. The common effect of these additions was the reduction of oxygen, although they may also have stimulated nitrogen fixation directly. The addition of glucose stimulates heterotrophic activity and respiration but some cyanobacteria are capable of photoheterotrophic metabolism. Photoheterotrophy may support diazotrophic growth. Sulfide may reduce oxygen chemically or biologically but can be used as electron donor in anoxygenic photosynthesis by a number of cyanobacteria (Cohen, 1989). Sulfide is also well-known as an inhibitor of oxygenic photosynthesis (Cohen et al., 1986). Ferrous iron reacts instantaneously with oxygen. Ferrous iron has also been reported as a possible electron donor for cyanobacterial photosynthesis (Cohen, 1989). In contrast, glucose (Figure 8b) did not greatly effect nitrogenase activity in the Bassin d'Arcachon, while sulfide (Figure 9b) and ferrous iron (Figure 10b) appeared to be inhibitory. Reduction of oxygen does not stimulate nitrogen fixation in heterocystous cyanobacteria. The inhibitory effect of iron may be explained by the fact that the indigenous iron content in this sediment was already very high and that with the extra addition the tolerance limits for this metal may have been exceeded. The inhibitory effect of sulfide is more easily explained by the overall toxic effect this substance has.

These experiments account for the occurrence of the different types of nitrogen-fixing cyanobacteria in the two lagoons. Heterocystous cyanobacteria are clearly best adapted for diazotrophic growth. The reason why such organisms do not occur in Étang du Prévost may be explained by the fact that in this lagoon considerable concentrations of free sulfide may be present. Heterocystous cyanobacteria are much more sensitive towards sulfide than non-heterocystous species (Howsley & Pearson, 1979) (Table 7). In contrast, the large amount of iron present in the Bassin d'Arcachon prevents the production of free sulfide, whereas this will not be the case in Étang du Prévost (Table 4).

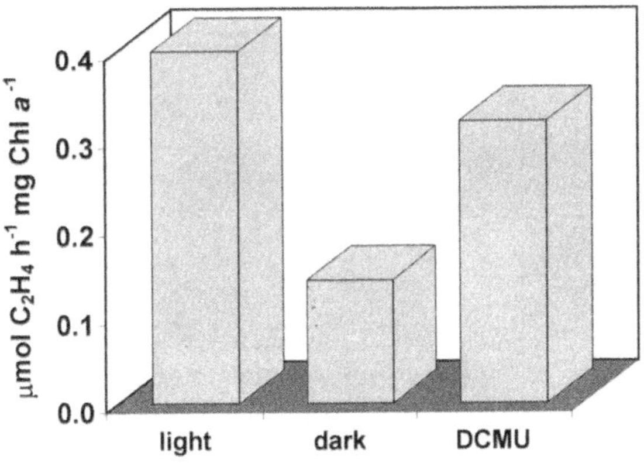


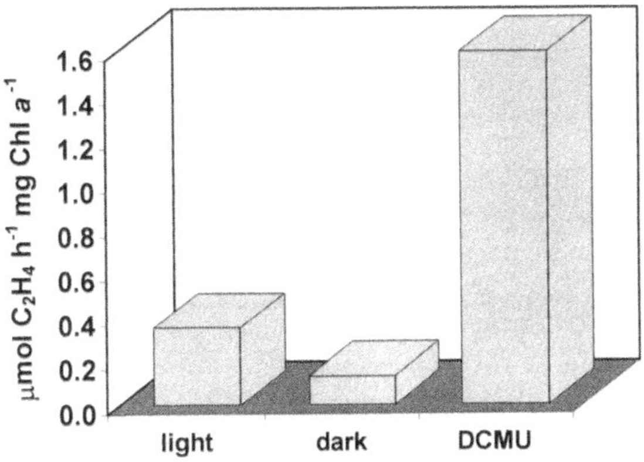


*Figure 7.* Effect of light and $10^{-5}$ M 3-(3,4-dichlorophenyl)1,1-dimethylurea (DCMU) on nitrogenase activity in the Bassin d'Arcachon (Station C) (A) and Étang du Prévost (Station 12) (B). Data from September 1993.

## Conclusions

From the results of the investigations on the biogeochemistry of the sediments of the two coastal lagoons Bassin d'Arcachon and Étang du Prévost a number of important conclusions can be drawn. From this study

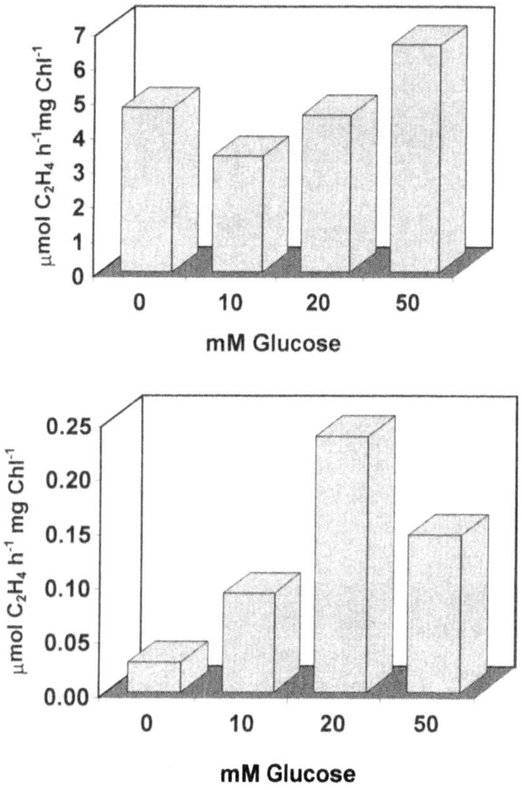


*Figure 8.* The effect of the addition of glucose on nitrogenase activity (acetylene reduction) in sediment samples from the Bassin d'Arcachon (Station C) (A) and Étang du Prévost (Station 12) (B). Incubation was done under ambient light.

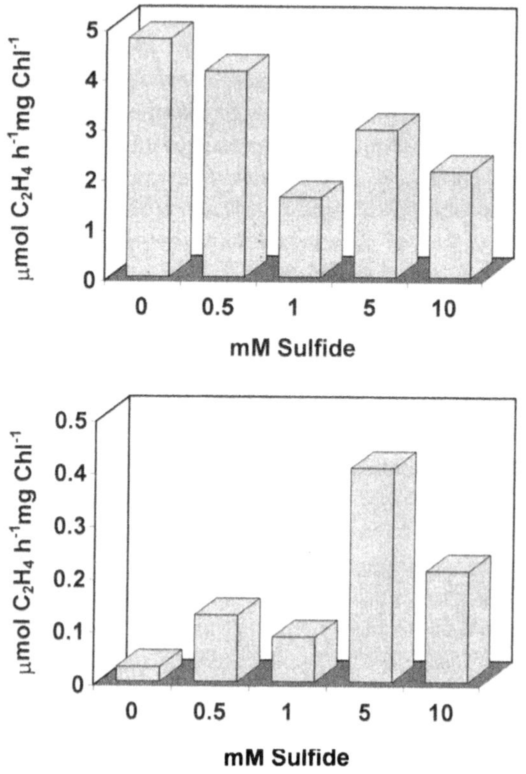


*Figure 9.* The effect of the addition of sulfide on nitrogenase activity (acetylene reduction) in sediment samples collected from the Bassin d'Arcachon (Station C) (A) and Étang du Prévost (Station 12) (B). Incubation was done under ambient light.

it is evident that iron plays a key role in the biogeochemical processes occurring in these lagoons. The Bassin d'Arcachon receives a large amount of iron which accumulates in the sediments and there is a strong seasonal fluctuation in sediment iron content. The regulating mechanisms of iron (im)mobilization in this lagoon are not precisely known and need further investigation. Oxidation of ferrous iron may lead to insoluble precipitates of ferric oxides. Also precipitation of FeS or pyrite ($FeS_2$) is important for iron immobilization. Benthic cyanobacteria may also be involved in binding of iron to their extracellular polysaccharide sheaths (Decho, 1990). Thus, immobilization of iron may be largely regulated by primary productivity and sulfate reduction (Howarth & Merkel, 1984). This explains why iron accumulation in these sediments seemed to occur predominantly during summer, when primary productivity and sulfate reduction were important. It is unlikely that the load of iron into the basin was higher in summer than in other seasons. The results also clearly demonstrate the mobilization of iron. Mobilization of iron may proceed simultaneously with re-oxidation of sulfide and the liberation of phosphate. In the Bassin d'Arcachon iron was in large excess of acid volatile sulfide. This means that any sulfide that is produced will precipitate as FeS and free sulfide will never reach toxic concentrations. In addition, ferric iron was in excess of total extractable phosphate in the sediment and addition of phosphate resulted in instantaneous binding to the sediment. Although ferric iron was certainly involved, it was not the only mechanism of phosphate binding in these sediments. Calcium possibly may also be important in the binding of phosphate. Iron reduction did not necessarily result in phosphate liberation, which would have been expected if ferric iron was the only phosphate binding component in these sediments. Phosphate liberation was possible after enriching the sediment with glucose however, the addition of molybdate, which inhibits sulfate reduction, did not prevent

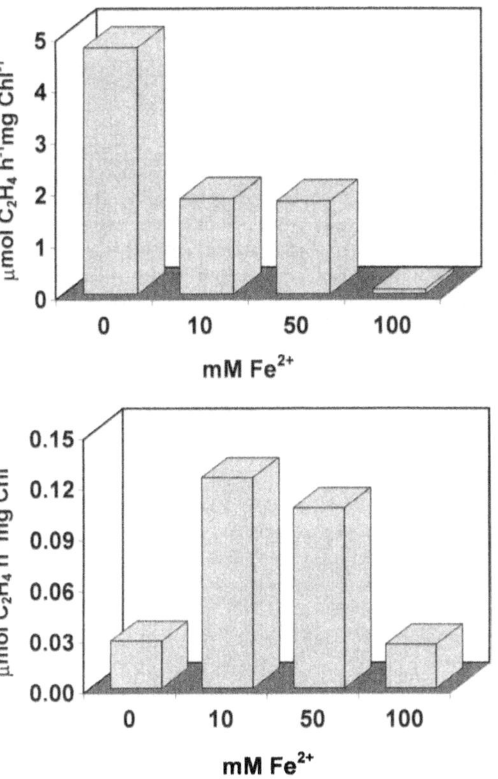


*Figure 10.* The effect of the addition of ferrous iron on nitrogenase activity (acetylene reduction) in sediment samples from the Bassin d'Arcachon (Station C) (A) and Étang du Prévost (Station 12) (B). Incubation was done under ambient light.

this. This indicates that sulfate reduction was not a prerequisite for phosphate liberation from these sediments and confirmed that ferric iron was only partially involved. These findings contrast with those of Caraco et al. (1989) but it is likely that different environments behave differently. Anaerobic degradation of organic matter, particularly in the absence of sulfate reduction, may lead to acidification which may dissolve calcium phosphate complexes (Pettersson et al., 1992; Robertson & Alexander, 1992). Nevertheless, together with a high iron content, the Bassin d'Arcachon had a high phosphate binding capacity. The situation in the Étang du Prévost is totally different. The iron content in this lagoon was much lower. Although there was a similar seasonal difference of iron content this was much less pronounced. In contrast to the Bassin d'Arcachon, the Étang du Prévost has almost no tidal influence and therefore the water exchange was much less. This would explain the small seasonal difference in iron content of Étang du Prévost. The lower iron content allowed the formation of free sulfide. Although in this lagoon there was some uncertainty about the role of iron in phosphate binding, its capacity was clearly lower than in the Bassin d'Arcachon. This suggests that in the Étang du Prévost primary productivity was probably not phosphate limited but nitrogen limited. Nitrogen fixation in the cyanobacterial mats is therefore an important process. These mats consisted of filamentous non-heterocystous cyanobacteria of the genus *Oscillatoria* sp. Nitrogen fixation by these organisms was clearly impaired by oxygen. Inhibition of oxygenic photosynthesis by 3,-(3,4-dichlorophenyl)-1,1-dimethylurea (DCMU) stimulated nitrogenase activity several fold. The addition of glucose, ferrous iron and sulfide were also stimulatory to nitrogenase activity. While sulfide is a potent inhibitor of oxygenic photosynthesis, all of these compounds may potentially serve as electron donors. Isolated strains of *Oscillatoria* from this site were only capable of anaerobic nitrogen fixation. Thus, although diazotrophy was obvious it was still affected by the presence of oxygen and there is no doubt that specialized heterocystous cyanobacteria would have been more appropriate. However, heterocystous cyanobacteria are rarely found in environments where high levels of free sulfide occur (Stal et al., 1994). This might explain why heterocystous species were present in the Bassin d'Arcachon where free sulfide was unlikely to occur due to the excess iron in the sediment.

## References


Caraco, N. F., J. J. Cole & G. E. Likens, 1989. Evidence for sulphate-controlled phosphorus release from sediments of aquatic systems. Nature 341: 316–318.

Caumette, P., 1986. Phototrophic sulfur bacteria and sulfate-reducing bacteria causing red waters in a shallow brackish coastal lagoon (Prévost Lagoon, France). FEMS Microbiol. Ecol. 38: 113–124.

Cohen, Y., 1989. Photosynthesis in cyanobacterial mats and its relation to the sulfur cycle: a model for microbial sulfur interactions. In Y. Cohen & E. Rosenberg (eds) Microbial mats. Physiological ecology of benthic microbial communities. ASM, Washington: 22–36.

Cohen, Y., B. B. Jorgensen, N. P. Revsbech & R. Poplawski, 1986. Adaptation to hydrogen sulfide of oxygenic and anoxygenic photosynthesis among cyanobacteria. Appl. Envir. Microbiol. 51: 398–407.

Coleman, M. L., D. B. Hedrick, D. R. Lovley, D. C. White & K. Pye, 1993. Reduction of Fe(III) in sediments by sulphate-reducing bacteria. Nature 361: 436–438.

Decho, A. W., 1990. Microbial exopolymer secretions in ocean environments: their role(s) in food webs and marine processes. Oceanogr. Mar. Biol. Annu. Rev. 28: 73–153.

Danenlouwerse, H., L. Lijklema & M. Coenraats, 1993. Iron content of sediment and phosphate adsorption properties. Hydrobiologia 253: 311–317.

Ehrlich, H. L., 1990. Geomicrobiology. Marcel Dekker, New York, 646 pp.

Fay, P., 1992. Oxygen relations of nitrogen fixation in cyanobacteria. Microbiol. Rev. 56: 340–373.

Gallon, J. R., 1992. Reconciling the incompatible: $N_2$ fixation and $O_2$. New Phytol. 122: 571–609.

Gallon, J. R. & L. J. Stal, 1992. $N_2$ fixation in non-heterocystous cyanobacteria: an overview. In E. J. Carpenter, D. G. Capone & J. G. Rueter (eds) Marine pelagic cyanobacteria: *Trichodesmium* and other diazotrophs. Kluwer Academic Publishers, Dordrecht, NATO ASI Series C. Vol. 362: 115–139.

Haselkorn, R., 1978. Heterocysts. Annu. Rev. Plant Physiol. 29: 319–344.

Hirschler, A., J. Lucas & J.-C. Hubert, 1990. Bacterial involvement in apatite genesis. FEMS Microbiol. Ecol. 73: 211–220.

Howarth, R. W. & S. Merkel, 1984. Pyrite formation and the measurement of sulfate reduction in salt marsh sediments. Limnol. Oceanogr. 29: 598–608.

Howsley, R. & H. W. Pearson, 1979. pH dependent sulphide toxicity to oxygenic photosynthesis in cyanobacteria. FEMS Microbiol. Lett. 6: 287–292.

Lovley, D. R., 1987. Organic matter mineralization with the reduction of ferric iron: a review. Geomicrobiol. J. 5: 375–399.

Lovley, D. R., 1991. Dissimilatory Fe(III) and Mn(IV) reduction. Microbiol. Rev. 55: 259–287.

Lovley, D. R. & E. J. P. Phillips, 1986. Organic matter mineralization with reduction of ferric iron in anaerobic sediments. Appl. Envir. Microbiol. 51: 683–689.

Lovley, D. R. & E. J. P. Phillips, 1987a. Rapid assay for microbially reducible ferric iron in aquatic sediments. Appl. Envir. Microbiol. 53: 1536–1540.

Lovley, D. R. & E. J. Phillips, 1987b. Competitive mechanisms for inhibition of sulfate reduction and methane production in the zone of ferric iron reduction in sediments. Appl. Envir. Microbiol. 53: 2636–2641.

Lovley, D. R. & E. J. P. Phillips, 1988. Novel mode of microbial energy metabolism: organic carbon oxidation coupled to dissimilatory reduction of iron or manganese. Appl. Envir. Microbiol. 54: 1472–1480.

Lovley, D. R., E. J. P. Phillips & F. Caccavo, 1992. Acetate oxidation by dissimilatory Fe(III) reducers. Appl. Envir. Microbiol. 58: 3205–3206.

Oremland, R. S. & D. G. Capone, 1988. Use of 'specific' inhibitors in biogeochemistry and microbial ecology. Adv. Microb. Ecol. 10: 285–383.

Oremland, R. S. & S. Polcin, 1982. Methanogenesis and sulfate reduction: competitive and noncompetitive substrates in estuarine sediments. Appl. Envir. Microbiol. 44: 1270–1276.

Pettersson, K., B. Bostrom & O.-S. Jacobsen, 1988. Phosphorus in sediments – speciation and analysis. Hydrobiologia 170: 91–101.

Phillips, E. J. P., D. R. Lovley & E. E. Roden, 1993. Composition of non-microbially reducible Fe(III) in aquatic sediments. Appl. Envir. Microbiol. 59: 2727–2729.

Pugnetti, A., P. Viaroli & I. Ferrari, 1992. Processes leading to dystrophy in a Po River delta lagoon (Sacca Di Goro): phytoplankton-macroalgae interactions. Science Tot. Envir. Suppl. 1992: 445–456.

Reynolds, A. & A. E. Walsby, 1975. Water-blooms. Biol. Rev. 50: 437–481.

Riegman, R., A. A. M. Noordeloos & G. C. Cadee, 1992. *Phaeocystis* blooms and eutrophication of the continental coastal zones of the North Sea. Mar. Biol. 112: 479–484.

Robertson, B. K. & M. Alexander, 1992. Influence of calcium, iron, and pH on phosphate availability for microbial mineralization of organic chemicals. Appl. Envir. Microbiol. 58: 38–41.

Skyring, G. W., 1987. Sulfate reduction in coastal ecosystems. Geomicrobiol. J. 5: 295–374.

Stal, 1994. Microbial mats in coastal environments. In L. J. Stal & P. Caumette (eds), Microbial mats. Structure, development and environmental significance. NATO ASI Series 35 Springer Verlag, Heidelberg: 21–32.

Stal, 1995. Physiological ecology of cyanobacteria in microbial mats and other communities. New Phytol. 131: 1–32.

Stal, L. J., H. W. Paerl, B. Bebout & M. Villbrandt, 1994. Heterocystous versus non- -heterocystous cyanobacteria in microbial mats. In L. J. Stal & P. Caumette (eds), Microbial mats. Structure, development and environmental significance. NATO ASI Series 35 Springer Verlag, Heidelberg: 403–414.

Sukenik, A., W. Schroeder, J. Lauer, G. Shelef & C. J. Soeder, 1985. Coprecipitation of microalgal biomass with calcium and phosphate ions. Water Res. 19: 127–129.

*Hydrobiologia* **329**: 199–210, 1996.
*P. Caumette, J. Castel & R. Herbert (eds), Coastal Lagoon Eutrophication and ANaerobic Processes (C.L.E.AN.).*



# Sulfur bacteria in sediments of two coastal ecosystems: the Bassin d'Arcachon and the Etang du Prévost, France

Bartholomeus E. M. Schaub[2] & Hans van Gemerden[1]
*Department of Microbiology, University of Groningen, Kerklaan 30, NL-9751 NN Haren, The Netherlands*
*[1]Corresponding author [2]Present address: Laboratory of Microbiology, University of Amsterdam, Nieuwe Achtergracht 127, 1018 WS Amsterdam, the Netherlands.*

*Key words:* Sulfur bacteria, MPN counts, vertical distribution, marine sediments, oxygen, sulfide, eutrophication

## Abstract

Enumeration of the functional groups of sulfur bacteria was performed in the sediments in the Bassin d'Arcachon, a mesotidal lagoon with strong tidal currents and dominant populations of seagrass (*Zostera noltii*), and in the Etang du Prévost, a shallow lagoon with moderate tidal fluctuations and dominant populations of floating seaweed (*Ulva* sp.). In addition, data were collected on the distribution of oxygen and sulfide at the water-sediment interface during diel cycles. Bacterial enumeration studies revealed highest numbers in the top two cm of the sediments for three functional groups of sulfur bacteria, these being the sulfate-reducing bacteria (SRB), the colorless sulfur bacteria (CSB), and the phototrophic sulfur bacteria (PSB). In both systems high numbers of SRB were encountered, suggesting ample availability of organic matter. A comparison between different sites in each ecosystem showed that sediments overlain by more stagnant water were dominated by PSB, whereas those overlain by more oxygenated water were dominated by CSB. Important factors are the physical forces induced by tidal currents and the degree of daily exchange of water between the lagoons and the sea. These factors may explain the differences observed between the two systems with regard to the development of anoxic conditions, more so than the level of eutrophication. It appears that rooted plants play an important role in the introduction of oxygen into the sediments, thus enhancing the competitive position of CSB compared to PSB. Mini-electrodes studies revealed high concentrations of free sulfide at the inner site of the Etang du Prévost but very low concentrations at the inner station of the Bassin d'Arcachon, which may be explained by the high iron input of the latter, rather than by differences in the rate of sulfide production.

## Introduction

In the present study aspects of the sulfur cycle in two eutrophic ecosystems were compared. In the Bassin d'Arcachon, a mesotidal lagoon (i.e. lagoon with a maximum tidal amplitude of 4 m), situated on the French Atlantic coast, tidal currents are strong, resulting in a substantial daily interchange of water with the ocean. On the intertidal flats in this system the eelgrasses *Zostera noltii* Hornem. and *Ruppia* sp. are present in abundance, while the green algae *Monostroma* sp is increasingly colonizing the basin. Decaying seagrass and algae, and faecal products from the abundantly present fauna, provide a rich supply of organic matter to the sediments.

In the second ecosystem studied, the Etang du Prévost, a different situation is encountered. Due to the low tidal amplitude and the relatively small channel (Grau) between the Etang and the Mediterranean the exchange of water is extremely small compared to that observed between the Bassin d'Arcachon and the Atlantic Ocean. The Etang du Prévost receives a heavy load of nutrients from agricultural run-off and sewage. The dominant flora comprises the marine macro alga *Ulva lactuca*, which develops profusely during the growing season. In the water column anoxic conditions occur frequently during this period, when oxygen-consuming processes in the water column exceed the supply of oxygen by production and diffusion. This often coincides with the presence of sulfide in the water column, resulting in suspended elemental sulfur ('white water'). During these periods, blooms of phototrophic sulfur bacteria in the water column are frequently observed (Caumette, 1987). In terms of eutrophication, the Etang du Prévost is considered to be more eutrophic than the Bassin d'Arcachon.

In coastal environments, when high mineralization and respiration rates prevail, oxygen often becomes limiting. Under these conditions a shift to alternative electron acceptors is observed of which sulfate, due to its abundance in the marine environment, is the most important. In anoxic coastal sediments a considerable part of the organic matter (>50%) is mineralized via sulfate reduction by sulfate-reducing bacteria (SRB)(Jørgensen, 1978; Thode Andersen & Jørgensen, 1989). The $H_2S$ produced is extremely toxic for most aerobic organisms; abiotically it can be trapped by iron and precipitated as ferrous sulfide or $FeS_2$ (pyrite). Sulfide can also be oxidized by two functional groups of sulfur bacteria: the colorless sulfur bacteria (CSB) and the phototrophic sulfur bacteria (PSB). The availability of electron acceptors, such as oxygen or nitrate, is important for the growth of CSB, whereas phototrophic growth of PSB depends on the availability of light. Alternatively, PSB may use oxygen during chemotrophic growth as has been demonstrated for *Thiocapsa roseopersicina* (Bogorov, 1974; De Wit & Van Gemerden, 1987), and other Chromatiaceae (Kämpf & Pfennig, 1980, 1986; Overmann & Pfennig, 1992). Similarly the okenone-containing *Thiocapsa* strain 5811, isolated from the Etang du Prévost has the potential for chemotrophic growth (Caumette et al., 1985).

The abundance of these groups is largely determined by the environmental factors described above. Sulfide oxidation by CSB and PSB is considered to play an important role in these environments, in addition to sulfide removal by FeS formation.

As part of the research programme for the Coastal Lagoon Eutrophication and ANaerobic processes (C.L.E.AN.), a project on the abundance of three functional groups of sulfur bacteria (SRB, CSB, and PSB) in the sediments of the two lagoons was carried. The hypothesis was put forward that in the Etang du Prévost these bacterial groups are more important than in the Bassin d'Arcachon, on the basis of the occurrence of 'white water' and blooms of phototrophic sulfur bacteria. Data were collected on the vertical distribution of oxygen and sulfide in both systems in the sediment and at the water-sediment interface during a full diel cycle. Special attention was paid to the impact of *Zostera* beds on the distribution of oxygen in the sediments and the influence there of on the vertical distribution of CSB.

## Materials and methods

### *Description of systems and sites*

The systems under study are both situated in France, the Arcachon Basin (Bassin d'Arcachon) is situated 80 km south-west of Bordeaux at the Atlantic coast, the Prévost lagoon (Etang du Prévost) is situated 40 km south-west of Montpellier on the Mediterranean.

The Bassin d'Arcachon is a triangular-shaped mesotidal lagoon with a total area of 156 $km^2$. At low tide only 40 $km^2$ of the basin is covered by water, from which an area of 10 $km^2$ is used for oyster farming. In the system, the tidal amplitude ranges from 3.90 at spring tide to 2.10 m at neap tide. The Etang du Prévost is a relatively shallow lagoon covering an area of 4.8 $km^2$ with a water depth ranging from 1.5 to 2 m. In each system two locations were chosen, one relatively close to the channel through which water exchange takes place (the outer station), and one distant from the channel (the inner station). In the Bassin d'Arcachon, the outer station (Station A) was situated on an intertidal flat in the center of the basin where coastal water was well mixed with basin water during each tidal cycle. The sediment at this site consisted of muddy sand. Station B was located in the intertidal area in the eastern part of the basin where little water exchange with coastal waters occurs. Sediment composition at this site was predominantly black mud mixed with some sand. The sampling sites A and B were both situated in eelgrass beds.

In the Etang du Prévost samples were taken at Stations X and 11. Station X was situated on a tidal flat near the channel (Grau) to the Mediterranean. This station was devoid of a well developed flora. The sediment consisted of grey-colored coarse sand. The second sampling site (Station 11) was situated in the western part of lagoon where the exchange of water is minimal. In these sediments (silt mixed with sand) a significant infauna was present, revealed by the large quantities of shells. At this station a dense population of floating algae was encountered, consisting predominantly of *Ulva* sp. Station 11 was continuously submersed and situated at a depth of approximately 50 cm below the water surface.

More detailed information on the different sampling stations is given in the introductory chapter of this volume. For data on seasonal variations the reader is referred to Bourguès et al. (1995) and De Wit et al. (1994).

*Enumeration of functional groups, MPN methodology*

Data on the vertical distribution of CSB, PSB and SRB were collected in the Bassin d'Arcachon in June and September 1993, and June 1994. The Etang du Prévost was sampled only during the two field campaigns in 1993.

Sediment samples were taken using 3.5 cm diameter stainless steel cores. These cores were either taken directly to the laboratory and processed immediately upon arrival, or kept refrigerated and processed within one day. Based on introductory experiments, cores were sliced in six consecutive layers, corresponding to depth layers of 0–5 mm, 5–10 mm, 10–15 mm, 15–20 mm, 20–30 mm and 30–50 mm. Each slice was suspended in sterilized, filtered and oxygen-free aged seawater (1:4 vol/vol) and subsequently subjected to mild ultrasonic treatment (15 Hz for 30 s, Branson B3 bath) in order to detach the cells from sediment particles and to loosen clumps. The resulting slurry, which was used for the enumeration of all three functional groups of sulfur bacteria, was directly pipetted in the first row of three microtiter plates (12 rows of eight wells, each maximally containing 250 $\mu$l). The remaining wells were previously filled with 200 $\mu$l of growth medium for the group to be enumerated. Ten-fold dilution steps were used. Microtiter plates for PSB and SRB were incubated in an oxygen-free environment in gas-tight plastic bags (Anaerocult system, Merck) and incubated at room temperature in incandescent light and in darkness, respectively. Medium composition for the enumeration of the different functional groups was as described previously (Visscher et al., 1992a). Growth of PSB was assessed by pigmentation of the wells, growth of CSB was judged from acid formation and turbidity, whereas for the presence of SRB blackening of the growth medium due to FeS precipitation was taken as the criterion. Growth was checked frequently and final scores were obtained after six to ten weeks of incubation. The most probable numbers and 95% confidence limits were calculated according to the method of Klee (1993).

*Vertical distribution of oxygen and sulfide*

In 1994, data were collected during two field campaigns. During the June campaign attention was focussed on the possibility of sulfide re-oxidation in *Zostera* beds at station B in the Bassin d'Arcachon, while during the August campaign most attention was paid to the measurement of oxygen and sulfide profiles during full diel cycles.

The small-scale sensors for the measurement of oxygen and sulfide were sheathed in stainless steel needles (outer diameter 0.9 mm) in order to prevent damage to the sensors by shells and other small objects. Particularly in the sediments of Station 11 in the Etang du Prévost are shells abundantly present. Details of the construction and calibration of the mini-electrodes are given in Van Gemerden et al. (1989) and Visscher et al. (1991).

During the June 1994 campaign, a platform was installed at station B in an area dominated by eelgrass. At each side of the platform a 0.75 × 2 m area was selected for sampling: in one part *Zostera* leaves were carefully removed without removing the roots, while in the other part *Zostera* plants were left intact. During periods with no overlying water, cores were taken at regular intervals at each side of the platform. Oxygen- and sulfide profiles were recorded at depth intervals of 0.1–1 mm within 15 minutes after sampling, using a custom made nanoAmp- and mV-meter (Electronic Workshop, Biology Department, University of Groningen).

During the August campaign electrode studies were performed using a computer controlled submersible device developed for *in situ* measurements in shallow marine environments (Mechanical and Electronic Workshops, Biology Department, University of Groningen). The software for the interactive control was developed by TiePie Engineering, Jornwerd, The Netherlands. A brief description follows, details of the system will be published separately. On a 1-m high stainless steel frame, an electrode block was mounted, which can be moved horizontally, along an X-and an Y-axis for a distance of 50 cm. Vertically, the maximum range was 30 cm with a minimal resolution of 0.1 mm. Profiles were recorded after a depth sensor, mounted on the electrode block, has detected the sediment surface. Needle electrodes for oxygen, sulfide, pH, redox and temperature (four each) were mounted on the electrode block. Amplifiers and nanoAmp-mV converters were mounted directly on top of the electrodes and millivolt signals sent to an on-land computer for further analysis.

During a diel cycle, measurements were repeated each hour and the horizontal spacing between measurements was 2.5 cm.

## Results

### *Bassin d'Arcachon*

*Station A*

Data on the vertical distributions of sulfur bacteria in the sediments of the Bassin d'Arcachon are presented in Figure 1.

At station A (Figure 1, left-hand panels) CSB were present in relatively high numbers with cell densities up to $10^9$ cells $cm^{-3}$ sediment in the upper 5 mm of the sediment. The average cell count in the top two cm was $6 \times 10^8$ $cm^{-3}$, which accounted for 99.5% of the total population of CSB in the 0–5 cm depth layer. In the 30–50 mm layer maximum counts were $2 \times 10^6$ cells $cm^{-3}$ sediment. PSB were detected in all depth layers from 0–5 cm, however, numbers were low when compared to CSB. Maximum population densities were found in the 10–15 mm depth layer ($3 \times 10^5$ cells $cm^{-3}$), cell densities in the other depth layers were two orders of magnitude lower. The average count in the top two cm of the sediment was $8 \times 10^4$ cells $cm^{-3}$, accounting for 98% of the population from 0–5 cm. At this site the CSB clearly outnumbered the PSB. The same was found during the campaigns in June 1993 and June 1994. For a proper evaluation of functional groups, the average individual cell size of each groups has to be taken into account, for PSB this value is approximately 1 0-times higher than for CSB (Visscher & Van Gemerden, 1993; Van den Ende et al., 1995). At Station A, the total biomass of CSB thus was approximately 100-fold higher than that of the PSB.

Highest population densities of SRB also were found in the 0–2 cm layer ($2 \times 10^7$ cells $cm^{-3}$), representing 90% of the population present in the layer of 0–5 cm. In the deeper layers cell numbers decreased to $10^6$ $cm^{-3}$.

*Station B*

The vertical distribution, as well as the abundance, of the three groups of sulfur bacteria was comparable to Station A, i.e. highest population densities were found in the upper two cm of the sediments (Figure 1, right-hand panels). Maximum cell densities for CSB were found in the 5–10 mm depth layer ($7 \times 10^8$ cells $cm^{-3}$), but counts were still $10^6$ $cm^{-3}$ in the 30–50 mm layers.

As at Station A, low numbers of PSB were encountered. The average population density in the 0–2 cm layer, containing 70% of the cells in the 0–5 cm layer, was $3 \times 10^3$ cells $cm^{-3}$. As at Station A, CSB dominated over the PSB.

For the SRB, maximum counts were obtained in the 10–15 mm depth layer ($6 \times 10^7$ cells $cm^{-3}$). As in Station A, 90% of the population of this group revealed to be present in the 0–2 cm depth layer, numbers close to $10^6$ $cm^{-3}$ were observed in the deeper layers.

In June 1994 a comparison was made between the numbers of CSB and PSB inside and outside the *Zostera* beds. In both conditions, the majority of CSB and PSB were present in the upper 20 mm of the sediments, for CSB being 96% and 93%, respectively, and for PSB being 84% and 98%, respectively. For CSB, the average count for the top 20 mm inside the *Zostera* bed was $4.6 \times 10^6$ $cm^{-3}$, and outside the seagrass bed $30.4 \times 10^6$ $cm^{-3}$. For the PSB the corresponding numbers were $2.1 \times 10^3$ $cm^{-3}$ and $2.0 \times 10^6$ $cm^{-3}$, respectively. The impact of the seagrass on the abundance of sulfide-oxidizing bacteria is elegantly illustrated by comparing the ratio CSB/PSB on a biomass basis. Outside the *Zostera* beds the contribution of CSB to the total biovolume of the populations of CSB and PSB was 60.8%, inside the *Zostera* beds the corresponding figure was 99.5%. It thus appears that the competitive position of colorless sulfur bacteria was enhanced in seagrass beds.

During the June campaign 1994 data were collected on the profiles of oxygen and sulfide in sediment covered with *Zostera* sp. (Figure 2, left-hand panels) and in an area from which the leaves of *Zostera* were removed while the root system was left intact (Figure 2, right-hand panels). Cores were taken at 2-h intervals, except when the sampling site was submerged. The differences observed between four different electrodes used for each parameter were minimal, except for 18:50 h when one electrode showed much deeper penetration of oxygen than the others. At the surface of the *Zostera*-covered area, oxygen concentrations were maximal in the late afternoon (200–300 $\mu$mol $l^{-1}$), but otherwise showed little variation with time. The depth of oxygen penetration was <5 mm during the night and up to 10 mm during the day. The removal of the *Zostera* leaves resulted in lower maximum oxygen concentrations in the superficial layers, and in much shallower penetration. No clear differences were observed

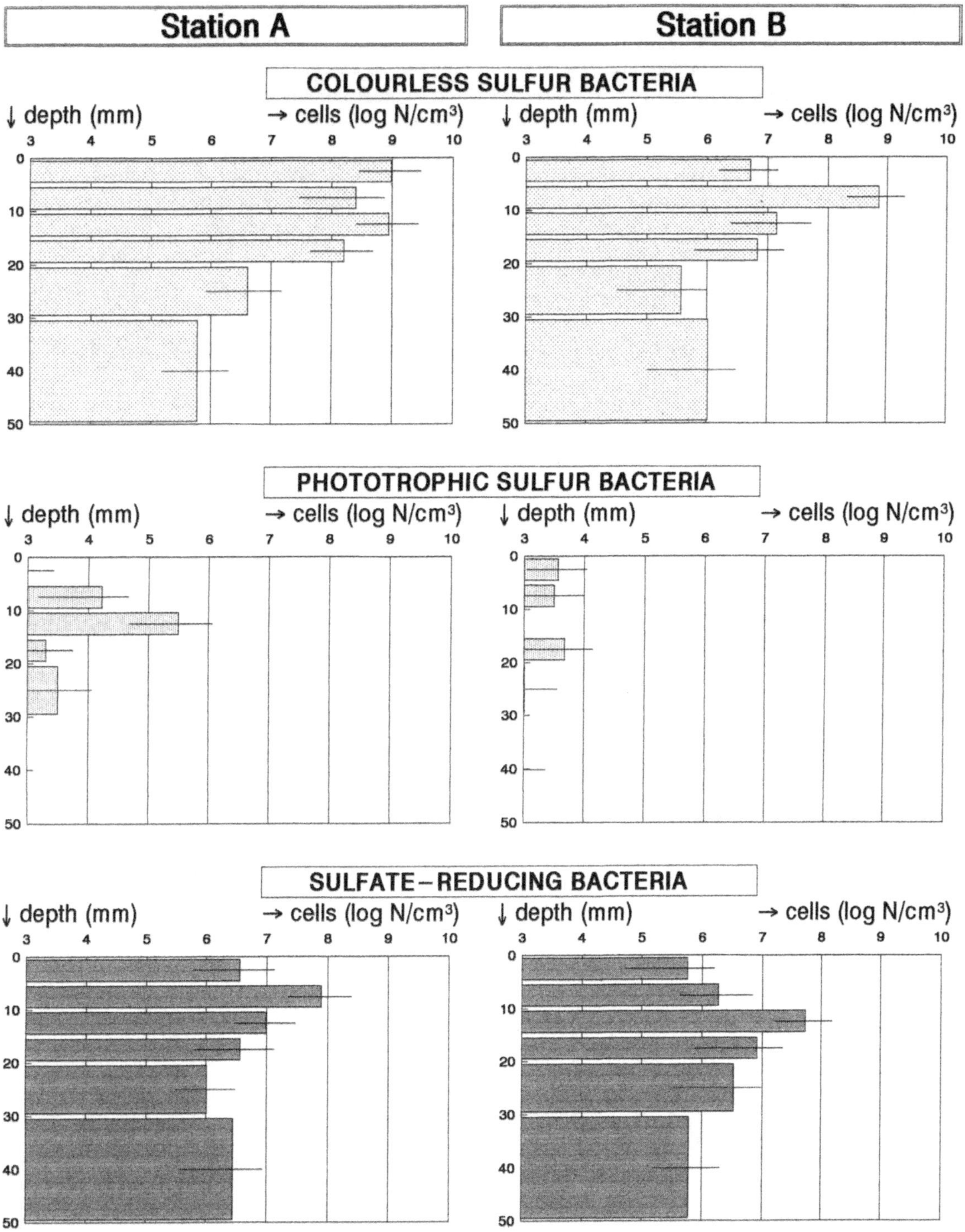


*Figure 1.* Vertical distribution of sulfur bacteria counted by the MPN-method in the top 50 mm at the outer station (Station A) and the inner station (Station B) in the Bassin d'Arcachon. Cores were collected on September $3^{rd}$ 1993. Sulfate-reducing bacteria were enumerated with lactate and acetate; colorless bacteria with thiosulfate and anoxigenic phototrophic sulfur bacteria with sulfide and thiosulfate. Error bars indicate 95% confidence intervals after Klee (1993).

at different times, indicating that oxygen production was primarily due to *Zostera* together with the epiphytes removed with the leaves of the latter. It thus appears that sediments with intact *Zostera* plant are more oxygenated than those from which the leaves were removed. Free sulfide could not be detected, neither underneath the *Zostera* beds, nor in the sediment from which the *Zostera* leaves were removed. The mV-signal of the sulfide electrode did not allow calculation of concentrations, nevertheless the trend observed was that sediments without *Zostera* leaves were more reduced than the sediments without *Zostera* leaves (Figure 2).

In August 1994, data were collected at 1-h intervals on the distribution of oxygen and sulfide in the *Zostera* beds during a full diel cycle. Oxygen concentrations, in general, were lower than in June (data not shown). The oxygen concentration in the 10 mm water phase just above the sediments and in between the *Zostera* leaves ranged from 25 to 150 $\mu$mol $l^{-1}$, indicating under-saturation. Maximum oxygen concentrations were observed in the late afternoon, while minimum concentrations were found between midnight and early morning. At all times, concentrations in the sediments gradually decreased to zero. Again, oxygen penetration depth was highly variable and ranged from only a few mm to more then 10 mm. On one occasion, concentrations up to 50 $\mu$mol $l^{-1}$ $O_2$ were observed in the deeper depth layers (8–30 mm) while oxygen in the 6–8 mm was absent. This phenomena, is probably due to the presence of *Zostera* roots. In contrast to measurements performed in September 1993 and June 1994, sulfide could be detected in the sediments of Station B, although at very low concentrations (<6 $\mu$mol $l^{-1}$ $S^{2-}$).

*Etang du Prévost*

*Station X*

The vertical distribution of sulfur bacteria at the outer Station X is shown in Figure 3 (left-hand panels).

CSB attained maximum population densities in the upper 3 cm of the sediment with an average density of $10^7$ cells $cm^{-3}$, whereas in the 30–50 mm layer the population density was two orders of magnitude lower. Of the entire population in the top 50 mm, 75% was found in the top 20 mm.

The population of PSB exceeded that observed in the Bassin d'Arcachon and was rather uniformly ($\approx 5 \times 10^4$ cells $cm^{-3}$ sediment) spread over the entire depth layer analyzed.

Although maximum numbers of SRB were found in the upper cm of the sediment ($10^6$ cells $cm^{-3}$), a substantial number of viable cells was present in the deeper layers, resulting in an average of $3 \times 10^5$ cells $cm^{-3}$ over the entire depth horizon from 0–50 mm.

In August 1994 depth profile data were collected on the distribution of oxygen and sulfide. Oxygen concentrations in the water phase just above the sediment ranged from 75 $\mu$mol $l^{-1}$ to 250 $\mu$mol $l^{-1}$. Concentrations were low in the early morning and progressively increased during the day to become maximal in the afternoon. The oxygen penetration depth into the sediments was fairly constant over time. In general, during the day anoxic conditions were observed at depths exceeding 5 mm, during the night oxygen penetration was less. The maximum oxygen concentrations at this site never exceeded 100% saturation levels. During the 24-h period sulfide was detected incidentally at very low concentrations.

*Station 11*

At Station 11 (Figure 3, right-hand panels), $8 \times 10^7$ cells $cm^{-3}$ of CSB were found in the top 5 mm of the sediment. The average number in the top 20 mm was $2 \times 10^7$ cells $cm^{-3}$ sediment, accounting for 90% of the viable population of CSB from 0–5 cm.

Relatively high PSB numbers ($5 \times 10^7$ $cm^{-3}$) were found in the top 10 mm of the sediment, and 96% of the viable population was enumerated in the top 20 mm. At this location, PSB were more numerous than CSB. When the numbers were transformed to biomass PSB dominated the CSB by a factor of 11. Highest cell densities for the SRB likewise were encountered in the top layers, with a maximum of $6 \times 10^8$ cells $cm^{-3}$ sediment in the upper 5 mm. The numbers of SRB showed a remarkable decrease of two to three orders of magnitude between the upper 5 mm and the 5–10 mm layer. The high numbers of SRB in the surface layers of the sediment may explain the formation of 'white water' observed in the days before sampling. If so, it may reflect a faster response of SRB to an increase in substrate availability when compared to the response of CSB and PSB to increased sulfide concentrations.

Free sulfide was clearly present in Station 11, the maximum concentration observed was 6 mmol $l^{-1}$. The data on oxygen concentrations in the water phase were comparable to those measured at Station X in the Etang du Prévost, however the pattern in the diel cycle

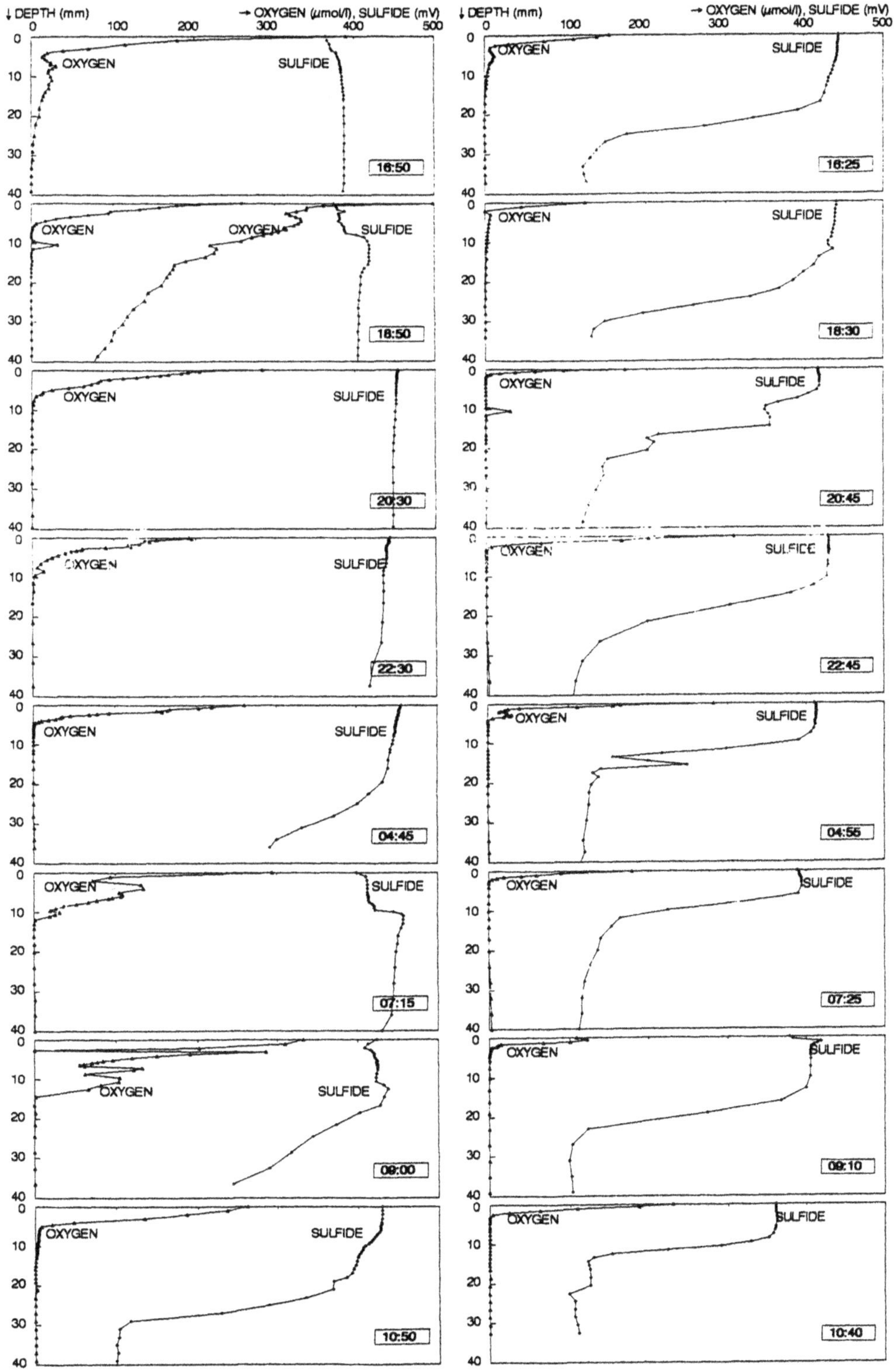


*Figure 2.* Concentration of oxygen and signal of sulfide needle sensor in cores taken in a *Zostera* bed at different times at Station B in the Bassin d'Arcachon. Left column: intact *Zostera* plants; right column: leaves of *Zostera* cut off on June 2 1994 at 15–16 h. Data were collected on 2–3 June 1994. High tides were at 0:23 h and 13:05 at June 2, and at 1:33 and 14:18 on June 3. Low tides were at 6:19 and 18:49 at June 2, and at 7:23 and 19:54 on June 3.

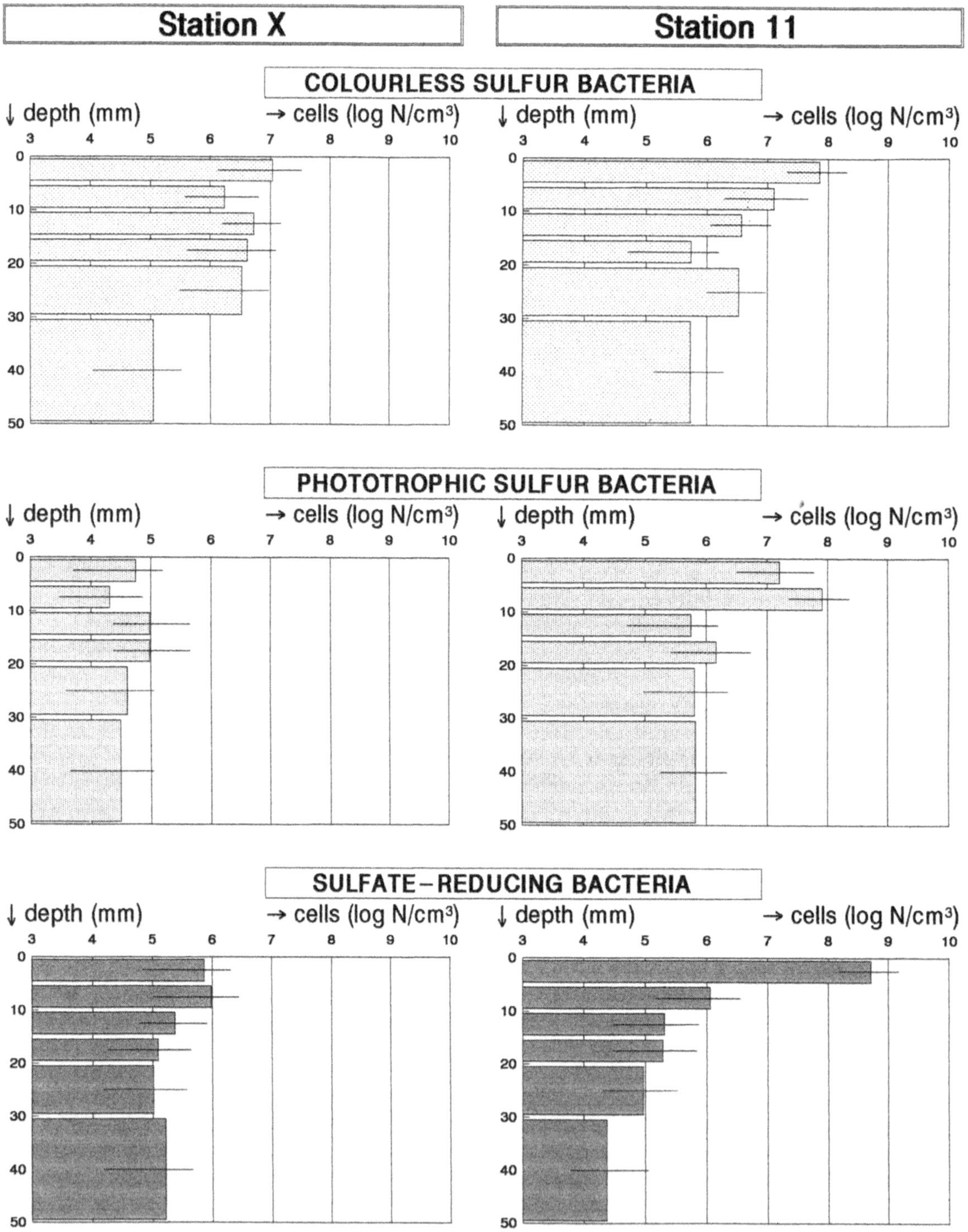


*Figure 3.* Vertical distribution of sulfur bacteria counted by the MPN-method in the top 50 mm at the outer station (Station X) and the inner station (Station 11) in the Etang du Prévost. Cores were collected on September $6^{th}$ 1993. Sulfate-reducing bacteria were enumerated with lactate and acetate; colorless bacteria with thiosulfate and anoxigenic phototrophic sulfur bacteria with sulfide and thiosulfate. Error bars indicate 95% confidence intervals after Klee (1993).

was more pronounced. During the night concentrations decreased progressively to reach lowest values in the very early morning ($\approx 10$ $\mu$mol $l^{-1}$ at 10 mm above the water-sediment interface), while highest oxygen concentrations were observed between 18 and 20 h.

## Discussion

### *Vertical distribution and abundance of sulfur bacteria*

Two principal conclusions can be drawn from the data on the vertical distribution of sulfur bacteria. Firstly, the substantial numbers observed for the colorless sulfur bacteria (CSB), the phototrophic sulfur bacteria (PSB), and the sulfate-reducing bacteria (SRB), demonstrate that the sulfur cycle is of major importance both in the Bassin d'Arcachon and the Etang du Prévost. When cell counts were transposed to biomass, it was generally observed that the sum of the biomass of CSB and PSB was roughly proportional to the biomass of the SRB. Secondly, the fact that the bulk of the population of all three groups were found in the surficial layers of the sediments, indicates that the processes primarily take place in the top layers of the sediments.

The high population density of SRB measured at the continuously submerged Station 11 (Etang du Prévost), probably was the consequence of the massive decay of macroalgae in the days before sampling. The observed 'white water' could be the result of sulfur particles, either produced biotically by CSB and/or PSB, or abiotically. However, carbonate precipitation may also have contributed to the formation of 'white water' (Caumette & Baleux, 1980).

The population densities of sulfur bacteria observed in general are high compared to other observations. Caumette (1987) reported maximum numbers of SRB and PSB at an inner station in Etang du Prévost in the order of $10^6$ $cm^{-3}$ and $10^5$ $cm^{-3}$, whereas in the inner station of Bassin d'Arcachon PSB counts maximally were $3 \times 10^4$ $ml^{-1}$, respectively. In the Ems-Dollard Estuary maximum numbers of SRB were present just below the oxic/anoxic and accounted to $10^4$ colony-forming-unit (Laanbroek & Pfennig, 1981), whereas in the Kattegat the populations of SRB were reported to be no higher than $10^6$ $cm^{-3}$ (Jørgensen & Bak, 1991). However, higher numbers were observed in a well developed *Microcoleus*-dominated microbial mat on the island of Texel (The Netherlands), in which the population densities of CSB, PSB, and SRB respectively were $2 \times 10^9$ $cm^{-3}$, $5 \times 10^7$ $cm^{-3}$, and $5 \times 10^3$ $cm^{-3}$ (Visscher et al., 1992a; Visscher & Van Gemerden, 1993).

The MPN counts obtained, particularly those of SRB, are generally assumed to be gross underestimates due to the use of highly selective media (Herbert, 1985). Recovery percentages as low as 0.011% have been reported (Gibson et al., 1987), however, a careful choice of conditions, in particular with regard to the composition of the growth medium, resulted in much higher recoveries (Bak & Pfennig, 1991). Further improvements were obtained by using water from the sampling site in the preparation of media (Visscher et al., 1992a).

High numbers of SRB were not only found at the inner station in the Etang du Prévost, but also at the sampling sites in the Bassin d'Arcachon. This may be regarded as indicative for the importance of the input of organic matter. The prominent role of SRB in the mineralization of organic matter is demonstrated by the high rates of sulfate reduction observed in the zones colonized by *Zostera* sp. and *Ruppia* sp. (Finster et al., 1994). No significant accumulation of reduced sulfur compounds was observed, suggesting that most of the sulfide produced by SRB was re-oxidized, either biotically or abiotically. However, the rate of oxygen uptake, as measured by the diffusive flux of oxygen into the sediments, was too low to account for the re-oxidation. In order to explain this discrepancy, Finster et al. (1994) proposed that rooted plants play an important role in the influx of oxygen in these 'anoxic' sediments. In the sediments of stations devoid of rooted plants (e.g. Station 11, Etang du Prévost), bioturbation by bivalves may act as the major way to introduce oxygen into the sediment.

The hypothesis proposed by Finster et al. (1994) may also explain the high numbers of CSB observed at sites colonized by eelgrass (Figure 1), since oxygen is a key environmental factor for the development of CSB.

The simultaneous presence of light and reduced sulfur compounds is favourable for the development of phototrophic sulfur bacteria. The presence of PSB in cell densities upto $8 \times 10^7$ $cm^{-3}$ and $3 \times 10^6$ $cm^{-3}$ in the sediments of the Etang du Prévost and the Bassin d'Arcachon, respectively, may be explained in this way. The re-oxidation of sulfide without the availability of sufficient oxygen (Finster et al., 1994), could be explained by the presence of phototrophically growing PSB, although growth and activities of PSB in these ecosystems likely are restricted by the limited availability of light. Tidal movements, resulting in resuspension of sediment particles create a shadowing effect. Also the small particle size of the sediment and the presence of high concentrations of FeS (data not shown; Stal et al. 1994) are likely to reduce the light intensity in deeper layers.Light penetration in sediments is restricted to the upper few mm (Jørgensen, 1989; Lassen et al., 1992; Lassen & De Wit, 1994;). At station B, the shadowing effects of dense *Zostera* coverage, results in low photosynthetically available radiation (PAR) in the surface layers of the sediments, however, near infrared radiation was less affected (Lassen et al., 1994). The significance of the very low numbers of PSB (1% of the surface abundance) detected in the deeper layers of the sediments, is marginal. Little is known about the survival characteristics of PSB under the prevailing environmental conditions. Possibly, bioturbation could be responsible for the presence of phototrophic bacteria in deep and dark depth horizons.

*Profiles of oxygen and sulfide*

Oxygen and sulfide are parameters of crucial importance, either stimulating or inhibiting the development of CSB and PSB. In general, oxygen is required for the proliferation of CSB. Nitrate, the alternative electron acceptor for these organisms, is only present at very low concentrations and, consequently, nitrate plays a minor role (Herbert & Welsh, 1994; Sloth et al., 1994). In addition to oxygen, CSB require sulfide, or another form of reduced sulfur, as electron donor.

CSB and PSB, although competitors for reduced sulfur compounds, might not be expected to thrive in the same habitat since CSB require oxygen, whereas PSB primarily are anoxygenic phototrophs whose pigment synthesis is inhibited by oxygen (De Wit & Van Gemerden 1990a, b). However, when oxygen concentrations show (diel) fluctuations and/or are limiting co-existence is feasible. The co-occurrence of CSB and PSB in these sediments can be explained by the phenomenon that PSB can oxidize intermediate forms of sulfur – such as thiosulfate, elemental sulfur, and poly-sulfide – which are produced by CSB from sulfide upon oxygen limitation (Van Den Ende and Van Gemerden, 1993, 1994). It is evident from the data presented in Figures 1 and 2 that, the conditions required by CSB for growth were met in sediments covered with *Zostera* (Station A and B, Bassin d'Arcachon). These data infer that oxygen was introduced into the sediments via the eelgrass roots.

The oxygen concentrations observed in June 1994 at the surface of the sediments in Station B, were higher then those recorded in August. The community at this site was less healthy in August compared to June, when oxygen saturating conditions were measured at the sediment surface. At the end of the *Zostera* growing season (August–September), oxygen concentrations in the sediment between the *Zostera* leaves were lower, which can be explained by a higher oxygen demand resulting from the accumulation of organic debris and decomposition the *Zostera* leaves. In August–September low oxygen concentrations or even anoxia may be expected, and red colored patches of phototrophic sulfur bacteria have been observed between decaying *Zostera* leaves (P. J. Labourg, pers. comm.).

At Station 11, oxygen concentrations in the water phase were below 25 $\mu$mol $l^{-1}$ during the night. At this site the pelagic alga *Ulva*, whose standing crop may exceed 500 g dry weight $m^{-2}$ (Viaroli et al., 1994) constitutes an important source of organic matter, and respiration and mineralization results in a high oxygen demand. Comparable low oxygen concentrations were observed in 1993, and appear to be the result of increased respiration rates, and not decreased rates of primary production (Viaroli et al., 1994).

At station X, variations in oxygen concentration during the diel cycle were less pronounced. The organic matter content of the sediments was low in comparison to site 11 (De Wit et al., 1994). The absence of a dense *Ulva* community and the location adjacent to the exchange channel may be important factors in this respect.

In general, oxygen penetration into the sediments of the Etang du Prévost was limited to the upper 10 mm of the sediments. However, oxygen penetration may extend to depths of 20 mm or more. Deeper oxygen penetration may result from the presence of roots (e.g. *Zostera*), animal burrows, or sea water seeping downwards through the sediments due to tidal movements. Although tidal differences are small in the Etang du Prévost (about 20 cm), the coarse sediment will result in exchange phenomena. In the Etang du Prévost animal burrows may be the most likely explanation. Furthermore the surface structure of the sediments is highly variable in these systems particularly at Station 11 (Etang du Prévost).

The hypothesis put forward by Finster et al. (1994), that *Zostera* plants enrich the superficial layers of sediments with oxygen, is substantiated by the vertical distribution of CSB inside and outside the *Zostera* beds.

Removal of *Zostera* leaves, which prevents transport of oxygen from the leaves to the roots, was found to coincide with reduced concentrations of oxygen in the surface sediment layers, and, more importantly, to result in reduced depth of penetration of oxygen (Figure 2). Although numbers of CSB inside and outside *Zostera* beds were not substantially different, the ratio CSB/PSB was higher inside compared to outside *Zostera* beds. CSB were found to dominate ecosystems covered with *Zostera* (Stations A and B), whereas PSB dominated systems devoid of *Zostera* Beds (Station B and 11).

The high oxygen-uptake rates of these systems (Sloth et al., 1994) are in agreement with the dense populations of CSB, although aerobic heterotrophic organisms can be expected to have contributed. Since the complete oxidation of the sulfide formed by the high sulfate-reduction rates observed did not correspond to the uptake of oxygen, Finster et al. (1994) postulated that sulfide was only partially oxidized, resulting in the temporary accumulation of reduced intermediates.

In this study the hypothesis that cutting of the *Zostera* leaves might limit aerobic respiration by eliminating an important way of introducing oxygen into the sediment was tested. Although the sediment became more reduced free sulfide did not accumulate under these conditions.

Free sulfide, in considerable concentrations, was only observed at Station 11 in the Etang du Prévost, and not in the sediments of Station B in the Bassin d'Arcachon. While little organic matter was found at Station X, Station B was characterized by high organic matter content and high rates of sulfate reduction (De Wit et al., 1994; Finster et al., 1994). At Station 11 (Etang du Prévost) the sulfide concentrations in the sediments were undoubtedly related to the dense *Ulva* populations. Laboratory experiments showed that the mineralization of 1 g dried *Ulva* resulted in the production of approximately 10 mmol sulfide (unpublished).

*Comparison between the two ecosystems*

On the basis of the population densities of the functional groups of sulfur bacteria, one might expect the inner stations of the Bassin d'Arcachon and the Etang du Prévost (Stations B and 11, respectively) to be comparable with respect to eutrophication levels. Assuming that the number of SRB reflects the availability of substrates for these bacteria, which is substantiated by the numbers of CSB and/or PSB, it can be inferred that the apparent absence of free sulfide in the sediments of the Bassin d'Arcachon cannot be explained by a lower level of eutrophication of the Bassin compared to that of the Etang du Prévost.

A number of explanations can be invoked to explain the differences between the Bassin d'Arcachon and the Etang du Prévost, but it appears that two factors are of major significance. Firstly the strong currents in the Bassin d'Arcachon, due to the large tidal difference between low and high water, as compared to the Etang du Prévost, ensure a thorough mixing of the water column, and as a consequence increased oxygen availability. Secondly, the higher iron concentrations in the Bassin d'Arcachon as compared to the Etang du Prévost, which effectively and efficiently removes free sulfide (Stal et al., 1994) and thus may limit the activities of PSB and CSB.

## Acknowledgments

The authors would like to acknowledge Frank P. van den Ende, Henk M. Jonkers, Jutta Meyer and Marcel van der Meer for help and stimulating discussions. We particularly like to thank Wim J. Beukema, Joost H. Roede (Electronic Workshop Biology, Haren), Karel Visser, E. Joop Wittenaar and Bert L. A. Cazemier (Mechanical Workshop Biology, Haren) for the precision and craftsmanship during the development and construction of the submersible multiple-electrode measuring device. This study was financed by a joint EU project on 'Coastal Lagoons, Eutrophication and ANaerobic processes (CLEAN)', contract number EV5V-CT92-0080.

## References

Bak, F. & N. Pfennig, 1991. Sulfate-reducing bacteria in littoral sediment of Lake Constance. FEMS Microbiol. Ecol. 85: 43–52.

Bogorov, L. V., 1974. About the properties of *Thiocapsa roseopersicina* strain BBS, isolated from the estuary of the White sea. Microbiology (Transl.) 43: 275–2801.

Bourguès, S., O. Pringault & R. De Wit, 1995. Spatial and temporal variations of oxygen uptake, sulfide production and population of sulfur cycle bacteria at the C.L.E.AN. stations. C.L.E.AN. Proc. Prog. Rep. 11: 109–132, E.U., Environment, Brussels.

Caumette, P., 1987. Rôle des bactéries phototrophes et des bactéries sulfato-réductrices dans les milieux lagunaires. Editions de l'ORSTOM., Paris: 1–305.

Caumette, P., K. Schmidt, H. Biebl & N. Pfennig, 1985. Characterization of a *Thiocapsa* strain containing okenone as major carotenoid. System. appl. Microbiol. 6: 132–136.

Caumette, P. & B. Baleux, 1980. Etude des eaux rouges dues à la prolifération des bactéries photosynthétiques sulfo-oxydantes dans l'étang du Prévost, lagune saumâtre méditerranéenne. Mar. Biol. 56: 183–194.

De Wit, R. & H. van Gemerden, 1987. Chemolithotrophic growth of the purple sulfur bacterium *Thiocapsa roseopersicina*. FEMS Microbiol. Ecol. 45: 117–126.

De Wit, R. & H. van Gemerden, 1990a. Growth of the phototrophic sulfur bacterium *Thiocapsa roseopersicina* under oxic/anoxic regimens in the light. FEMS Microbiol. Ecol. 73: 69–76.

De Wit, R. & H. van Gemerden, 1990b. Growth and metabolism of the purple sulfur bacterium *Thiocapsa roseopersicina* under combined light/dark and oxic/anoxic regimens. Arch. Microbiol. 154: 459–464.

De Wit R., E. Buffan, S. Bourguès & H. Etcheber, 1994. Seasonal and diel variations of the physicochemical parameters at the C.L.E.AN. sampling station in Bassin d'Arcachon, Etang du Prévost and Domaine de Certes. C.L.E.AN. Proc. Prog. Rep. II: 15–21, E.U., Environment, DG XII, Brussels.

Finster, K., A. Gammelgard, N. Risgård-Petersen, S. Rysgård, S. Pelgri & N. P. Revbech, 1994. Sulfate-reduction in eelgrass-covered sediments of the Arcachon Basin (France). C.L.E.AN Proc. Progr. Rep II: 364–379, E.U., environment, DG XII, Brussels.

Gibson, G. R., R. J. Parkes & R. A. Herbert, 1987. Evaluation of viable counting procedures for the enumeration of sulfate-reducing bacteria in estuarine sediments. J. Microbiol. Meth. 7: 201–210.

Herbert, R. A., 1985. Development of mass blooms of photosynthetic bacteria on sheltered beaches in Scapa Flow, Orkney Islands. Proc. R. Soc. of Edinburgh. 87B: 15–25.

Herbert, R. A. & D. Welsh 1994. Nitrifying and denitrifying bacteria in the Arcachon basin and Prevost lagoon: abundance, activities and diversity. C.L.E.AN Proc. Prog. Rep. II: 199–229, E.U. environment, DG XII, Brussels.

Jørgensen, B. B., 1978. A comparison of methods for the quantification of bacterial sulfate reduction in coastal marine sediments. III Estimation from chemical and bacteriological field data. Biogeochem. J. 1: 49–64.

Jørgensen, B. B., 1989. Light penetration, absorption, and action spectra in cyanobacterial mats. In Y. Cohen & E. Rosenberg (eds) Physiological ecology of benthic microbial communities, A.S.M., Washington: 123–137.

Jørgensen, B. B. & F. Bak, 1991. Pathways and microbiology of thiosulfate transformation and sulfate reduction in a marine sediment (Kattegat, Denmark). Appl. Envir. Microbiol. 57: 847–856.

Kämpf, C. & N. Pfennig, 1980. Capacity of Chromatiaceae for chemotrophic growth. Specific respiration rates of *Thiocystis violacea* and *Chromatium vinosum*. Arch. Microbiol. 127: 125–135.

Kämpf, C. & N. Pfennig, 1986. Isolation and characterization of some chemoautotrophic Chromatiaceae. J. Basic. Microbiol. 26: 507–515.

Klee, A. J., 1993. A computer program for the determination of most probable number and its confidence limits. J. Microbiol. Meth. 18: 91–98.

Laanbroek, H. J. & N. Pfennig, 1981. Oxidation of short-chain fatty acids by sulfate-reducing bacteria in fresh water and in marine sediments. Arch. Microbiol. 128: 330–335.

Lassen, C., H. Ploug & B. B. Jørgensen, 1992. A fibre-optic scalar irradiance microsensor: application for spectral light measurements in sediments. FEMS Microbiol. Ecol 12: 247–254.

Lassen, C., R. de Wit, L. Lorentzen & N. P. Revsbech, 1994. Light penetration in the sediments of the Arcachon basin and Prevost lagoon as measured by optic fibre microsensor. C.L.E.AN Proc. Prog. Rep. II: 23–40, E.U., Environment, CG XII, Brussels.

Overmann, J. & N. Pfennig 1992. Continuous chemotrophic growth and respiration of Chromatiaceae species at low oxygen concentrations. Arch. Microbiol. 158: 59–67.

Sloth, N. P., N. Risgaard-Petersen, S. Rysgaard & S. P. Pelegri, 1994. Nitrification, denitrification, and nitrate ammonification in sediments of two coastal lagoons in southern France. C.L.E.AN. Proc. Prog. Rep. II, E.U. Environment, DG XII, Brussels.

Stal, L., S. B. Behrens & F. Kruyning, 1994. Interrelations of nitrogen, sulfur, and iron in benthic cyanobacterial communities in two eutrophic marine lagoons. C.L.E.AN Proc. Prog. Rep. II: 231–253, E.U. Environment, DG XII, Brussels.

Thode Andersen, S. & B. B. Jørgensen, 1989. Sulfate reduction and the formation of $^{35}$S-labelled FeS, $FeS_2$, and S° in coastal marine sediments. Limnol. Oceanogr. 34: 793–806.

Van den Ende, F. P. & H. van Gemerden, 1993. Sulfide oxidation under oxygen limitation by a *Thiobacillus thioparus* isolated from a marine microbial mat. FEMS Microbiol. Ecol. 13: 69–78.

Van den Ende, F. P. & H. van Gemerden, 1994. Relationships between functional groups of organisms in microbial mats. In L. J. Stal & P. Caumette (eds), Microbial Mats: Structure, Development and Environmental Significance, NATO ASI Series vol G35. Springer-Verlag, Heidelberg: 339–352.

Van Gemerden, H., C. S. Tughan, R. de Wit & R. A. Herbert, 1989. Laminated microbial ecosystems on sheltered beaches in Scapa Flow, Orkney Islands. FEMS Microbiol. Ecol. 62: 87–102.

Viaroli, P., M. Bartoli, C. Bondavalli, M. Cattadori, G. Giordani, M. Naldi & I. Ferrari, 1994. Macroalgae and macrophytes, oxygen and nutrient fluxes at the community level. C.L.E.AN. Proc. Prog. Rep. II: 255–279, E.U., Environment, DG XII, Brussels.

Visscher, P. T., J. Beukema & H. van Gemerden, 1991. In situ characterization of sediments: measurements of oxygen and sulfide profiles with a novel combined needle electrode. Limnol. Oceanogr. 36: 1476–1480.

Visscher, P. T., R. P. Prins & H. van Gemerden, 1992a. Rates of sulfate reduction and thiosulfate consumption in a marine microbial mat. FEMS Microbiol. Ecol. 86: 283–294.

Visscher, P. T., F. van den Ende, B. E. M. Schaub & H. van Gemerden, 1992b. Competition between anoxygenic phototrophic bacteria and colorless sulfur bacteria in a microbial mat. FEMS Microbiol. Ecol. 101: 51–58.

Visscher, P. T. & H. van Gemerden, 1993. Sulfur cycling in laminated marine microbial ecosystems. In Biogeochemistry of global Change: Radiatively Active Trace Gases. R. S. Oremland (ed.) Chapman and Hall, New York: 672–690.

*Hydrobiologia* **329**: 211–222, 1996.
P. Caumette, J. Castel & R. Herbert (eds), *Coastal Lagoon Eutrophication and ANaerobic Processes (C.L.E.AN.).*



# Sulphide release from anoxic sediments in relation to iron availability and organic matter recalcitrance and its effects on inorganic phosphorus recycling

Gianmarco Giordani, Marco Bartoli, Matteo Cattadori & Pierluigi Viaroli
*Dipartimento di Scienze Ambientali, Università di Parma, 43100 Parma, Italy*

*Key words:* Sulphide, orthophosphates, iron, organic matter, buffer capacity

## Abstract

This research aims to analyse the sediment capacity to buffer free sulphide release in three coastal lagoons which differ in terms of eutrophication level, tide influence and primary producer communities. A preliminary estimate of soluble reactive phosphorus (SRP) regeneration coupled with sulphide fluxes is also made. Sediment profiles of ferrous and ferric iron and reduced sulphur pools were determined in three stations in the Bassin d'Arcachon (South West France), in one site in the Etang du Prévost lagoon (Southern France), and in three stations in the Sacca di Goro lagoon (Northern Italy). Laboratory experiments were also conducted by incubating sediment slurries. Slurries from the French lagoons were also enriched with about 2% d.w. of organic detritus obtained from the dominant macrophytes of each site, namely *Zostera noltii* and *Ruppia cirrhosa* (Bassin d'Arcachon), and *Ulva rigida* (Etang du Prévost). In the Sacca di Goro, slurry experiments were conducted at two sites with different salinity range, sediment composition and hydrodynamics.

Field data showed that concentrations of available iron (Fe(II)+Fe(III)) ranged from a minimum of 28.5 $\mu$mol cm$^{-3}$ (Etang du Prévost) to a maximum of 275.7 $\mu$mol cm$^{-3}$ (Sacca di Goro). Moreover, in the French lagoons, acid volatile sulphide (AVS) accumulation in the superficial sediment was related to ferrous iron concentrations. Laboratory experiments showed that in spite of strong reducing conditions, sulphide and SRP release was weaker in iron-rich sediments and in those enriched with the most refractory organic matter. The highest fluxes were detected in sediment slurries from the Etang du Prévost, which had the lowest iron content, supplied by 2% of the labile detritus from *Ulva rigida*. In this case, SRP release was directly related to sulphide production.

Two factors seem significant to evaluate the buffer capacity against free sulphide and SRP release from anoxic sediment: organic matter biodegradability, which forces sediment toward reducing conditions, and iron availability, which can affect sulphide mobility as well as the ironhydroxide-phosphate-sulphide system.

## Introduction

Most of the lagoons along the Northern Mediterranean coast suffer high trophic levels, with the production of excess macroalgal biomass, mostly in the warmer months. The decomposition of such large amounts of organic matter, coinciding with high water temperature and slow water renewal, renders the superficial sediment completely anoxic and causes so-called dystrophic crises. Anoxia and dystrophic crises, have been described as particularly severe in the Etang du Thau (De Casabianca, 1993), the Etang du Prévost (Caumette, 1986), the lagoons of Orbetello and Valli di Comacchio (Izzo & Hull, 1991), the lagoon of Venice (Sfriso et al., 1992) and the Sacca di Goro (Viaroli et al., 1992). Although rarer and significantly less intense, dystrophic events have also been observed in some sheltered lagoons and impoundments along the French Atlantic coast (Labourg, 1975). However, in most of this coastal area lower temperature and larger tidal amplitudes prevent the accumulation of organic matter and keep the sediment oxidised.

Under the high summer temperature, organic matter delivered in the sediment undergoes rapid decomposition with sulphate reduction and associated fermentative reactions exceeding oxic respiration (Jørgensen, 1983; Giblin & Wieder, 1992). In the anoxic sediment, sulphides produced by sulphate reduction are rapidly precipitated, for the most part as iron sulphides (Canfield, 1989; Howarth & Stewart, 1992). The buffering capacity exerted by iron against sulphide keeps free-sulphides at very low levels minimising their toxic effects, since free-sulphides are highly toxic to aquatic life (Koch et al., 1990). However, in sheltered lagoons with excess macroalgal production (up to $10^3$ g $m^{-2}$ as dry weight), when water renewal decreases, the sulphide produced from biomass decomposition can exceed the buffering capacity of the system. Thus, a considerable amount of free-sulphides can diffuse upwards to the overlying water leading to the ultimate stage of dystrophic crisis.

During dystrophic events, noticeable fluxes of soluble reactive phosphorus (SRP) have also been observed as the result of detritus decomposition and strong anoxia (Viaroli et al., 1992). Even though most organic phosphate compounds are readily metabolised by microbes, their mobility and availability appear to be controlled by the formation of complexes with metal ions and hydroxides (De Groot, 1991; De Groot & Golterman, 1993). Sediment can accumulate phosphate throughout adsorption on ferric hydroxides, precipitation with calcium ions and adsorption on clay particles (Ponnamperuma, 1972; Golterman, 1984, 1995). Among these complexing compounds, ferric oxides and hydroxides have been considered to play a key role in marine environments, where iron reactivity is strictly linked to sulphide production (Coleman et al., 1993; Jensen & Thamdrup, 1993; Golterman, 1995). Thus phosphorus cycling can be viewed as strictly related to reduced sulphur metabolism.

Data presented in this paper describes free-sulphide fluxes in three coastal environments which differ in terms of eutrophication level, tide influence and primary producer communities. A preliminary estimate of soluble reactive phosphorus (SRP) regeneration coupled with sulphide fluxes is also described.

Sediment profiles of ferrous and ferric iron and reduced sulphur pools were determined in three stations in the Bassin d'Arcachon (South West France), in one site in Etang du Prévost lagoon, (Southern France), and in three stations in the Sacca di Goro lagoon (Northern Italy).

Laboratory experiments were also conducted by incubating slurries of sediment collected from all sites in the French lagoons and at two sites in the Sacca di Goro.

## Study area

The Bassin d'Arcachon is located on the French Atlantic coast (Figure 1, see also Castel et al., 1996). Sampling station A lies in the mesotidal centre of the lagoon and station B in the inner tidal mud-flat. In both stations, the phanerogam *Zostera noltii* Hornem. is forming large permanent meadows. The third site (station C1) is located along the eastern shore of the Bay, in the artificial impoundments of Certes covered by large stands of *Ruppia cirrhosa* (Pet.) Grande.

The Etang du Prévost is a sheltered lagoon located on the Mediterranean coast of France. The chosen station (station 11) lies in the western area of the lagoon, where high production of the seaweed *Ulva rigida* C. Agardh occurs and since water renewal is limited results in the accumulation of large quantities of macroalgal detritus (see also Caumette, 1986; Viaroli et al., 1996).

The Sacca di Goro is a microtidal lagoon in the southern part of the Po Delta (Northern Italy). Fresh water flowing into the lagoon from the Po di Volano Canal and from the Po di Goro deltaic branch determine the salinity gradient. The zonation of the lagoon also depends on *Ulva rigida* growth. We considered four sites located in different zones of the lagoon. Station G is close to the Po di Volano mouth where mean depth is about 50 cm, salinity range is high and macrophyte cover is negligible. Station 4 (150 cm deep), located in the central area of the lagoon, is affected mainly by *Gracilaria* spp. but also by *Ulva rigida* growth. Station 8 is located in the shallowest (about 30 cm deep) and most sheltered area of the lagoon. Station 7, along the southern sand bank, is about 50 cm deep and is covered in spring and early summer by *Ulva rigida*. The latter site was considered only for slurry experiments.

## Materials and methods

### *Sediment profiles*

In the Sacca di Goro triplicate sediment cores (inner diameter 50 mm, length 300 mm) were collected at stations G, 4 and 8 with a hand corer. Samples were collected on the 16 March, 15 July and 12 December 1994. In late summer 1994, in the French lagoons

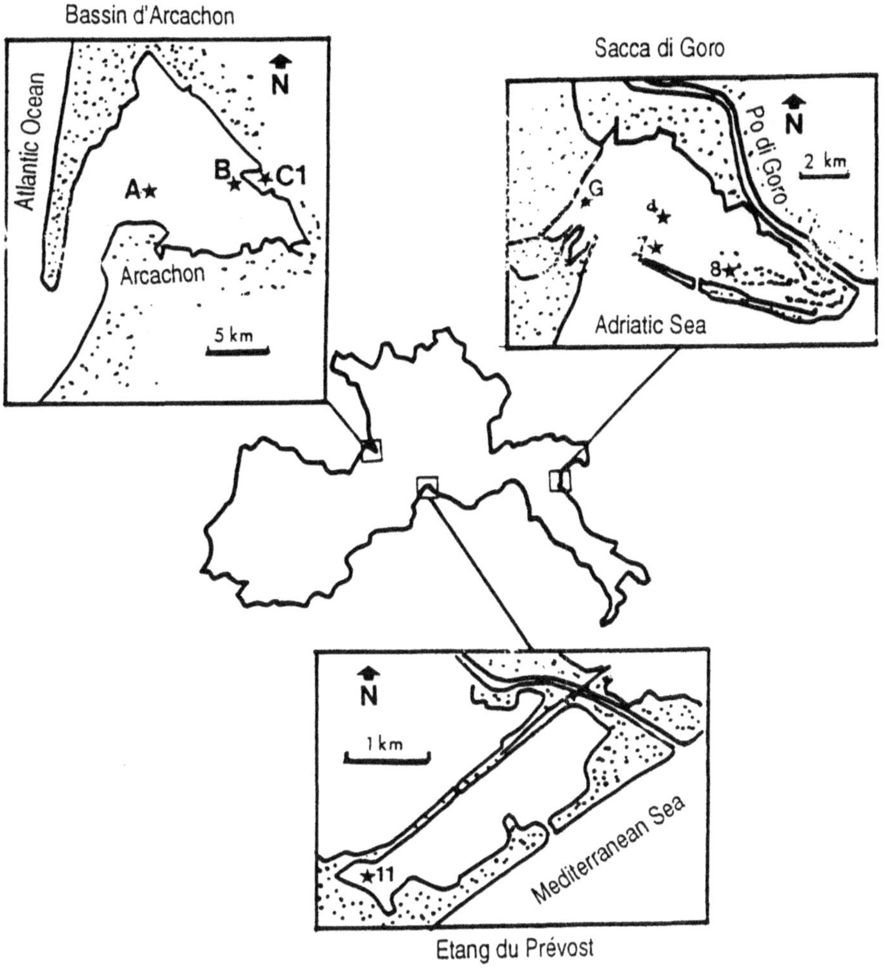


*Figure 1.* Map of the sampling stations.

triplicate sediment cores (inner diameter 50 mm, length 300 mm) were collected by hand at low tide on 18 August at station 11, on 29 August at station C1, on 30 August at station A and on 1 September at station B.

For each station one core was sliced every two cm from 0 to 10 cm; each sediment section was immediately fixed in 10 ml of 20% w/v Zn acetate and kept frozen at −20 °C until the determination of reduced inorganic sulphur pools. In parallel, one core was sliced in the same way and immediately after the slicing, 1 $cm^3$ of sediment was fixed with 5 $cm^3$ of 0.5 M HCl for the determination of ferric and ferrous iron. For the Sacca di Goro, cores for iron analyses were frozen after collection and sliced within 1 month. Finally, the remaining core was sliced in the same way to determine sediment density and porosity.

### *Slurry experiments*

In late August 1994, ten sediment cores (i.d. 5 cm, height 30 cm) were collected at station 11 (19 August), at station C1 (26 August) and at stations A and B (30 August). The upper 5 cm of sediment were removed from each core collected at each station, mixed and screened through a 1 mm sieve. The material obtained was mixed with water collected at the corresponding site to form a slurry of about 30% (by wet sediment). Slurries were stirred and aerated for 1–2 days until they reached positive Eh values. Then, 1 litre dark bottles were filled with 750 $cm^3$ of the slurry. Half of the bottles were left untreated and considered as blanks, while the remaining ones were treated by the addition of a fixed amount of organic matter (Table 1). Organic matter was obtained from the dominant primary producer of the corresponding site and was added as a finely ground powder: *Ulva* was added to the slurry from station 11, *Ruppia* and epiphytes (50% each component) to the slurries from station C1, and *Zostera* to the slurries from stations A and B.

On 5 July 1994, similar experiments were conducted with slurries obtained from stations G and 7 in the Sacca di Goro. Sediment and slurries were treated identically to those of French lagoons except for the organic matter treatment which was omitted. Incubation times were also shorter than those adopted for experiments carried out in the French sites.

Sediment manipulation and organic matter (O.M.) addition resulted in small variations of O.M. concentrations and redox of the slurries in comparison with those normally observed in the field.

After the collection of the initial samples, the bottles were incubated in the dark under constant $N_2$ flux and stirring for about 2 days in a water bath at controlled temperature. All the sulphide produced in the slurry was carried in $N_2$ gas flow and trapped in 100 ml of 2% w/v Zn acetate. At selected time intervals during the experiment, 40 $cm^3$ of slurry were withdrawn with a syringe from a valve located in the $N_2$ inlet tube. These samples were taken more frequently at the beginning than at the end of the incubation period. Immediately after collection, the Eh was determined and 3 $cm^3$ of slurry were fixed with 2% w/v zinc acetate for the analysis of reduced sulphur pools. The remaining material was centrifuged within a few minutes and the supernatant was filtered through GF/C filters (Whatman) for porewater analysis (pH, $NH_4^+$ and SRP). In parallel, the zinc acetate trap was vigorously shaken and a subsample was taken for free sulphides analysis. Density, dry weight, organic matter, total phosphorus and ferrous iron as well as the above mentioned variables were determined on the initial and final samples.

### *Analytical determinations*

Eh was measured with a pH/mV-meter (TIM 90, Radiometer) by using a platinum electrode (P101, Radiometer) and saturated calomel reference (K701, Radiometer). pH was measured with a combined electrode (GK2401C, Radiometer) and a temperature probe (T801, Radiometer) for temperature compensation. Dissolved oxygen was determined by the Winkler method (A.P.H.A., 1975). Soluble reactive phosphorus (SRP) was measured by means of the ascorbic acid method (Valderrama, 1977), and ammonium by the indophenol blue method (Koroleff, 1970). Sulphide concentrations were determined with the methylene blue method (Cline, 1969) after fixation with 2% w/v zinc acetate.

Reduced sulphur pools, namely acid volatile sulphide (AVS) and chromium reducible sulphur (CRS), were determined by means of two sequential distillations under anoxic conditions (Fossing & Jørgensen, 1989). AVS was detected as $H_2S$ released by the addition of 10 $cm^3$ of 5 M HCl to about 1 g of sediment fixed in 2% w/v zinc acetate. CRS, which represents the remaining reduced inorganic sulphur, was extracted with 10 $cm^3$ of 1 M $CrCl_2$ in 0.5 M HCl. The latter extraction was carried out at room temperature for 20 minutes and at boiling temperature for 40 minutes. At each step, the sulphides produced were removed in a $N_2$ gas stream, fixed in 2% w/v Zn-acetate and then measured by the methylene blue method.

Ferrous iron was measured after fixation of 1 $cm^3$ of sediment in 5 ml of 0.5 M HCl with the phenanthroline method (A.P.H.A., 1975). Reactive ferric iron was determined as ferrous iron after reduction with 10% ascorbic acid.

Dry weight was determined at 70 °C until a constant weight was reached. Organic matter content (O.M.) was measured as weight loss after ignition of 2–3 g of finely powered dry sample at 550 °C for 24 hours.

Total phosphorus was determined spectrophotometrically after extraction with 20 $cm^3$ of 1 M HCl from 1 g of sediment ash (Aspila et al., 1976).

## Results

### *Sediment profiles of iron and reduced sulphur pools*

Measurements conducted in the French lagoons are restricted to short periods of observation. However, they cover the transitional phase from active growth towards biomass decay of the dominant macrophytes. We thus assumed that this step would be representative of the most critical conditions for the biogeochemistry of inorganic sulphur and iron.

In the Bassin d'Arcachon and in the Etang du Prévost, the first 10 cm of the sediment horizon were characterized by a wide range of iron concentrations (Fe(II) and Fe(III)) (Figure 2). Maximum values were found at station C1 (106.5 to 152.7 $\mu$mol $cm^{-3}$) and to a lesser degree in the uppermost sediment layer (0–6 cm) of station B (93.9 to 130.6 $\mu$mol $cm^{-3}$). Noticeable differences were found between station B, in the inner mud-flat, and station A, located in the central sand bank. The lowest concentrations (28.5

*Table 1*. Experimental design of the slurry experiments. Initial values of organic matter (O.M.), total phosphorus (TP) and Fe(II) are reported.

| St. | O.M. added % d.w. | g dry sed. per litre of slurry | O.M. % d.w. | TP $\mu$mol cm$^{-3}$ | Fe(II) $\mu$mol cm$^{-3}$ |
|---|---|---|---|---|---|
| A | – | 120 | 10.0 | 12.2 | 33.0 |
| A | +2.6% *Zostera* | 120 | 11.9 | 13.2 | 33.0 |
| B | – | 140 | 13.0 | 9.8 | 58.8 |
| B | +2.2% *Zostera* | 140 | 14.7 | 10.6 | 58.8 |
| C1 | – | 116 | 18.0 | 21.7 | 67.2 |
| C1 | +2.2% *Ruppia* + epiphytes | 116 | 18.8 | 22.6 | 67.2 |
| 11 | – | 120 | 17.0 | 11.7 | 29.2 |
| 11 | +2.8% *Ulva* | 120 | 19.2 | 13.1 | 29.2 |
| G | – | 60 | 7.1 | 23.0 | 43.4 |
| 7 | – | 100 | 2.7 | 9.3 | 7.5 |

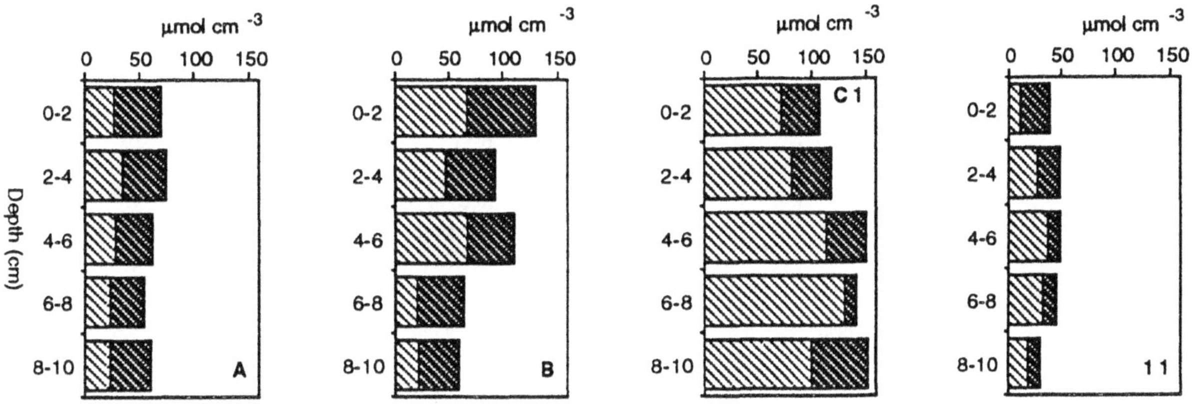


*Figure 2*. Depth profiles of ferrous (light) and ferric (dark) iron concentrations in the sediments collected from stations A, B and C1 in the Bassin d'Arcachon, and station 11 in the Etang du Prévost.

to 48.6 $\mu$mol cm$^{-3}$) were detected in the dystrophic Etang du Prévost, which was already known to have a very low iron content (Stal et al., 1994). Nevertheless, at this site the limited iron supply was coupled to AVS concentration up to 33 $\mu$mol cm$^{-3}$ (Figure 3). In the Bassin d'Arcachon, AVS concentrations in the first 10 cm of sediment followed a similar pattern to iron distribution. Maximum AVS concentrations (60.5 to 79.9 $\mu$mol cm$^{-3}$) were detected in the uppermost sediment layer (0–6 cm) at station C1, while minimum concentrations occurred at station B (less than 10 $\mu$mol cm$^{-3}$) and A (less than 20 $\mu$mol cm$^{-3}$). At these sites, even though sulphate reduction rates reached the same order of magnitude as at station 11 (Bourguès et al., 1994), AVS accumulation was low in the superficial sediment. This may be due to AVS reoxidation which occurs at low tide when the sediment surface is exposed to the air. Station C1 is different because of the high O.M. content in the surface sediment and the limited water renewal. In late August 1994, settling of large amounts epiphytes growing on *Ruppia cirrhosa*, increased the O.M. content in the surface sediment to 24.6 ± 1.4% d.w. (Bartoli et al., 1996). Such an increase in O.M. availability would stimulate sulphate reducing bacteria activity which would account for the AVS accumulation (see also Viaroli et al., 1996).

Low iron content in the surficial sediment at sites A and 11 was also coupled to low CRS concentrations compared with those detected at stations C1 and B. At the latter site, CRS levels exceeding 500 $\mu$mol cm$^{-3}$ could be achieved due to high iron availability and the reoxidation of reduced sulphur induced by the sediment surface drying out at low tide and by oxygen diffusion from the plant roots (Canfield, 1989; Finster et al., 1994).

The relationship between iron and reduced sulphur pools was also analysed in the Sacca di Goro where three sites located in different zones of the lagoon were

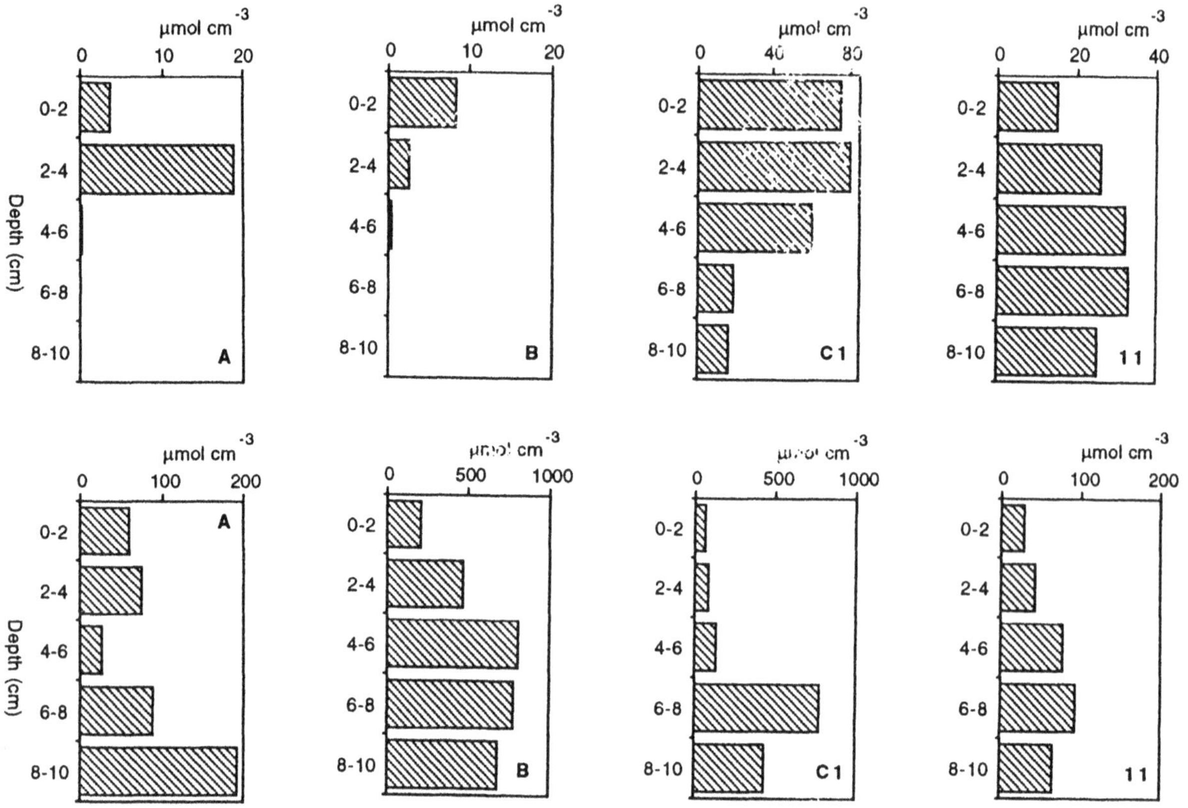


*Figure 3.* Depth profiles of AVS (upper panels) and CRS (lower panels) concentrations in the sediments collected from stations A, B and C1 in the Bassin d'Arcachon, and station 11 in the Etang du Prévost.

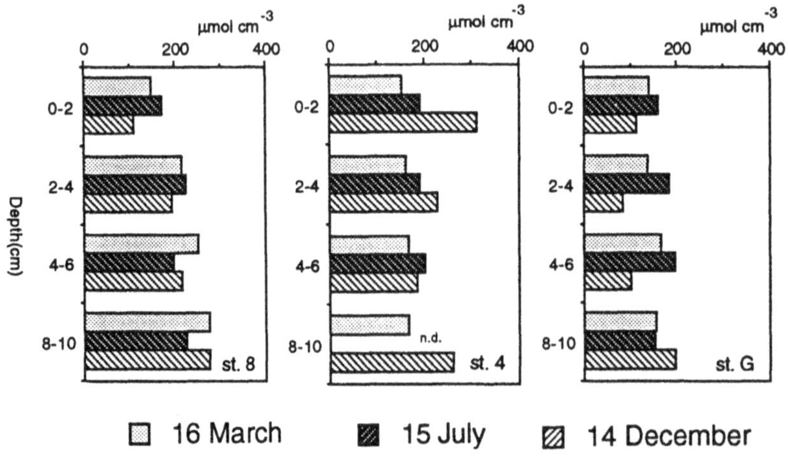


*Figure 4.* Depth profiles of total available iron (Fe(II) + Fe(III)) concentrations in the sediments collected from Stations 8, 4 and G in the Sacca di Goro in 1994.

studied: station G near the freshwater outlet from the Po di Volano Canal, station 4 affected by tidal currents and station 8 in the sheltered part of the lagoon where both freshwater inflow from the Po di Goro and marine inputs determine the salinity regime. Total available iron (Fe(II) + Fe(III)) concentrations were on average

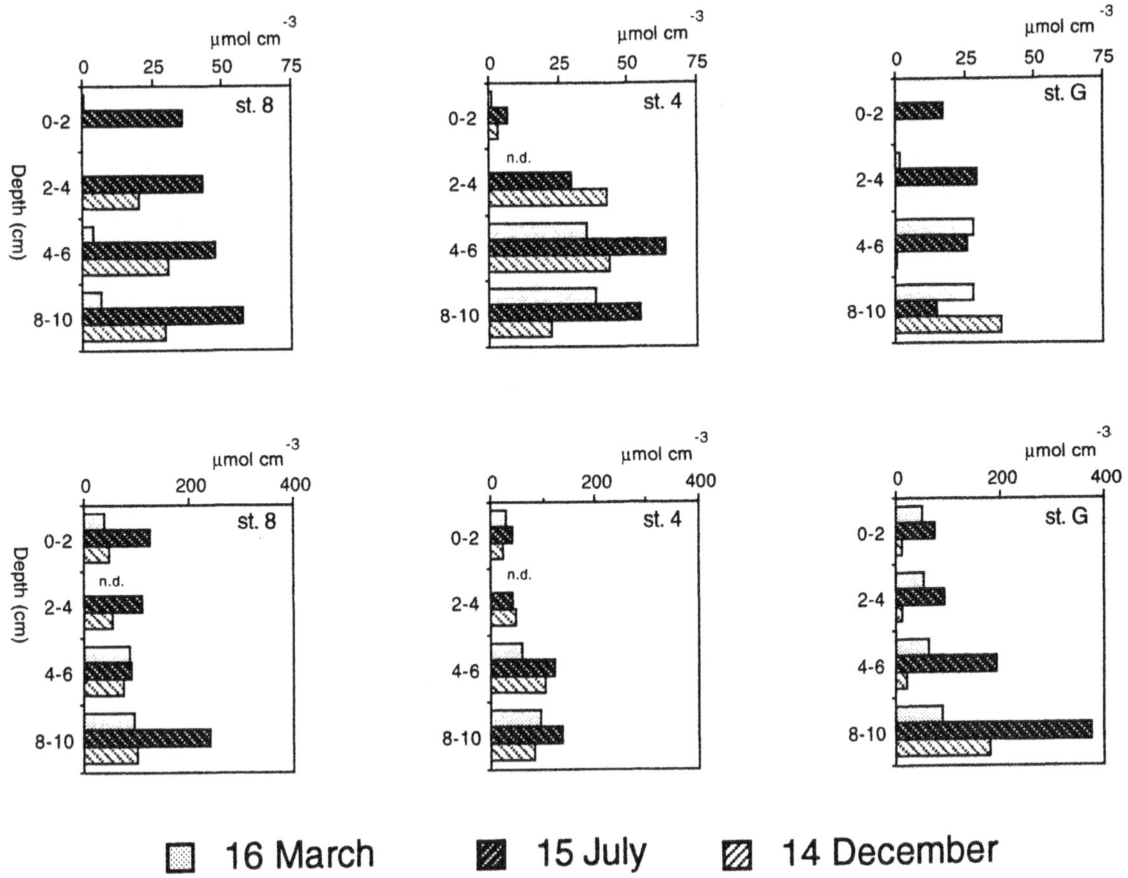


*Figure 5.* Depth profiles of AVS (upper panels) and CRS (lower panels) concentrations in the sediments collected from stations 8, 4 and G in the Sacca di Goro in 1994.

higher than those determined in the French lagoons, with minimum values ranging from 100 to 110 $\mu$mol cm$^{-3}$ and maximum values of about 310 $\mu$mol cm$^{-3}$ (Figure 4). AVS concentrations showed wide seasonal variations, with maximum values in July. CRS followed similar patterns with maximum values in the warmer summer months and minimum concentrations in December (Figure 5). This clear seasonal trend suggests that reduced sulphur pools are controlled by 'iron availability' as well as reoxidation processes. However, the parallel decrease of AVS and CRS in December appears to show that most of the reduced sulphur is reoxidised, minimising sulphur burial through pyrite deposition.

*Sediment slurry incubation*

The sediment capacity to buffer free sulphide release was tested by incubating sediment slurries both with and without organic matter (O.M.) added. During the course of the experiment, the Eh decreased to highly negative values, mostly in slurries with O.M. added (Table 2). In these experiments, wide differences between sites was also found, with the most negative value at station 11 (−315 mV). In the case of station C1, O.M. addition resulted in only a slight decrease in Eh in comparison with the untreated slurry. Moreover, slurries from station C1 showed the least negative Eh values under the experimental conditions tested. The differences indicated by the Eh values, conform to those shown by trends of free sulphide, AVS and SRP (Figure 6). For all these variables, differences between treated and untreated slurries were in the range of approximately one order of magnitude, except for the free sulphide and SRP at station C1, which did not appears to be affected by O.M. addition.

Sulphide production increased exponentially in the treated slurry for station 11, where the final concentration of 4.3 $\mu$mol cm$^{-3}$ was reached. Considerable release of sulphide was also detected in the treated slur-

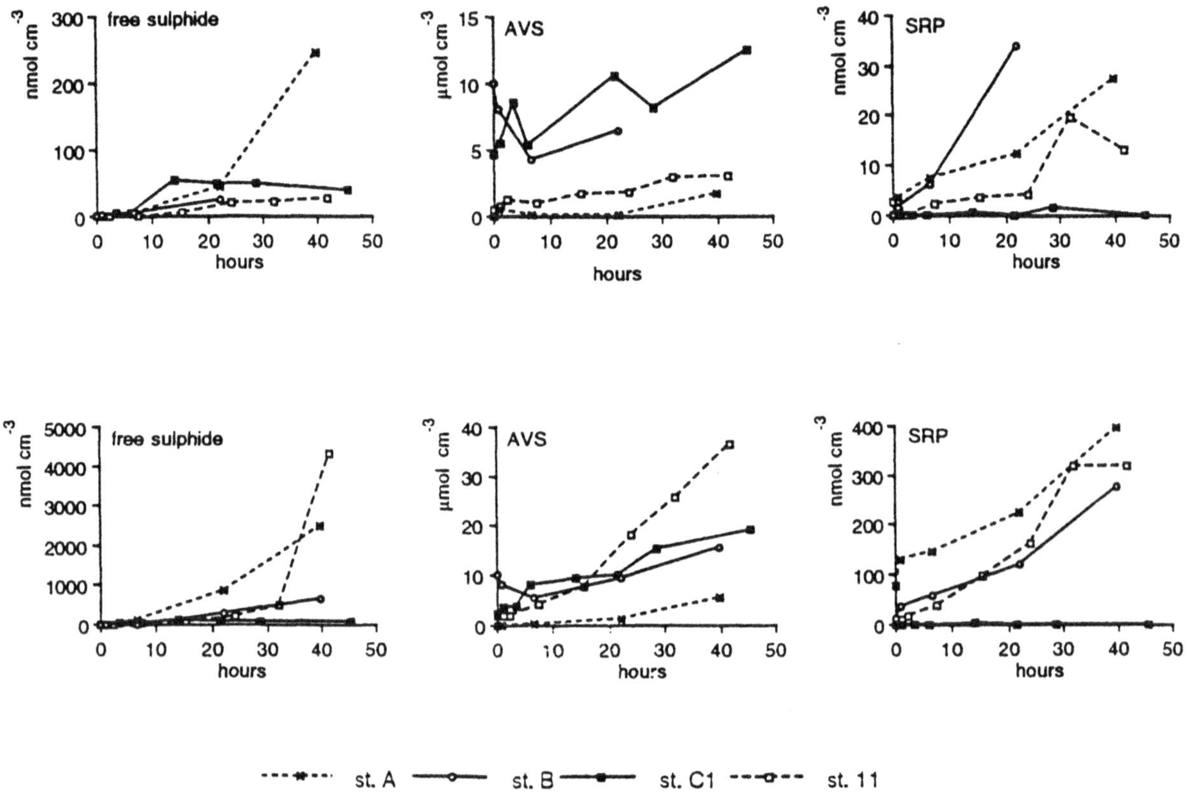


*Figure 6.* Changes of free sulphide (nmol $cm^{-3}$), AVS ($\mu$mol $cm^{-3}$) and SRP (nmol $cm^{-3}$) concentrations during the incubation of sediment slurries from stations A, B, C1 and 11 without (upper panels) and with (lower panels) organic matter added. X-axis: incubation duration (hours), Y-axis: concentrations.

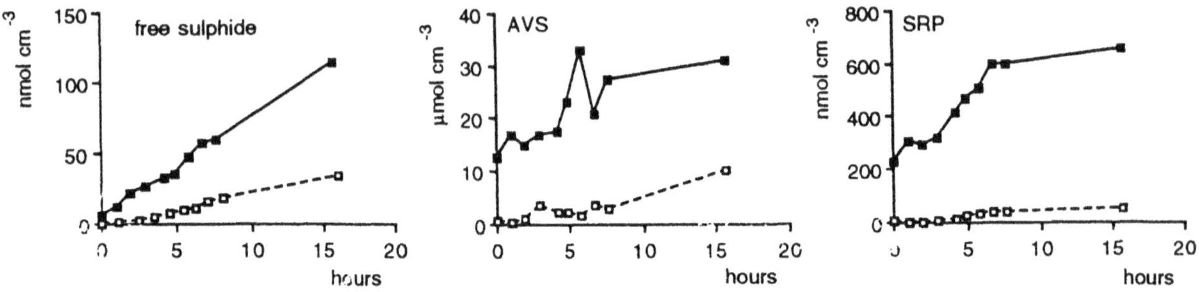


*Figure 7.* Changes of free sulphide (nmol $cm^{-3}$), AVS ($\mu$mol $cm^{-3}$) and SRP (nmol $cm^{-3}$) concentrations during the incubation of sediment slurries of stations 7 (open symbols) and G (black symbols). X-axis: incubation duration (hours), Y-axis: concentrations.

ries from stations A (2.5 $\mu$mol $cm^{-3}$) and B (0.6 $\mu$mol $cm^{-3}$).

AVS concentrations followed an almost linear increase under all the conditions investigated, showing a clear gradient with maximum values in the treated slurry from station 11.

Free sulphide and AVS release was coincident with significant SRP-fluxes, except for station C1, which did not show appreciable release of SRP.

Preliminary experiments were also conducted in the Sacca di Goro by incubating slurries of sediment collected from two distinct sites: the sandy station 7 and the muddy station G. Changes of free sulphide, AVS and SRP concentrations conformed to those observed

*Table 2.* Main characteristics of slurries without (–) and with (+) organic matter added at the end of incubation period. Data were adjusted to 40 hours of incubation for stations A, B, C1 and 11 and to 15 hours for stations G and 7. Final values of Eh and CRS mean value of the whole incubation period were reported.

| Station | | Eh mV | $Fe^{2+}$ $\mu$mol cm$^{-3}$ | $S^{2-}$ $\mu$mol cm$^{-3}$ | AVS $\mu$mol cm$^{-3}$ | CRS $\mu$mol cm$^{-3}$ | SRP nmol cm$^{-3}$ |
|---|---|---|---|---|---|---|---|
| A | – | –201 | 46.0 | 0.247 | 1.65 | 123.3± 9.9 | 27.5 |
| B | – | –184 | 64.2 | 0.045 | 6.47 | 308.7±12.3 | 33.7 |
| C1 | – | –171 | 82.6 | 0.034 | 12.58 | 174.3± 6.3 | 1.3 |
| 11 | – | –219 | 43.1 | 0.025 | 2.90 | 78.2± 1.7 | 12.4 |
| A | + | –264 | 50.2 | 2.490 | 5.54 | 111.6± 3.6 | 399.3 |
| B | + | –238 | 77.4 | 0.633 | 15.69 | 328.6±28.2 | 276.6 |
| C1 | + | –203 | 83.7 | 0.085 | 17.03 | 145.6± 9.1 | 2.2 |
| 11 | + | –312 | 43.1 | 4.141 | 34.49 | 72.3± 6.8 | 236.9 |
| G | | n.d. | 72.4 | 0.114 | 30.9 | 90.6±13.2 | 655.9 |
| 7 | | n.d. | 53.2 | 0.034 | 10.0 | 6.3± 3.2 | 49.8 |

in the French lagoons, with a significant relationship between AVS and SRP (Figure 7).

## Discussion

Several studies have demonstrated that in coastal marine environments most of the organic matter is mineralised by means of sulphate reduction (Giblin & Wieder, 1992). Nevertheless, sulphide produced by sulphate reduction remains at very low levels, except in a number of dystrophic lagoons which are characterized either by limited water renewal or abnormal primary production.

The sulphide pathway is based on ferric iron reduction by sulphide, which in turn is oxidised to zero-valent sulphur. This step is followed by the reaction of the produced ferrous iron with another sulphide to form ferrous monosulphide (FeS). This sequence of reactions which either reoxidises or precipitates free sulphide in effect functions as a buffer against sulphide mobilisation (Canfield, 1989). Therefore, the reactions which detoxify free sulphides are for the most part based on iron availability. Likewise, the latter would be considered as an indicator of lagoon resistance against the harmful effects of sulphide, e.g. the ultimate stage of dystrophy.

As for the sediment profiles determined in the French lagoons we found a significant relationship between AVS concentration and AVS/Fe(II) ratio (Figure 8). We have been tentatively able to establish that the ratio AVS/Fe(II)=1 represents the threshold level of the sediment capacity to precipitate free sulphide. This level can also be inferred from the fact that sulphide is precipitated mainly as insoluble monosulphuric ferrous iron (Canfield, 1989). Thus we can show that at station 11, the sediment capacity to take up sulphide was exceeded at very low concentrations (around 30 $\mu$mol $S^{2-}$ cm$^{-3}$), while at station C1 the saturation threshold was reached approximately at 80 $\mu$mol $S^{2-}$ cm$^{-3}$. Moreover, in the uppermost sediment layers (0–4 cm) of station 11, the AVS/Fe(II) ratio, was 1.4, which suggests that available ferrous iron was oversaturated by sulphide. At station C1, the iron system appears to be only saturated in the most superficial layers, where sulphate reduction was probably high (see Viaroli et al., 1996). At stations A and B, and in the deeper sediment layer of station C1 AVS/Fe(II) ratio

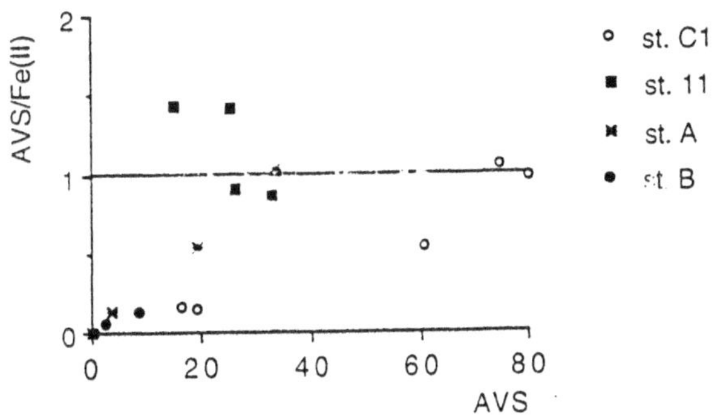


*Figure 8.* Relationship between AVS concentration ($\mu$mol cm$^{-3}$) and the AVS/Fe(II) molar ratio. The horizontal line refers to the theoretical ratio of iron monosulphide.

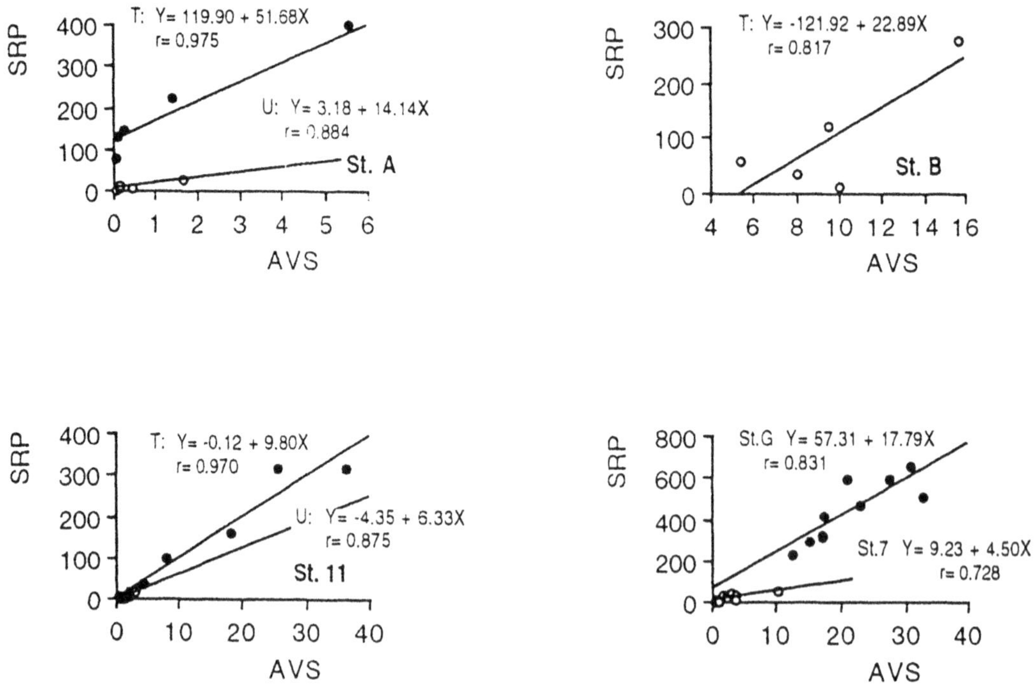


*Figure 9.* Relationship between AVS ($\mu$mol cm$^{-3}$) and SRP (nmol cm$^{-3}$) concentrations during slurry incubation.

was less than 0.5. The latter value might indicate that the buffering capacity of ferrous iron was greatly in excess of sulphide production. However, at stations A and B, sulphide mobility as well as AVS precipitation may not be simply related to iron availability since the effects of tidal currents are also involved. In fact, drying out of the sediment at low tide allows oxidation of the superficial sediment, thus enabling sulphide to be reoxidised. The recorded sediment profiles do not conform completely to those obtained with sediment slurry incubations (Table 2). In the sediment slurries without O.M. addition, we found that station C1 accumulated AVS, with an AVS/Fe(II) ratio = 0.15 resembling those of the sediment profiles. Unexpected significant production of free sulphide was found at station A. This was probably due to the high sulphate reduction rates (SRR) (Bourguès et al., 1994) as well as sediment porosity, which would make free sulphide stripping by the $N_2$ gas stream easier. Moreover, free sulphide removal might stimulate SRR, since this is controlled by sulphide concentration. Organic matter addition (approximately 1–2% by final concentration) enhanced microbial processes as evidenced by the abrupt decrease in Eh. However, station C1 still maintained a good capacity to control sulphide release, with an almost balanced AVS/Fe(II) ratio and negligible free sulphide fluxes. A significant sulphide flux without relevant AVS accumulation was found in the treated slurry from station A, similar to that observed in the untreated slurry. In the case of station 11, sulphide production greatly exceeded the sediment capacity to precipitate AVS, leading to an AVS/Fe(II) ratio of 1.2 and to 4.14 $\mu$mol cm$^{-3}$ of free sulphide concentration.

This wide range of AVS and free sulphide concentrations can be explained to some extent by considering ferrous iron availability as well as organic detritus quality. Detritus from *Ulva* can undergo more rapid decomposition than that from *Zostera* and *Ruppia*, due chiefly to the higher content of soluble carbohydrates (Lahaye & Jegou, 1993; Viaroli et al., 1996).

With regard to the slurry experiments we found a clear relationship between AVS and SRP concentration (Figure 9). This aspect may be of considerable significance in the analysis of lagoon eutrophication, since it can be considered as a form of positive feedback due to significant SRP recycling. Inorganic phosphate is usually assumed to be present in the sediment as iron and calcium bound phosphate (Golterman, 1995). However, for a given sediment, the distribution of SRP within the inorganic sedimentary pool is mostly controlled by the iron hydroxide-sulphide system (Davison and Heany, 1978; De Groot, 1991; Golterman, 1995). This seems to be the case for all the sites considered in this study, except for station C1. At the latter site, SRP concentrations were always negligible despite high AVS concentrations. Accordingly, we postulate that there, SRP fluxes are mainly controlled either by clay (Ponnamperuma, 1972; Golterman, 1984) or by the calcium-carbonate system (Golterman, 1984, 1995). The significant differences among treated and untreated slurries may also be dependent on direct SRP recycling from decomposition of the added detritus.

Although this investigation was restricted to a short period of observation, we can conclude that iron availability and O.M. quality can be considered as key indicators to predict the ability of a given lagoon to buffer against dystrophic events. For example, the dystrophic Etang du Prévost suffers from a severe iron limitation and receives a high input of readily decomposable detritus from recurring 'green tides' (Ulvacean blooms). In contrast, dystrophic events are unusual in the sheltered impoundments of the Bassin d'Arcachon, where iron concentrations are 2–3 times higher and Ulvacean blooms have a minor impact. However, to what extent these measurements enable us to predict more accurately dystrophic events remains to be seen.

## Acknowledgements

We are indebted to Mr Tony Zappata (Battello Hydra, Amministrazione Provinciale di Ferrara, Assessorato Ambiente) and Mr Vittorio Gaiani (Dipartimento di Biologia, Università di Ferrara) for the support during sampling in the Sacca di Goro. This study was financed by the joint EU project on 'Coastal Lagoon Eutrophication and Anaerobic processes (CLEAN)', contract number EV5V-CT92-0080.

## References

A.P.H.A., 1975. Standard Methods For the Examination of Water and Wastewater. 14th Edition. APHA, Washington, 1193 pp.

Aspila, K. I., H. Agemian & A. S. Y. Chau, 1976. A semiautomated method for the determination of inorganic, organic and total phosphate in sediments. Analyst 101: 187–197.

Bartoli, M., M. Cattadori, G. Giordani & P. Viaroli, 1996. Benthic oxygen respiration, ammonium and phosphorus regeneration in surficial sediments of the Sacca di Goro (Northern Italy) and two French coastal lagoons: a comparative study. Hydrobiologia 329 (Dev. Hydrobiol. 117): 143–159.

Bourguès, S., O. Pringault & R. de Wit, 1994. Spatial and temporal variation of oxygen uptake, sulphide production and population of sulphur cycle bacteria at the CLEAN stations. In P. Caumette (Coord.), C.L.E.A.N. Progress Report, European Commission: 109–132. European Community Environment Programme, DG XII, Brussels.

Canfield, D. E., 1989. Reactive iron in marine sediments. Geochim. Cosmochim. Acta 53: 619–632.

Castel, J., P. Caumette & R. Herbert, 1996. Eutrophication gradients in coastal lagoons as exemplified by the Bassin d'Arcachon and the Étang du Prévost. Hydrobiologia 329 (Dev. Hydrobiol. 117): xi–xxx.

Cline, J. D., 1969. Spectrophotometric determination of hydrogen sulphide in natural waters. Limnol. Oceanogr. 14: 454–459.

Coleman, M. L., D. B. Hedrick, D. R. Lovley, D. C. White & K. Pye, 1993. Reduction of Fe(III) in sediments by sulphate-reducing bacteria. Nature 361: 436–438.

Caumette, P., 1986. Phototrophic sulphur bacteria and sulphate reducing bacteria causing red waters in a shallow brackish coastal lagoon (Prévost Lagoon, France). FEMS Microbiol. Ecol. 38: 113–124.

Davison, W. & S. I. Heany, 1978. Ferrous iron-sulphide interactions in anoxic hypolimnetic waters. Limnol. Oceanogr. 23: 1194–1200.

De Casabianca, M. L., 1993. Macronutrients relative evolution in Mediterranean macrophyte system subject to dystrophic crises. In J. W. Rijstenbil & S. Haritonidis (eds), Macroalgae, eutrophication and trace metal cycling in estuaries and lagoons. Proceedings of the COST-48 Symposium of Sub Group III, Thessaloniki 24–26 September. EU-BRIDGE, 31–34.

De Groot, C. J., 1991. The influence of FeS on the inorganic phosphate system in sediments. Verh. int. Ver. Limnol. 24: 3029–3035.

De Groot, C. J. & H. L. Golterman, 1993. On the presence of organic phosphate in some Camargue sediments: evidence for the importance on phytate. Hydrobiologia 252: 117–126.

Finster, K., A. Gammerlaad & L. S. Hansen, 1994. Sulphate reduction rates, inorganic sulphur pools and the accumulation of volatile fatty acid in the Bassin d'Arcachon and the Prévost lagoon, southern France: a comparative study. In P. Caumette (Coord.), C.L.E.A.N. Progress Report, European Commission: 223–252. European Community Environment Programme, DG XII, Brussels.

Fossing, H. & B. B. Jørgensen, 1989. Measurement of bacterial sulphate reduction in sediment: evaluation of a single-step chromium reduction method. Biogeochemistry 8: 205–222.

Giblin, A. E. & R. K. Wieder, 1992. Sulphur cycling in marine and freshwater wetlands. In R. W. Howarth, J. W. B. Stewart & M. U. Ivanov (eds), Sulphur cycling on the continents: wetlands, terrestrial ecosystems and associated water bodies. SCOPE 33, J. Wiley & Sons, New York: 85–117.

Golterman, H. L., 1984. Sediments, modifying and equilibrating factors in the chemistry of freshwater. Verh. int. Ver. Limnol. 22: 23–59.

Golterman, H. L., 1995. The role of the iron hydroxide-phosphate-sulphide system in the phosphate exchange between sediments and water. Hydrobiologia 297: 43–54.

Howarth, R. W. & J. W. B. Stewart, 1992. The interactions of sulphur with other element cycles in ecosystems. In R. W. Howarth, J. W. B. Stewart & M. U. Ivanov (eds), Sulphur cycling on the continents: wetlands, terrestrial ecosystems and associated water bodies. SCOPE 33, J. Wiley & Sons, New York: 67–84.

Izzo, G. & V. Hull, 1991. The anoxic crises in dystrophic processes of coastal lagoons: an energetic explanation. In C. Rossi & E. Tiezzi (eds), Ecological Physical Chemistry. Elsevier Sci. Pbls. Amsterdam: 559–572.

Jensen, H. S. & B. Thamdrup, 1993. Iron-bound phosphorus in marine sediments as measured by bicarbonate-dithionite extraction. Hydrobiologia 253: 47–59.

Jørgensen, B. B., 1983. Processes at the water-sediment interface. In B. Bolin & R. B. Cook (eds), The major biogeochemical cycles and their interactions. SCOPE 21, J. Wiley & Sons, New York: 477–509.

Koch, M. S., I. A. Mendelssohn & K. L. McKee, 1990. Mechanism for the hydrogen sulphide growth limitation in wetland macrophytes. Limnol. Oceanogr. 35: 399–408.

Koroleff, F., 1970. Direct determination of ammonia in natural waters as indophenol blue. Information on techniques and methods for seawater analysis. I.C.E.S. Interlaboratory Rep. No. 3: 19–22.

Labourg, P. J., 1975. Contribution à l'hydrologie des étang saumâtres de la région d'Arcachon: description des phénomènes d'eaux blanches. Bull. soc. linn. Bordeaux 5: 3–8.

Lahaye, M. & D. Jegou, 1993. Chemical and physical-chemical characteristics of dietary fibres from *Ulva lactuca* (L.) Thuret and *Enteromorpha compressa* (L.) Grev. J. Appl. Phycol. 5: 195–200.

Ponnamperuma, F. N., 1972. The chemistry of submerged soils. Adv. Agron. 24: 29–96.

Sfriso, A., B. Pavoni, A. Marcomini & A. A. Orio, 1992. Macroalgae, nutrient cycles, and pollutants in the lagoon of Venice. Estuaries 15: 517–528.

Stal, L. J., F. Kruying, S. B. Behrens & M. Vilbrant, 1994. Spatial and seasonal variations of iron, phosphate and sulphide in the sediments of two eutrophic lagoon. In P. Caumette (Coord.), C.L.E.A.N. Progress Report, European Commission: 167–186. European Community Environment Programme, DG XII, Brussels.

Valderrama, J. C., 1977. Methods used by the Hydrographic Department of National Board of Fisheries, Sweden. In K. Grasshoff (ed.), Report of the Baltic Intercalibration Workshop. Annex, Interim Commission for the Protection of the Environment of the Baltic Sea: 14–34.

Viaroli, P., A. Pugnetti & I. Ferrari, 1992. *Ulva rigida* growth and decomposition processes and related effects on nitrogen and phosphorus cycles in a coastal lagoon (Sacca di Goro, Po River Delta). In G. Colombo, I. Ferrari, V. U. Ceccherelli and R. Rossi (eds), Marine eutrophication and population dynamics. Olsen & Olsen, Fredensborg: 77–84.

Viaroli, P., M. Bartoli, C. Bondavalli, R. R. Christian, G. Giordani & M. Naldi, 1996. Macrophytes communities and their impact on benthic fluxes of oxygen, sulphide and nutrients in shallow eutrophic environments. Hydrobiologia 329 (Dev. Hydrobiol. 117): 105–119.

# Subject index

Page numbers refer to the first page of an article in which the entry is discussed. As regards biological organisms, only genera and, when appropriate, higher taxonomic categories are indexed.

GPSR Compliance

*The European Union's (EU) General Product Safety Regulation (GPSR) is a set of rules that requires consumer products to be safe and our obligations to ensure this.*

*If you have any concerns about our products, you can contact us on ProductSafety@springernature.com*

In case Publisher is established outside the EU, the EU authorized representative is:

Springer Nature Customer Service Center GmbH
Europaplatz 3
69115 Heidelberg, Germany

**Batch number: 10809834**

Printed by Printforce, the Netherlands